The Participatory Role of Indigenous Women

Nilakshi Moran · Pardeep Singh ·
Kshitij Kumar Singh
Editors

The Participatory Role of Indigenous Women

Environmental Conservation in the Himalaya Region, Volume 2

Springer

Editors
Nilakshi Moran
Department of English, Satyawati College (Evening)
University of Delhi
New Delhi, Delhi, India

Pardeep Singh
Department of Environmental Studies
PGDAV College
University of Delhi
New Delhi, Delhi, India

Kshitij Kumar Singh
Faculty of Law
University of Delhi
New Delhi, Delhi, India

ISBN 978-3-032-24058-3 ISBN 978-3-032-24059-0 (eBook)
https://doi.org/10.1007/978-3-032-24059-0

This Springer imprint is published by the registered company Springer Nature Switzerland AG
The registered company address is: Gewerbestrasse 11, 6330 Cham, Switzerland

Preface

The Himalayas, long regarded as an ecologically sensitive and culturally diverse region, offer a critical vantage point for understanding the interplay of gender, environment, and sustainability. Building on the foundation laid in The Participatory Role of Indigenous Women: Conservation in the Himalaya Region (Volume I), this second volume advances the discussion by interrogating Indigenous women's evolving roles in conservation through the lenses of forest governance, policy frameworks, globalization, and socio-economic transformation. While Volume I emphasized agricultural practices, biodiversity stewardship, and community resilience, Volume II situates women's agency within broader institutional, historical, and global contexts. The chapters underscore that Indigenous women are not passive beneficiaries but central actors in shaping ecological futures. The volume opens with The Unseen Guardian: Role of Native Women in the Conservation of the Himalayan Ecosystem, foregrounding women's invisible yet indispensable contributions as ecological custodians. This is followed by Climate-Smart Agroforestry as a Nature-Based Solution for Forest Landscape Restoration in Uttarakhand, situating women's practices within the emerging discourse of climate-smart and nature-based solutions. The legacies of grassroots environmental activism are revisited in Aftermath of Chipko: Women are the Undervalued Protagonists of Forest Management, which critically re-examines women's enduring yet undervalued contributions in the wake of one of South Asia's most iconic ecological movements. Community-based conservation forms a recurring theme. Women as Conservators: Protecting Forests through Community Action and Empowering Change: The Role of Women in Reviving Indigenous Environmental Governance demonstrates how women's collective initiatives contribute to participatory forest management and biodiversity conservation. Extending the focus beyond the Western Himalaya, Empowering Indigenous Women in Conservation Efforts through Participatory Roles and Livelihood Generations in India's Northeastern Region illustrates the integration of ecological stewardship with livelihood diversification, emphasizing how conservation and socio-economic empowerment are mutually reinforcing. The chapters Voices from the Himalayas: Indigenous Women in a Glocalized World and Policy Frameworks and Institutional Supports for

Indigenous Women in India expand the scope by examining the pressures and opportunities created by globalization and institutional interventions. These analyses make clear that while women's traditional ecological knowledge remains vital, its effectiveness depends on supportive policy environments and equitable institutional recognition. The gendered consequences of extractive industries are addressed in Protecting Water and Land: Indigenous Women and Mining Impacts, which interrogates the intersection of resource exploitation, gender inequality, and environmental justice. Cultural and spiritual dimensions of conservation are focused on in Eco-Custodians of the Himalayas: A Feminine Perspective on Sacred Natural Sites Conservation. Here, Indigenous women's custodianship of sacred ecologies highlights the interdependence of spiritual values and ecological governance. The final chapter, Beyond Essentialism: Rethinking the Women–Nature Nexus in an Age of Globalization, offers a critical theoretical intervention. It challenges the essentialist notion of women's inherent closeness to nature, instead situating their ecological roles within the shifting terrain of political economy, globalization, and socio-ecological change. Collectively, the contributions to this volume enrich contemporary debates within feminist political ecology, environmental governance, and sustainability studies. They provide empirical evidence and conceptual clarity to demonstrate that Indigenous women's ecological leadership must be recognized not only as an ancillary dimension of conservation but also as a central pillar of environmental resilience, resource governance, and cultural continuity in the Himalaya. This work with Volume I consolidates an interdisciplinary and comparative framework for understanding women's participatory roles in conservation. It speaks with confidence to scholars, policymakers, and practitioners alike: sustainable futures for the Himalayan region cannot be envisioned without Indigenous women's recognition, empowerment, and leadership.

New Delhi, India

Nilakshi Moran
Pardeep Singh
Kshitij Kumar Singh

Acknowledgements

We, the editors, express our sincere gratitude to the Indian Council of Social Science Research (ICSSR), New Delhi, for the financial support provided under the Research Project Grant ICSSR/RPD/MJ/2023-24/ST/98, which enabled us to undertake field research in the Indian Himalayan region. This book is an outcome of our academic engagement, field experiences, and research reflections in this ecologically fragile and culturally rich mountain region. We are deeply thankful to our respective institutions within the University of Delhi for their continuous academic and administrative support throughout this work. We also acknowledge with gratitude the cooperation and support received from individuals, communities, and local stakeholders who participated in the research process and generously shared their knowledge, experiences, and perspectives. Their valuable insights have significantly enriched the quality, depth, and relevance of this volume. We further extend our heartfelt thanks to all the contributing authors for their scholarly contributions to this book. We are equally grateful to the peer reviewers for their constructive comments, critical feedback, and valuable suggestions, which helped improve the quality of the chapters. We also acknowledge all those who directly or indirectly supported the successful completion of this edited volume.

Contents

Participatory Role of Indigenous Women in Environmental Sustainability, Management, and Governance in the Himalayan Region: A Critical Overview

Kshitij Kumar Singh, Nilakshi Moran, and Pardeep Singh

Abstract The Himalayas, as a global biodiversity hotspot, have been a critical source of ecosystems, including water resources and habitats for diverse indigenous communities that possess traditional ecological knowledge (TEK), thereby providing ecological security. The strategic importance of the Himalayan Region lies in its role as a source of water for many countries, its protective function against environmental disasters, and its rich bioresources, which boast phenomenal biodiversity. The grand environmental challenges, such as climate change, global warming and environmentally inconsistent development projects and human activities, disrupt the symbiotic relationship between indigenous people and the environment. The intrinsic insights of indigenous people include traditional ecological knowledge that helps them utilise the territory, land, and resources. TEK is helpful in environmental conservation and resource management. It is a cumulative body of knowledge embodying biological, spiritual, and cultural systems. The local cultures and environments reflect the sustainable practices of traditional wisdom. TEK is an evolving process and is unique to a region, where a pattern of living and resource use is established. All these patterns are time-tested, sustainable practices that do not harm the environment. It is instrumental in ensuring economic and social sustenance. Women, being the custodians of seeds, often attach their conservation efforts to their identity. They play a significant role in stewardship and environmental conservation.

K. K. Singh (✉)
Associate Professor, Campus Law Centre, Faculty of Law, University of Delhi, New Delhi, India
e-mail: singh.genetic@gmail.com

Research Affiliate to the Inter-CeBIL Programme at University of Copenhagen, Copenhagen, Denmark

N. Moran
Department of English, Satyawati College (Evening), University of Delhi, New Delhi, India

P. Singh
Department of Environmental Studies, PGDAV College, University of Delhi, New Delhi, India

N. Moran et al. (eds.), *The Participatory Role of Indigenous Women*,
https://doi.org/10.1007/978-3-032-24059-0_1

Keywords Indigenous women · Traditional ecological knowledge · Himalayan region · Environmental conservation and sustainability · Climate change

1 Introduction

The Himalayas, as a global biodiversity hotspot, have been a critical source of ecosystems, including water resources and habitats for diverse indigenous communities that possess traditional ecological knowledge (TEK), thereby providing ecological security.[1] The strategic importance of the Himalayan Region lies in its role as a source of water for many countries, its protective function against environmental disasters, and its rich bioresources, which boast phenomenal biodiversity. The grand environmental challenges, such as climate change, global warming and environmentally inconsistent development projects and human activities, disrupt the symbiotic relationship between indigenous people and the environment. The intrinsic insights of indigenous people include traditional ecological knowledge that helps them utilise the territory, land, and resources. TEK is helpful in environmental conservation and resource management. It is a cumulative body of knowledge embodying biological, spiritual, and cultural systems. The local cultures and environments reflect the sustainable practices of traditional wisdom. TEK is an evolving process and is unique to a region, where a pattern of living and resource use is established. All these patterns are time-tested, sustainable practices that do not harm the environment. It is instrumental in ensuring economic and social sustenance. Women, being the custodians of seeds, often attach their conservation efforts to their identity. They play a significant role in stewardship and environmental conservation.[2]

2 Role of Indigenous Women in Environmental Conservation and Management and Policymaking

Indigenous women in the Himalayan Region have been making significant contributions to environmental conservation and sustainability by adopting a harmonious approach in tune with nature, thanks to their in-depth understanding of sustainable practices in the region. They have a profound connection with the land and its resources and have the potential to contribute to environmental policy making. In addition to the environmental policy, the social policy also affects their role as these women often encounter socio-cultural exclusions due to the patriarchal colour of society and the male-dominated approach in the implementation of the laws and policies. In addition to adequate recognition of their participatory role, they require

[1] Chauhan (2023).

[2] Arthur and Eggerts (2024).

empowerment through targeted policy interventions. They can take a leadership role in environmental conservation and management.

3 Legal and Policy Gaps in Recognizing the Participatory Role of Indigenous Women

However, the discussion regarding the participatory role of women in environmental conservation and protection is often overlooked in formal governance and policy-making. Although international and national laws and policies are gradually recognising the role of indigenous women in environmental conservation, a significant gap remains between the international recognition of indigenous rights and their national implementation. Many national laws, though, recognise gender related issues; they fall short in effectively addressing the deep-rooted patriarchy, insecure land tenure, and disproportionate impact of climate change and environmental degradation on indigenous women.[3] Nevertheless, women-led initiatives brought about significant transformations, reflecting the power of collective action, social and economic empowerment, and environmental sustainability. Given the understanding of indigenous women of biological, spiritual, physical, and socio-cultural systems, there is a need to integrate their TEK in the legal and policy approaches to environmental conservation and management.[4]

Indigenous women, being the prime stakeholders in the Himalayan region, hold deep insights into the sustainable practices. These practices involve daily life activities such as farming, food production, seed conservation, and broader management of natural resources.[5] It can supplement the legal and policy measures applied for the conservation and management of the environment and biological resources.

4 Vulnerability of Indigenous Women in the Himalayan Region: A Human Rights Perspective

In addition to being a significant contributor to environmental sustainability and conservation, women are disproportionately vulnerable to 'the impacts of climate change and environmental degradation', which raises the issue of livelihood, food

[3] Dipika Adhikari, "Indigenous women's struggle for forest rights in India" *Crawford School of Public Policy* available at https://crawford.anu.edu.au/content-centre/article/news/indigenous-womens-struggle-forest-rights-india (last visited September 24, 2025).

[4] Sampson et al. (2024).

[5] Sally Jabiel, "Guardians of Nature: Indigenous Women and Biodiversity" *Sustainability for All* available at https://www.activesustainability.com/sustainable-development/women-indigenous-biodiversity/ (last visited September 24, 2025).

security, and economic sustenance to a critical ecological concern. Therefore, environmental conservation is not only an ecological issue, but it has become a human rights and social justice issue. Environmental degradation, such as water scarcity, extreme weather conditions, and deforestation, impacts indigenous women disproportionately as they face severe challenges to their livelihood, loss of income, food security, and enhanced workload.[6] They face multilayered discrimination, including indigenous identity, poverty, disability, or castes entrenched in power dynamics, and therefore need a multilayered and intersectional approach rather than a siloed approach—failure to integrate indigenous women's TEK and secure their land rights led to environmental injustice.[7] The marginalisation of indigenous women has a causal link to the environmental conservation outcomes. Therefore, it is an economic, social, and ecological imperative to the full and equitable inclusion of indigenous women.[8]

5 Rights of Indigenous Women: International Perspective

International legal instruments, though, provide a foundational recognition of the rights of indigenous people and indigenous women, yet the implementation of the same varies among nations. The United Nations Declaration on the Rights of Indigenous Peoples (UNDRIP) is the primary international legal instrument that provides a comprehensive, non-binding resolution establishing a universal framework of minimum standards that touches upon almost all aspects of indigenous peoples' lives, including their survival, well-being, and dignity worldwide.[9] Recognising their inherent rights, it affirms the right to self-determination, cultural rights, and collective and individual rights over their land's territories and resources. It also obliges states to establish and implement assistance programs for environmental conservation and to prevent "the storage of hazardous materials on indigenous lands without their "Free, Prior, and Informed Consent (FPIC)"".[10] It is particularly significant given the adverse impact of anthropogenic factors on the environment in the Himalayan region. Similarly, the International Labour Organisation (ILO) Convention No. "169 on the rights of Indigenous Peoples in Independent Countries" advances the rights of indigenous people by recognising the right to consultation and participation with

[6] Wafa Rasheeq, "Indigeneity and Environmental Security: A Gendered Analysis" *Centre for Studies of Plural Studies* available at https://cspsindia.org/indigeneity-and-environmental-security-a-gendered-analysis (last visited September 25, 2025).

[7] Krishna B. Bhattachan, Kamala Thapa et al., "Climate Justice for Indigenous Women: Urgency and Way Forward" available at https://lcipp.unfccc.int/sites/default/files/2022-04/AIPP%20Climate%20justice%20for%20indigenous%20women.pdf (last visited September 25, 2025).

[8] Ibid.

[9] The United Nations (2007).

[10] Ibid.

robust provisions relating to the administration of justice and indigenous customary law.[11]

Additionally, the Convention on Biological Diversity (1992) addresses biodiversity conservation and the rights of indigenous peoples over their traditional knowledge and resources. It recognises that the communities have the "right to be involved in the wider application of their knowledge and to participate in the equitable sharing of benefits arising from its utilisation".[12] To facilitate this, the Parties to the Convention have developed a set of Voluntary Guidelines for cultural, environmental, and social impact assessment regarding indigenous communities known as "the Akwé: Kon Guidelines".[13]

Given the interdependence of environmental, social, and economic concerns outlined in the sustainable development goals (SDGs), the provisions of other international and regional legal instruments relating to women's human rights become relevant. The Convention on the Elimination of All Forms of Discrimination against Women (CEDAW), 1979 stresses the disadvantaged position and peculiar needs of indigenous women, recognising them as particularly vulnerable and disadvantaged group. The Paris Agreement of 2015 also acknowledges human rights and the rights of indigenous peoples in climate action, emphasising the obligations of parties to respect these rights. Similarly, the Escazu Agreement 2018 as a regional agreement requests states to assist indigenous peoples in preparing requests for environmental information.[14]

---

[11] *The United Nations Human Rights Office of the High Commissioner*, "International Standards-Special Rapporteur on the rights of Indigenous Peoples" available at https://www.ohchr.org/en/special-procedures/sr-indigenous-peoples/international-standards (last visited September 26, 2025).

[12] See Art. 8(j) of the Convention on Biological Diversity available at https://www.cbd.int/traditional/default.shtml (last visited September 26, 2025).

[13] See Akwé: Kon Guidelines available at https://www.cbd.int/traditional/guidelines.shtml (last visited September 26, 2025).

[14] See Regional Agreement on Access to Information, Public Participation and Justice in Environmental Matters in Latin America and the Caribbean, available at https://treaties.un.org/pages/viewdetails.aspx?src=treaty&mtdsg_no=xxvii-18&chapter=27&clang=_en (last visited September 26, 2025).

6 Participatory Role of Indigenous Women in Himalayan Region: Legal and Policy Frameworks of India, Nepal and Bhutan

National laws of the nations in the Himalayan region, particularly those of India, Nepal, and Bhutan, either lack provisions like international legal instruments or have a superficial and slow implementation of these provisions.[15] It creates a critical governance challenge, where an apparent inconsistency persists between the theoretical framework of law and its implementation, leaving the vulnerabilities of these women unaddressed. Nations in the Himalayan region vary in their approaches to recognising the participatory role of women in environmental protection and conservation. Although they have made significant progress in empowering indigenous communities through legal and policy initiatives, the degree of protection and recognition varies.

India's The Scheduled Tribes and Other Traditional Forest Dwellers (Recognition of Forest Rights) Act 2006 formally recognises "the rights of indigenous peoples, often referred to as Adivasis and other forest dwellers, including land tenure, access to and use of minor forest products, and the right to manage and conserve forests".[16] The Act provisions for gender inclusivity in Gram Sabhas and Forest Rights Committees and recognises the joint land titles for married couples, as well as the land claims made by women in single-headed households. Women from the Van Gujjar community reportedly gained a sense of legal identity and inclusiveness through this Act.[17] However, the Act does not address the systemic barriers to equality and recognize the decision-making role of women. The Biodiversity Act 2002 also contains mechanisms for community and women's involvement, and the Biodiversity Rules, 2004 mandate a minimum of one-third representation of women in the total membership of Biodiversity Management Committees.[18] It also stresses the participation of marginalised communities, including women, in managing Biodiversity Heritage Sites.[19] Despite ensuring the numerical representation of women, it does

---

[15] Pratima Gurung, Penelope J.S. Stein et al., "Intersectionality, Indigeneity and Disability Climate Justice in Nepal" *The Petrie-Flom Center-Bill of Health* available at https://petrieflom.law.harvard.edu/2024/02/29/intersectionality-indigeneity-and-disability-climate-justice-in-nepal/ (last visited September 26, 2025).

[16] Dipika Adhikari, "Indigenous women's struggle for forest rights in India" *Crawford School of Public Policy* available at https://crawford.anu.edu.au/content-centre/article/news/indigenous-womens-struggle-forest-rights-india (last visited September 24, 2025).

[17] Raghwai and Rukmani (2024).

[18] See ABS Mechanism under the Biological Diversity Act, 2002 available at https://sbbsikkim.nic.in/pdf/abs/abs_guidance_manual.pdf (last visited September 26, 2025).

[19] See generally Kannaiyan (2007).

not provide specific provisions relating to indigenous women; instead, it only recognises the provisions for local people and communities or benefit claimers.[20] Similarly, the prime environmental legislation of India, the Environment (Protection) Act, 1986 also lack specific provisions relating to indigenous women, which leaves them unprotected against the disproportionate impacts of environmental degradation.[21]

Nepal has adopted a community-based forest management model, which has been highly successful in enabling indigenous peoples and local communities to actively leverage their traditional knowledge in protecting and managing forests.[22] The Community Forestry Guidelines 1995 mandate significant female representation. It requires that at least 50% of the members of the community forestry committees be women, and that women hold two key leadership roles (chairperson or secretary).[23] On similar lines, the Federation of Community Forest Users (FECOFUN) has formally prescribed a 50% women's leadership quota in its executive body since 2010.[24] Nepal also ensures the numerical representation of women in leadership roles. However, systemic barriers, lack of awareness, incompatible meeting timings and locations, and use of technical language make it superficial and not substantial.[25] The National Parks and Wildlife Conservation (NPWC) Act 1973, which governs protected areas that often overlap with ancestral and indigenous territories, lacks the provisions for Free, Prior, and Informed Consent (FPIC) and therefore falls short of its commitment under the Nepal ratified United Nations Declaration on the Rights of Indigenous Peoples (UNDRP).[26] Recent amendments in the NPWC Act in 2024 further deepen the vulnerability of indigenous people, especially women, by opening the conservation areas to private infrastructure projects, including hydropower plants and cable cars.[27] Despite Nepal's policies and plans to address climate change, which recognise the concerns of indigenous people and secure their access to participation in decision-making, they fall short in addressing the disproportionate impact of environmental degradation.

---

[20] See ABS Mechanism under the Biological Diversity Act, 2002 available at https://sbbsikkim.nic.in/pdf/abs/abs_guidance_manual.pdf (last visited September 26, 2025).

[21] See generally the Environment (Protection) Act, 1986 available at https://www.indiacode.nic.in/bitstream/123456789/4316/1/ep_act_1986.pdf (last visited September 26, 2025).

[22] *World Bank Group*, "Supporting the Indigenous Peoples and Local Communities in Nepal's Forest Sector" available at https://www.worldbank.org/en/news/feature/2024/08/08/supporting-the-indigenous-peoples-and-local-communities-in-nepal-s-forest-sector (last visited September 26, 2025).

[23] Niyogi (2025).

[24] "These women ran for office to protect Nepal's forests" *Rights and Resources Initiatives*, available at https://rightsandresources.org/women-fecofun/ (last visited September 28, 2025).

[25] Murer (2022).

[26] Pratima Gurung, Penelope J. S. Stein et al., "Intersectionality, Indigeneity and Disability Climate Justice in Nepal" *The Petrie-Flom Center-Bill of Health* available at https://petrieflom.law.harvard.edu/2024/02/29/intersectionality-indigeneity-and-disability-climate-justice-in-nepal/ (last visited September 26, 2025).

[27] Joshi (2024).

Bhutan approaches environmental conservation guided by its philosophy, rooted in the concept of Gross National Happiness, with a constitutional mandate to maintain a forest cover of at least 60%. The Forest and Nature Conservation Act (FNCA), 1995 recognises community rights to use and manage forests.[28] It uses gender-neutral language "to promote equitable access to forest resources and benefits".[29] The Biodiversity Action Plan 1997 and the National Biodiversity Policy 2014 recognise the crucial role of women in biodiversity conservation, knowledge sharing, food security, and forest governance. Gender mainstreaming efforts have been evident in various policies.[30] Nevertheless, despite these approaches and efforts of gender mainstreaming, the role of indigenous women in decision-making remains ambiguous. The gender-neutral approach does not adequately address the unique needs of indigenous women.[31] The Environmental Assessment Act, 2000 and the National Environment Protection Act, 2007 of Bhutan promote public participation. However, it does not provide special provisions for indigenous women and generally keeps them out of the loop of decision-making.[32] Although the Gross National Happiness has become an integral part of Bhutan's policy initiatives, backed by a constitutional mandate, it has still to achieve gender equality in environmental governance.[33]

7 Need for a Bottom-Up Approach to the Governance and Community Initiatives

Instead of the top-down model, a bottom-up approach with a more inclusive and participatory role for women makes environmental conservation efforts more impactful. It has been well reflected in the women-led community initiatives for environmental protection and the recognition of indigenous peoples' rights. Numerous women-driven community initiatives set best practices for environmental conservation. "The Chipko Movement, initiated in 1974 by Adivasi woman Gaura Devi in Uttarakhand, India, involved women who literally embraced trees to prevent deforestation."[34] Adivasi women's fight for the recognition of their forest rights in Rajasthan and Odisha's women-led forest protection committees set good examples of the participatory role of women in their fight for their rights. The Bishnoi community has also made significant contributions to environmental conservation

[28] Wangmo and Magry (2024).

[29] Ibid.

[30] Ibid.

[31] Ibid.

[32] See generally Environment Assessment Act, 2000 available at https://oag.gov.bt/wp-content/uploads/2010/05/Environmental-Assessment-Act-of-Bhutan-2000English-version.pdf; The National Environment Act, 2007 available at https://policy.asiapacificenergy.org/node/79 (last visited September 28, 2025).

[33] Zangmo et al. (2024).

[34] Sashidhar (2024).

and protection through a community-driven movement, including the participation of women.[35] In Nepal, Community Forest Users Groups and Chepang and Tharu communities set best practices for community-based forest management, adapting frugal use of the resources and effective fire control.[36] Similarly, Bhutan's eco-tourism and Wash Initiatives recognise the same, and projects like "Water for Women" exemplify the inclusive and participatory role of indigenous women, where trained women work as caretakers and health assistants.[37]

8 Indigenous Women and Sustainable Development Goals

The participatory role of indigenous women in the Himalayan region can be significant in achieving the Sustainable Development Goals (SDGs).[38] Given the interdependencies among the SDGs, empowering women to address their socio-economic needs and to be recognised culturally promotes sustainable practices in the Himalayan region. Indigenous women can foster agricultural productivity in the Himalayan region, focusing on SDG 2—Zero Hunger. Modern agricultural practices, backed by new technologies, can be applied in harmony with biodiversity by incorporating the long-practised illumination of indigenous farmers, which involves a significant portion of work by women. The policy, informed by the traditional wisdom of indigenous women, can help build climate-change resilience, for example, by promoting mixed cropping. SDG 5: Gender equality in the Himalayan region also needs a push, as the persistent conventional roles for women restrict access to resources and exclude them from decision-making. It limits their participation substantially and could be a barrier to giving effect to other SDGs. Therefore, recognising women's participatory role and extending it to decision-making and governance regarding the common resources is imperative.

SDG 6: Clean Water and Sanitation is another important goal, given that the Himalayas, a source of water for many other states, offer a strategic advantage; they require water management and forest conservation. Ensuring the participation of indigenous women provides sustainable insights for water management and supports forest conservation against the impacts of climate change, such as glacial melt. SDG13: Climate Action is relevant to recognising the special needs of indigenous women in the Himalayan region, who face disproportionate impacts of climate change. However, given their close association with nature and adaptability to change, their role as conservationists, biodiversity protectors, and forest restoration agents

[35] Ibid.

[36] *World Bank Group*, "Supporting the Indigenous Peoples and Local Communities in Nepal's Forest Sector" available at https://www.worldbank.org/en/news/feature/2024/08/08/supporting-the-indigenous-peoples-and-local-communities-in-nepal-s-forest-sector (last visited September 26, 2025).

[37] See "Water for Women-Bhutan- Strengthening climate resilience, coping mechanisms, and adaptive capacities for inclusive WASH services in rural Bhutan" available at https://www.waterforwomenfund.org/en/project/water-for-women%2D%2D-Bhutan.aspx (last visited September 28, 2025).

[38] Banerjee (2011).

is vital. They can help promote SDG15: Life on the Land by fostering coexistence with other living creatures and by guiding tourists and other human activities in tune with nature in a sustainable manner.[39] Recognising women's participatory role in achieving the SDGs and empowering them socially, culturally, and politically may promote holistic environmental initiatives.

9 Conclusion

Empowering indigenous women through policy interventions can reduce their dependency on nature and enable them to contribute meaningfully. There is a need to identify and address the specific issues affecting indigenous women in the Himalayan region, and to provide them with decision-making and leadership roles in environmental conservation, protection and management. Indigenous community-based initiatives help to redress the implementation challenges and develop best practices. The traditional wisdom and time-tested experience of indigenous women in the Himalayan region supplement the policymaking destined to provide sustainable solutions to the contemporary environmental challenges.

Acknowledgements Author PS expresses his sincere gratitude to the Indian Council of Social Science Research (ICSSR), New Delhi, for the financial support provided under the Research Project Grant ICSSR/RPD/MJ/2023-24/ST/98, which enabled him to undertake field research in the Indian Himalayan region. I am deeply thankful to PGDAV College, University of Delhi, for its continuous academic and administrative support throughout this study. I also extend my appreciation to all individuals and community members who participated in the research process and shared their invaluable insights. Their cooperation and contributions have enriched the quality and depth of this work.

References

Arthur, A., & Eggerts, E. (2024, October 29). How gender equality drives effective forest conservation. *UNDP Climate Promise*. Available at https://climatepromise.undp.org/news-and-stories/how-gender-equality-drives-effective-forest-conservation. Last visited September 24, 2025.

Banerjee, A. (2011). Role of mountain women towards sustainable development. *Salesian Journal of Humanities & Social Sciences, 2*(2), 45. Available at https://publications.salesiancollege.net/wp-content/uploads/2023/06/Role-of-Mountain-Women-towards-Sustainable-Development.pdf. Last visited October 12, 2025.

Chauhan, H. K. (2023). The Himalayas: A climate change laboratory. *Environmental Development,* 45:100814. Available at https://www.sciencedirect.com/getaccess/pii/S2211464523000143/purchase. Last visited September 24, 2025.

[39] "Women Empowerment in the Himalayas" *Himalayan by Heco insight* available at https://www.himalayanecotourism.com/women-empowerment-himalaya/#:~:text=Park%20–%20Tirthan%20Valley-,Women%20in%20conservation,intern)%20performing%20a%20creative%20activity (last visited October 12, 2025).

Joshi, A. R. (2024, September 27). Eminent Nepali conservation NGOs demand repeal of controversial law. *Mongabay*. Available at https://news.mongabay.com/2024/09/eminent-nepali-conservation-ngos-demand-repeal-of-controversial-law/. Last visited September 28, 2025.

Kannaiyan, S. (2007). An overview of Biological Diversity Act-2000. In *Special Lecture in the National Horticulture Research Conference, 2007*. Available at http://nbaindia.org/uploaded/docs/biological-diversityact-ii.pdf. Last visited September 26, 2025.

Murer, C. (2022). Affirmative policy in Nepal's community forestry: Does it make a difference in terms of social sustainability? *Sustainability, 14*(9), 5598. Available at https://www.mdpi.com/2071-1050/14/9/5598. Last visited September 28, 2025.

Niyogi, D. G. (2025, June 25). Nepal's forest user groups are fighting climate change from the ground up. *Down to Earth*. Available at https://www.downtoearth.org.in/forests/nepals-forest-user-groups-are-fighting-climate-change-from-the-ground-up. Last visited September 28, 2025.

Raghwai & Rukmani. (2024). Empowering through recognition: An analysis of women's participation in Forest Rights Act 2006 in Andhra Pradesh, Uttarakhand and Rajasthan. *Educational Administration, 30*(1), 6743–6747. Available at https://kuey.net/index.php/kuey/article/view/10003/7642. Last visited September 26, 2024.

Sampson, L. E., Appiah-Opku, S., et al. (2024). Indigenous ecological knowledge and wildlife conservation: A systematic review of the relevant links. *Ecology & Conservation Science: Open Access, 4*, 1. Available at https://juniperpublishers.com/ecoa/pdf/ECOA.MS.ID.555632.pdf. Last visited September 24, 2025.

Sashidhar, K. (2024, December 6). Indigenous advocacy: A historical overview of environmental movements in India. *Peertechz-Annals of Environmental Science and Toxicology*. Available at https://www.agriscigroup.us/articles/AEST-8-181.pdf. Last visited September 28, 2025.

The United Nations. (2007). *The United Nations declaration on the rights of indigenous people*. Available at https://www.un.org/development/desa/indigenouspeoples/wp-content/uploads/sites/19/2018/11/UNDRIP_E_web.pdf. Last visited September 25, 2025.

Wangmo, K., & Magry, M. A. (2024). Evaluating forest conservation policies of Bhutan: Sensitiveness and responsiveness to gender considerations. *Discover Conservation, 1*, 10. Available at https://d-nb.info/1351522876/34. Last visited September 28, 2025.

Zangmo, N., Hiroshima, T., et al. (2024). Participatory forest management and gender inclusiveness within the community forest management groups of Bhutan. *Journal of Geoscience and Environment Protection, 12*(4), 12–30. Available at https://www.scirp.org/journal/paperinformation?paperid=132554. Last visited September 28, 2025.

The Unseen Guardian: Role of the Native Women in Conservation of the Himalayas Ecosystem

Sabina Nazakat, Shujaul Mulk Khan, Abdullah Abdullah, Zeeshan Ahmad, and Shakil Ahmad Zeb

Abstract Women play important role in the environmental conservation addressing the mountain ecosystem and addressing resource-based problems. The significant convergence of the objective policies between the advancement of women and theme of conservation of environmentally sustainable development. It is an attempt to portray a realistic picture of ecofeminist prospective, which drawn upon objectification and oppression of both women and nature in Himalayan communities. A close scrutiny addressing concerns, Practice, policy implications, and ecological movements initiated by Himalayan women. Biophysical and cultural characteristics of mountains that merit special consideration in the matter of conservation and preservation. These include three-dimensional nature involving altitudinal belts and slopes of varying ecosystems, their different aspects, frequent characteristics of inaccessibility, remoteness, cultural diversity, and spirituality in a short distance. Emphasizing the importance of indigenous knowledge systems and sacred landscapes in the management of natural resources and their ethnocultural prospective in conservation efforts. Arguing the challenges and opportunities of traditional ecological knowledge, an inclusive approach and contemporary conservational strategies incorporate the indigenous communities. Highlight studies on indigenous women's participation in Pakistan, Afghanistan, Tibet, and China with the collaborative efforts of its people, the government, and non-governmental organizations for restoring and protecting forest resources. The International Union for Conservation of Nature (IUCN) research further recognizes women's contribution in the legal framework for environmental conservation. This chapter ultimately recognizes the multiple perspectives in achieving equitable and effective conservation outcomes, proposing a more holistic understanding of nature and women's role, cultural values, scientific approaches, land rights and community-driven efforts for long-term resilience of the Himalayas' communities.

S. Nazakat · S. M. Khan · A. Abdullah · Z. Ahmad · S. A. Zeb
Department of Plant Sciences, Quaid-i-Azam University Islamabad, Islamabad, Pakistan

S. M. Khan (✉)
Pakistan Academy of Sciences, Islamabad, Pakistan
e-mail: smkhan@qau.edu.pk

N. Moran et al. (eds.), *The Participatory Role of Indigenous Women*,
https://doi.org/10.1007/978-3-032-24059-0_2

Keywords Himalaya communities · Indigenous women · Environmental conservation · Sustainable development · Ecofeminist prospective · International union

1 Introduction: Reorienting Conservation Through the Lens of Indigenous Women in the Himalayas

The complex relationship between humans and nature underpins the concept of sustainable development (Khan et al., 2013) and framework of the integrating ecological resilience. This framework often overlooks the gendered dimensions of environmental knowledge, particularly the indispensable roles played by Indigenous women in sustaining ecosystems (Hariram et al., 2023a, 2023b; Pradhan et al., 2021a, 2021b). The Himalayan region of Pakisdtan, specifically the highland communities of Northern areas, women are not passive recipients of development; they are the frontline managers of biodiversity, stewards of the traditional ecological knowledge (TEK) (Teixidor-Toneu et al., 2025), and key agents of adaptation in the face of climate disruption (Fatima et al., 2020a, 2020b).

Although the international policy has acknowledged women's contributions in principle since the United Nations Decade for Women (1975–1985), practical integration into environmental governance remains limited. Globally it is declared in the Agenda 21 and Chapter 24 of the United Nations Conference on Environment and Development (UNCED) emphasized the ivolment of women's in the sustainable development. But unfortunately these policies mostly lacked the conservation efforts of women in the mountain ecosystem (Jones, 2014a, 2014b; UNEP, 2000). In developing and underdeveloped countries such as Pakistan and Afghanistan the role of women in conservation in the Himalaya region are less explored or not explored (Abdulsalam et al., 2024a, 2024b; Hopkins & McKeown, 2002).

The women of this region are involved local communities, engage daily in activities that directly affect conservation outcomes, they manage household resources, collection of medicinal plants, protect seed biodiversity, and maintain cultural rituals tied to seasonal and spiritual rhythms (Bhasin, 2011a, 2011b; Lama & Buchy, 2021). However, these contributions remain underpin, often classified as informal or non-expert labor. Social hierarchies, patriarchal norms, and land tenure systems further marginalize women's authority in environmental matters, despite their practical ecological expertise (Khatwani, 2016a, 2016b).

The urgency of re-centering native women in conservation discourse is amplified by mounting environmental threats in the Himalayan belt glacial retreat, deforestation, ecosystem fragmentation, and climate-induced disasters. These threats disproportionately impact women who rely on natural resources for food, fuel, fodder, and cultural survival (Sherpa, 2023a, 2023b). Still, research shows that where women are actively involved in environmental governance, conservation outcomes are more sustainable and equitable (Agarwal, 2021; Johnson-Pynn & Johnson, 2005).

In the current chapter, we explore the unseen role of women in the Hamayalan region in the northern Pakistan. We assess how their traditional knowledge and grassroots initiatives align with global environmental strategies and examine the socio-political barriers that hinder their fuller participation in conservation planning. Through a gendered, decolonial lens, we argue that any meaningful path toward sustainable environmental development in the Himalayas must begin by centering the voices, knowledge, and rights of native people specifically women.

2 Indigenous Women's Experiences and Contributions in the Himalaya

Himalaya is the third pole of the world which host a lot a tribial commun ities and ethnic groups. They have their own culture and tradition, and hence its traditional ecological were also different in the these communities. In Himalayan communities diverse populations exist, tribal women plays a central role in cultural and social development and play dynamic roles in all challenging situations (Chettri et al., 2012). The traditional culinary knowledge of women are indispensable in all societies included mountainous regions. It is considered as gatekeepers of culinary traditions as food is an essential constituent and its intergenerational meticulous knowledge transfer from mother to daughter. In this way, women not only select and use food but also conserve and maintain genetic and biodiversity of plants. Its paramount, though overlook in conservation that erosion of culinary traditions lost of incentives for plant diversity. Knowledge of culinary related to scientific knowledge, thus women's indigenous techniques are highly valued for using, fermenting, reduction in spoilage and for preserving plants. The majority of edibles are derived from foraging which is predominantly carried by females. Wild plants are mostly gathered by women who furnish medicines, soil enrichment, energy, countless material for building and crafting, fuel and food. Women remove unnecessary weed while harvesting encouraging habitat for new plants, they also tended the roots with digging sticks fostering the development of uncurved and lengthy rhizomes. Rhizomes left by women during harvesting would continue for future yield. In northern areas of Pakistan women are principal custodians of traditional ecological knowledge (Sher et al., 2015). Historically the women perform diverse responsibilities including agricultural work, domestic management, craft production, financial provision, procurement of resources and spiritual instruction. Mountainous women also strengthen and build social ties because their handicrafts served as social resource for community well-being and income for the generation. In villages traditional machine made by wood called spinning wheel used for making cotton operated by women for income generation. In textile industry cotton generation through spinning would foster a platform for social community building, counselling and knowledge transfer. Women also made products like blankets, cloth, jute, mats etc. were organic, more resistant, biodegradable and environment friendly. Most raw materials of plants used as fodder (Zeb et al.,

2022) that increase milk production of cattle, women from various regions use traditional techniques to recycle by-products of various plants that increase yield of meat and milk in cattle. In agriculture women's expertise and economic activities made a positive contribution to rural sustainability. In western Himalaya the aged women 60 or above possess deeper understanding of plants than men, highlighting their pivotal part in knowledge dissemination. In 2020 survey 37% of women adopt manifold ways of earning in agriculture labore force so their input is considerable, they represent half of small-scale livestock keepers (FAO, 2022). Despite this woman experience gender limitation that reduce their gains and restrict their involvement in conservation decision making (James et al., 2021). Across Asia cultural, legal, political and socio-economic landscape differ vastly, women hold crucial positions as economic agents in communities. Numerous rural inhabitants across Asian mountains participate in livestock raising in which women hold distinctive positions. Nonetheless their command over land and other natural resources are uneven compared to men. Community led conservation typically leans on patriarchal community hierarchies in directing key initiatives (Rytkönen & Hotakainen, 2020). In rural areas women are unrepresented in developmental and conservational endeavors unless these programs are specifically designed for them (Márquez-García et al., 2024). Consequently, conservation activities may unintentionally worsen gender related and social imbalances. Research also evidence that including women in conservational and environmental work can produce enhanced outcomes so, women should have equivalent voice in conservation decisions ensuring their specific concerns and hazards are tackled and allowing them to reap reward from current conservation program.

3 Indigenous Knowledge of Nature Conservation

In human development, indigenous knowledge and biodiversity are essential and complementary phenomena. The accelerating trend of loss of fauna and flora has a negitive impact on biodiversity so the global commitment towards the importance of conservation is no longer a matter of question. Numerous scientists, research centers, and international development agencies also adding their collective concern to resolve the issue of species loss. Indigenous knowledge include domesticated plants, animals, water and soil on which humans lives depends. In ninetieth century social science and colonialism in criticizing and ignoring "indigenous knowledge" has been documented extensively (Chambers & Gillespie, 2000; Seth, 2009; Whitt, 2009).

In colonial era the negative attitude about Indigenous knowledge has now change after understanding the significance of indigenous knowledge because they never understood it before. National and international agencies have now recognized the merits of participatory approaches for sustainable development of nature. Indigenous knowledge and sustainable development have strong relationship and ethnobiological indigenous knowledge provide basis for decision making for identification and solution for problems related to environmental sustainability. At policy level in the developmental process the role of indigenous knowledge is now began to review by developmental agencies. In 1990 the cost effectiveness by the addition of indigenous knowledge has been demonstrated by Titilola in 1989. The international agency of Canada has also completed two reports on indigenous knowledge. The world bank also held seminar on this topic for the development of agriculture (Clemens & Kremer, 2016).

The developmental agencies show greater weightage on the significant and stronger role of indigenous knowledge. The documentation and compilation of indigenous knowledge should be highly prioritized otherwise it is being lost at an exceptional rate and its preservation in the form of data must take place immediately. The global network of resource centers of indigenous knowledge can be most effective in storing and record indigenous knowledge systematically so it can facilitate the developmental efforts nationally and internationally. Now there are 10 establishment centers out of them three function globally, five work nationally and two work at the regional level for biodiversity conservation, and another 17 more centers are becoming established in future. The National resource centers for indigenous knowledge are useful in storing, recording, and screening of indigenous knowledge for economic use as well as for the distribution of this knowledge to other centers (Adeyemo & Adebayo, 2018; Christen, 2012; Mantyka-Pringle et al., 2017). In other parts of the globe other innovative techniques are used that work equally well in similar ecological zones (Coccia, 2015). Indigenous knowledge is also controlled and shared by national centers that help to facilitate this effort, and they serve as a vehicle to introduce this knowledge and its components from schools to university level, including training institutes (Barnhardt, 2014). The growing focus on indigenous knowledge and biodiversity indicates the interest of the world in biological and cultural biodiversity.

4 Mountain Biodiversity and Conservation

The Himalaya vaunts a wealth of rich biodiversity in terms of flora and fauna (Shaheen et al., 2011; Khan et al., 2014). The indigenous knowledge is crucial for the resilience of these communities under socio-cultural and environmental change. In the western Himalaya, the forest has been recognized as global top priority for conservation of biodiversity (Zeb et al., 2025). Over centuries, the indigenous people of the mountaholdolds vast knowledge about natural resources and ecology based on their interaction with diverse ecosystems and their proximity to nature for livelihood (Din et al., 2020). Traditionally, they have been engaged in biodiversity conservation for health systems and food security in many ways such as budding, grafting, domesticating various species from the environment, storing propagules (seeds), and exchanging seeds among and outside the village (Abdullah & Khan, 2023). Their knowledge is transmitted through a long series of observations from generation to generation. They are aware of the significance of biodiversity conservation in generating resources for their livelihood as well as for ecological services (Schroth & Mcneely 2011). Their knowledge based on series of practices of "Rule of thumb" determined over long period of time through "Trial and error" method (Norström, 2011).

The implementation of their knowledge is indefinite and sometimes involves an intimate link with their belief system, such knowledge is diverse from the Western world of modern sciences. It is imperative to recognize fully the value of the knowledge-belief-practice complex of indigenous people relating to biodiversity conservation if biodiversity and ecosystems are sustainably managed (Kosoe et al., 2020).

5 Sustainable Biodiversity and the Role of Indigenous Women in Northern Pakistan

The Northern Areas of Pakistan, encompassing parts of the Himalayan and Karakoram Mountain ranges, play a pivotal role in maintaining national biodiversity and ecological stability. Although geographically limited in area, these regions serve as vital catchments for the Indus River, which supports Pakistan's hydroelectric infrastructure and irrigated agriculture (Khan et al., 2012a, 2012b). The region is also home to rich forests, endemic flora and fauna, mineral reserves, and cultural heritage sites. Women from Indigenous communities in Gilgit-Baltistan, Chitral, and Kohistan have long served as the custodians of traditional ecological knowledge (TEK). Their daily responsibilities such as seed preservation, livestock herding, herbal medicine preparation, and forest product collection are not just survival strategies but also conservation practices (Nazakat et al., 2021).

Despite their significant contributions, the Northern Areas face growing biodiversity pressures due to deforestation, tourism, road construction like the Karakoram Highway, population expansion, and unsustainable harvesting of wild resources (Abbas et al., 2021a, 2021b; Ullah et al., 2022a, 2022b). These pressures are compounded by structural inequalities that exclude women from formal conservation planning, despite their ecological expertise. For instance, many of the native plant species used by women for food, medicine, and ceremonies such as *Rheum emodi* and *Berberis lyceum* are increasingly threatened due to overharvesting and land degradation (Okorondu et al., 2022a, 2022b; Chakravarty et al., 2012).

In response, both international and local actors have initiated strategies for sustainable development in the region. The Northern Areas Strategy for Sustainable Development (NASSD), launched in 1999 with support from the Swiss Agency for Development and Cooperation and the World Conservation Union, proposed an integrated framework addressing environmental, economic, and social goals (Hemmati, 2012a, 2012b). However, gendered perspectivesparticularly the role of Indigenous womenremain underrepresented in such frameworks.

National conservation policies, including the designation of protected areas like Ayubia National Park and Khunjerab National Park, aim to safeguard biodiversity under the Wildlife Act of 1975. Yet, their effectiveness is undermined by limited local engagement and failure to incorporate women's ecological knowledge in governance (Pimbert & Pretty, 2013). Sustainable biodiversity conservation must move beyond protectionist models and embrace community-based, gender-inclusive strategies (Chen et al., 2023a, 2023b).

The erosion of biodiversity in these regions also carries spiritual and cultural consequences. Sacred landscapes, burial sites, and medicinal plant trails are intimately tied to Indigenous women's identities. The disappearance of these spaces not only depletes biodiversity but also disrupts intergenerational knowledge transmission (Sherpa, 2023a, 2023b). According to the IUCN Red List, 45 out of 2000 threatened species occur in Pakistan, and many are found in the Northern Areas (Hilton-Taylor, 2000a, 2000b).

To address these challenges, collaborative approaches are essential. NGOs such as the Aga Khan Rural Support Program (AKRSP), the Himalayan Wildlife Foundation (HWF), and IUCN Pakistan have initiated participatory conservation models that show promise. However, these must explicitly empower Indigenous women as planners, not just beneficiaries. Integrated, landscape-level strategies involving local women, forestry departments, agricultural bodies, and environmental NGOs are play significant role in the conserving the biodiversity loss (Boiral & Heras-Saizarbitoria, 2017; Niesenbaum, 2019a, 2019b) (Fig. 1).

Empowering women through land tenure recognition, inclusion in biodiversity councils, and training in ecological monitoring would significantly strengthen conservation outcomes. The future of Pakistan's Himalayan biodiversity depends not only on laws and reserves, but on the everyday practices and wisdom of Indigenous women whose lives are inseparable from these landscapes.

Fig. 1 **a** and **b** Scenic view of the Northern Himalaya

6 Development of Conservation in China

China spread over 9.6 million km^2 of land form of hills, plateaus and mountains (MEE, 2018). China is among 12 countries known for its rich biodiversity but under influence of climatic change crisis. In past centuries rise in temperature from 0.9 to 1.5 ° C above average in China have significant effects on rain, precipitation, snowfall, sunshine etc. as well as on climate that badly impact on its environmental, social, and economic condition (Chen et al., 2018).

China has devoted its full attention in addressing problem of changes in climate by making strategy to promote low-carbon and green development for its economic and social development. In 2007 various policies were documented such as the National Climate Change Program (NCCP) and the National Strategy for Climate Change Adaptation (NCCA) it was given in 2013, and in 2015 policies has adopted for ecosystem restoration and biodiversity conservation. Meanwhile, their government launched "National biodiversity strategy and action plan from (2011–2030) in which at least 28 plans prepared for biodiversity conservation and 10 plans for agriculture and food security. It showed institutional frame work administrative centers for effective implementation of their policies for addressing issue of climatic change (Chmutina et al., 2012).

China's biodiversity conservation approach has shifted from state controlled to community driven models acknowledging indigenous vital contributions, notably indigenous women whose environmental understanding enhance conservation successes (Ma et al., 2022). Previously indigenous women were excluded from official management of conservation but now become pivotal troupes in biodiversity restoration, seed autonomy and climatic resilience. In southwestern province of China their contributions shine brightest, such as in Qinghai, Sichuan, Yunan and Guangxi. For example, in "Jiabli" indigenous women leverage customary regulations and revered ecological methos to safeguard forests, manage communal resource use and guide biodiversity program (Xu et al., 2022).

Additionally, in Guangxi, women's seed cooperatives have revived conventional duck ice farming which has better soil fertility, higher harvests and strengthens biodiversity.

In Zhang Xiuyun a women bred 30 varieties of maize crop designed to tolerate rainfall shifts and severe climatic conditions, her novel approach resolve food security and climatic shifts. These instances demonstrate that conservation in China is more than simple scientific management, it transformed into community-led, gender aware process. Programs like Indigenous and community conserved area (ICCAs) and Stewardship pilot conservation program are expanding community-led environmental management and involvement of indigenous actors, specifically women. Methods like seed banking, growing medicinal herbs, and safeguarding sacred sites are crucial for the conservation of biodiversity, in which women were engaged their indigenous knowledge is ecological capital. They improve forest protection, sustainable land use and specie monitoring (Song & Jiggins, 2023).

To truly grasp the development of conservation in China, one can must recognize the essential part played by indigenous women, their inclusion in conservation improves biodiversity, advance gender equality, cultural preservation and local resilience. Indigenous women not only more recipients of the conservation strategy, but they are also its designers, pioneers and visionaries. Emphasizing their role redefines and moves conservation into a more inclusive, equitable and sustainable future (Gao & Xu, 2024).

7 Case Study Tibetan Rangelands Sustainable Conservation Development

Geographically the Tibetan highland is a unique zone that cover 2.57–106 km^2. It is considered as the third pole of the earth (Zhang et al. 2002). The rangelands of Tibet serve central role in keeping the livelihood and natural atmosphere, which together support overall social legacy of the Tibetan land (Miller 1999; Richard 2002). The grassland of Tibetan land serves biological functions, include water, carbon and soil maintenance that ensure stability of ice-covered soil as well as biodiversity (Xu et al. 2004). Tibetan highlands are the source of major Asian rivers, and it has an influence on native communities. According to a case study this plateau is noteworthy for sustainable developmental management due to its ecological functions and supporting livelihood of the population, but critical challenge is balancing the conflict for foliage resources. To resolve this conflict, considerable investment in resources and labor takes place over years but have failed. Numerous strategies based on research are presented for livestock production, its progress and for sustainable development of biodiversity. The Tibetan pastoral nomads have intricated indigenous knowledge that is strategic for environmental stability (Long, 2003). Any modification in rangelands livestock will influence the economic and social structure of Tibetan land (Shen et al., 2009) thus, its ecological maintenance and resilient response to global transformations are the key tasks for sustainable development of biodiversity (Swart et al., 2003). The energy flow and biogeochemical cycles occurring during harsh environmental conditions in Tibetan plateau have negative impact on ecosystem (Shang et al., 2006). Rangelands and livestock are core elements for sustainable development of Tibetan plateau (Brown et al., 2009). The last three decades have been spending for remarkable technical and scientific development of Tibetan plateau, the major focus is on livestock and grass lands (Du, 2006) and more additional investment has been encouraging on these sectors by government. In Rangelands advanced genetic technologies have been introduced for best breeds of sheep and yak and supply more feed during harsh cold seasons and farmers also store more fodder (Ji et al., 2010). Most issues arise in Tibet rangelands due to degradation of grass lands that has become worse that increase the risk of natural disasters and problems like soil erosion, land sliding and flooding that altimetry cause loss of habitat and biodiversity (Zhou et al., 2011). Many other factors impact production of livestock include lack of funding, marketing skills, policies and management of livestock etc. Thus, successful sustainable management to overcome these crises there should be multidisciplinary strategy that incorporate various methods and approaches for rising livestock (Kremsa, 2021). In Tibetan regions increasing demand of food due to population growth while less percentage of populates involve in agriculture that pose more pressure socioeconomically (Shen et al., 2009) and it also limit grass land resources. Some researchers also reveal that other than grazing pressure "Global Warming" is the other major cause for grassland degradation (Wang et al., 2012). Imbalance in livestock number and natural ecological space also spread degradation according to latest consensus so, it can be only resolved by using non pastoral areas. It has been assessed that 42% extra grazing by livestock than its grass land capacity was recorded in Northern areas of Tibet recorded during winter season (Wang et al., 2004). Through satellite remote sensing it has been recorded that 76% of change in grass land in Tibetan region lies within limits of natural variation, only 23% of grass land degraded in the years between 1981 and 2004. Most of grass land have high vegetation cover of poisonous weeds. Thus, there is an immediate need to resolve shortage of forage by lessening the livestock number for sustainable management of livestock in grass lands of Tibet. Harris (2010), concluded that "indigenous pastoralism" that was absent from literature review from past 5–6 years contributing to grass land degradation because the herdsmen consider herds size as a symbol of his wealth. Over stocking of livestock limits land that pose severe damage to grass lands.

It is highly recommended that latest policies and planning must be prepared on the basis of social and economic background status of Tibetan rangelands. In strategies most attention is needed to modern technologies and genetic resources for successful sustainable development of biodiversity. Additionally, further novel approaches should practice that retain nomadic pastoral herding while expanding spatial and temporal scope of rotational livestock to decrease degradation. Ensuring sustainable development on the Tibetan rangelands by building safe ecological shelter zones and resolve conflict of grassland and livestock.

8 Indigenous Biodiversity and Conservation in Afghanistan

Afghanistan's unique geographic location at the crossroads of Central Asia, South Asia, and the Middle East has made it a cradle of diverse ecosystems and endemic biodiversity. The country's topography ranging from alpine tundra in the Hindu Kush to dry steppes and river valleys supports a remarkable array of flora and fauna, including globally threatened species such as the snow leopard (*Panthera uncia*), Marco Polo sheep (*Ovis ammon polii*), and the endangered Asiatic black bear (*Ursus thibetanus*) (Kremen & Merenlender, 2018). However, decades of political instability, conflict, and environmental mismanagement have severely endangered Afghanistan's ecological integrity.

Among the most overlooked contributors to biodiversity conservation in Afghanistan are Indigenous women, especially those from nomadic and highland rural communities. These women have traditionally played an essential role in managing local ecosystems through sustainable herding, wild plant harvesting, and preservation of native seed varieties (Ullah et al., 2024). Their ecological knowledge often transmitted orally and embedded in spiritual or seasonal customs has sustained diverse agro-pastoral systems even in extreme environments (FAO, 2011). For instance, in Wakhan and Badakhshan, pastoralist women manage rotational grazing to prevent overgrazing of highland meadows, demonstrating a deep understanding of ecological thresholds and species resilience (Kazemi & Sadat, 2020).

Yet these women are largely excluded from national conservation discourse due to gendered power structures, low female literacy, and the ongoing conflict that limits women's mobility and agency. Biodiversity conservation programs in Afghanistan, such as those initiated under the National Environmental Protection Agency (NEPA), often lack a gender-inclusive lens. When women are not formally consulted or empowered, conservation strategies fail to capture critical ground-level insights into changing resource patterns, climate adaptation practices, and habitat shifts (Adil, 2000).

Afghanistan's environmental degradation is also linked to poverty and survival-driven exploitation of resources. Widespread deforestation for fuelwood, illegal poaching of wildlife for trade, and expansion of unsustainable agriculture have led to serious ecosystem fragmentation. These pressures are exacerbated by the absence of enforceable environmental laws and effective institutional frameworks. While over 80% of the population relies directly on land and biodiversity for survival, environmental protection remains secondary to humanitarian crises and governance instability (UNEP, 2020).

Despite these challenges, promising examples of community-led conservationoften spearheaded by womenare emerging. In Bamyan and Daykundi provinces, women's cooperatives have begun cultivating and conserving native medicinal plants such as *Ferula* and *Ephedra* species, both valuable for health and ecological functions (ICIMOD, 2021). These initiatives not only help restore native vegetation but also generate income and strengthen local stewardship. Moreover, women in Hazara communities are known to protect alpine forests through customary restrictions and storytelling traditions, reinforcing ethical relationships with nature (Tariq & Mahboob, 2019).

Sustainable biodiversity conservation in Afghanistan must be rooted in inclusive environmental governance that recognizes women as key knowledge holders and decision-makers. Policymakers should collaborate with local councils, religious leaders, and NGOs to ensure that Indigenous women are not only consulted but actively empowered in all stages of conservation planning. Programs should include gender-sensitive ecological education, access to land rights, and financial support for women-led conservation cooperatives (Wright et al., 2023a, 2023b). Strengthening the ecological agency of Afghan women is not only a gender justice issue but a practical requirement for the preservation of Afghanistan's rapidly vanishing biodiversity.

8.1 Strategic Measures for Biodiversity Conservation

Afghanistan's environmental challenges require not only policy responses but a grounded understanding of Indigenous ecological knowledge, particularly from women who have historically managed biodiversity at the household and community level. Effective biodiversity conservation in Afghanistan's fragile mountain and rangeland ecosystems calls for the establishment and management of legally recognized protected areas, along with participatory ecological planning that centers the roles of Indigenous communities (UNEP, 2020).

High-priority regions, such as the eastern forests and alpine pastures of Badakhshan and Nuristan, must undergo comprehensive ecological surveys to assess the viability of protected areas. These assessments should inform both national and transboundary conservation strategies, especially given the region's vulnerability to cross-border poaching plant species (Kremen & Merenlender, 2018). In this regard, the International Union for Conservation of Nature (IUCN) supports the development of regional Red Lists and species threat assessments, which guide conservation prioritization while advocating for inclusive environmental policies (IUCN, 2023).

Equally important is the revival of traditional ecological knowledge held primarily by Indigenous women that sustains rare crop varieties, native livestock breeds, and endemic plant species. For example, the conservation of *Pistacia atlantica* (wild pistachio) and juniper forests in eastern Afghanistan depends not only on formal reforestation programs but also on women's customary practices of rotational harvesting, seed saving, and habitat monitoring (FAO, 2022). A gender-responsive National Biodiversity Action Plan (NBAP) is essential to integrate these local knowledge systems with national and international biodiversity frameworks (ICIMOD, 2021).

Additionally, empowering Afghan scholars including women ecologiststo document local flora, medicinal plants, and agroecological systems will foster locally informed environmental governance. Deforestation, particularly in monsoon and riverine forests, must be addressed through holistic policy frameworks that combine scientific forestry with Indigenous conservation practices. Enforcement of environmental laws should include local participation, enabling women and community elders to serve as environmental monitors and knowledge transmitters (Wright et al., 2023a, 2023b).

8.2 Institutional and Community Participation

Institutional reform and inclusive governance are foundational for successful biodiversity conservation. In recent years, a number of Afghan national institutions and grassroots organizations have emerged as key actors in this space. Community-based associations such as Forest Management Associations (FMAs) and Rangeland Management Associations (RMAs) have facilitated bottom-up restoration and protection of ecologically sensitive landscapes (Tariq & Mahboob, 2019).

Indigenous women in highland areas frequently lead these efforts, whether by organizing collective tree planting campaigns, preserving heirloom seed banks, or mediating access to seasonal grazing lands. Their leadership is often informal but deeply rooted in cultural norms of stewardship and reciprocity with nature (Kazemi & Sadat, 2020). These initiatives exemplify how traditional knowledge systems can be harmonized with modern conservation science to create culturally relevant and ecologically effective solutions.

To scale up these efforts, development agencies and NGOs must invest in training programs that equip women with ecological monitoring tools, legal literacy, and access to financial incentives for sustainable land management. Multi-stakeholder platforms should ensure that local voices particularly those of Indigenous women are central in biodiversity planning and benefit-sharing arrangements (Wright et al., 2023a, 2023b; UNEP, 2020).

Long-term success in biodiversity conservation will depend on institutions that not only acknowledge Indigenous expertise but co-create governance mechanisms with local communities. Women's inclusion is not optional it is imperative for building resilient landscapes and equitable environmental futures in Afghanistan.

9 Gender Dynamics and Social Movements

The intersection of gender dynamics and social movements has become an increasingly critical area of scholarly inquiry, particularly within the fields of environmental justice, conservation, and sustainability. While feminist movements have historically focused on gender-based inequalities, contemporary sociological perspectives increasingly view gender not merely as a subject of activism but as a structuring force in how movements are formed, sustained, and mobilized (Rowbotham, 2013a, 2013b).

Over the past few decades, this intersection has been relatively underexplored in mainstream social sciences. However, emerging scholarships have begun to reshape how gender is conceptualized within collective action. Kindsey (2020) argues that traditional models of social movements often fail to account for the fluid and relational nature of gender identities and instead treat gender as a fixed category or marginal variable. This limited view overlooks how gender hierarchies are continuously produced and contested within movements themselves, including those oriented towards environmental conservation and climate justice.

9.1 Gender as a Structuring Logic in Movements

Gender functions not merely as a demographic trait but as a system of power relations that shapes participation, visibility, and leadership within movements. Parker et al. (2017a, 2017b) emphasize that the framing of protests, organizing strategies, and coalition-building are deeply influenced by the intersection of gender with other identity markers such as class, race, ethnicity, and sexuality. These dynamics are particularly evident in grassroots environmental movements, where they lived experiences of indigenous women often inform the very strategies and goals of resistance.

In many environmental and indigenous struggles, women are not only active participants but also serve as symbolic anchors, cultural mediators, and strategic leaders. Their roles extend beyond activism to include knowledge preservation, caregiving, and intergenerational teaching. Gender thus operates both as a lived experience and an organizing framework in community-based resistance.

9.2 Challenging Reinforcement: Gender, Resistance, and Transformation

Classical theories of social movements tend to emphasize how movements reinforce traditional gender roles, portraying women as either victims or supporting actors. However, more recent analyses highlight how gender also acts as a source of transformation, innovation, and disruption of dominant systems (Avelino, 2021a, 2021b).

9.3 Gendered Movements and Biodiversity Conservation

The link between gendered social movements and biodiversity conservation is both direct and profound. Although data from Afghanistan remains limited due to conflict and institutional barriers, early reports from provinces like Badakhshan and Bamyan suggest that women's informal organizing around water, forests, and seed systems plays a crucial role in biodiversity stewardship.

Gender dynamics are foundational to understanding social and environmental movements. Rather than viewing women as peripheral actors in conservation, it is vital to recognize them as knowledge holders, organizers, and agents of change. Their roles in shaping collective resistance, fostering ecological resilience, and challenging power structures highlight the need for gender-aware conservation frameworks. Integrating gender analysis into environmental policy is not only a matter of social justice but also a practical necessity for achieving sustainable biodiversity outcomes.

10 Women, Social Moments and Environmental Stability

It needs to be understood that in gender dynamics we typically outline division of labor, property, and power rather than solely to the notion that women are close to nature than men supported by prospect of ecofeminism which gives them a specific stake in the protection of environment (Mackintosh, 2023).

The emerging interest in gender, development and environment has taken hold due to rapid structuring of bionetworks, economics, culture, and political spheres from global to local levels. Globally, due to environmental, economic, and political shifts have affect both women and men as central participants in management of environment, resource use, allocation and establishment of environmental ethics related to well-being of society (Rocheleau et al., 2013). Conversely, others see the major difference in biology due to gender specific environmental realities. These differences are derived from the social constructs based on gender and sociological understanding of biology which vary by class, culture, place and race governed individual and social transformation (Lindsey, 2020).

10.1 Women Leadership and Intersectionality

In understanding the complex experience of women in social movements it is decisive to know the concept of intersectionality. Aulia et al. (2024) pinpoints the women leadership role in progressing movements related to gender equality. Research reveals major factors helpful in sustaining successful leadership includes: encouragement for policies based on gender equality and inclusive decision making. In leadership women also face challenges of institutional barriers and gender stereotypes (Thelma & Ngulube, 2024). In 2024 Sonia Kelly Uche also concluded that the historical role of women has transformed antagonized patriarchal structure and political scenarios.

10.2 Gender Legislative Tactics and Organizational Structure

The social movements organizational structures and strategies are gender based and are greatly influenced by systems of gender inequalities which affect forms, constraints and opportunities of activism (Hurwitz & Crossley, 2018). This viewpoint acknowledges that gender not only shapes the goals of social movements but also contributes to their external policies and internal dynamics.

11 Mountain Women and Concerns

Women play a predominant role in the conservation of biodiversity in mountainous areas. At country level it is clear that "women" have greater responsibility for the sustainable management of natural resources, and they are the keepers of indigenous knowledge (Aluko, 2018).

Strategies on smaller scale are necessary for fostering a deep profound knowledge of ecosystem and drive women to adopt multiple roles in a community (Balabantaray, 2023).

Women's informal groups are commonly engaged in management of collective resources helped by collective travel to seasonal areas. Being the key role in providing essential needs women are generally motivated by long term future prospects of their families (Duguid and Weber, 2019).

They equally participate with their men in livestock as well as agricultural responsibilities and spend added 4–5 h on work for fundamental domestic requirements. In poor mountain community's women bear greater work burden as compared to men, they are primary users of natural resources and endure great suffering if these resources are degraded (Dumas et al., 2018).

12 Women's Role in Food Security

In mountain ecosystems ongoing stresses affecting intensely the women status and their capacity to provide food for their livelihood.

The burden of women further increased due to habitat loss and decline of traditional ways of living and community support structure. In rapidly expanding economy it is exceptionally hard for women in obtaining new opportunities, there is likely to be less power of making decisions in the families (Shiva, 2014).

Traditionally, food crops were cultivated by females for their families, but these may displace by cash crops that can hinder the participation of women for conservational practices and the income from these cash crops also controlled by males (Ibnouf, 2011). Women need extra revenue for their families, they engaged in varieties of small earning activities such as food processing and trading thus, they manage income. Women contribute more than fifty percent of income from selling handicrafts, livestock and agricultural products even during crisis they increase their working hours and contribute up to 75% income for their families (Rais et al., 2013).

13 The Contribution of Women in Environmental Conservation

Women play vital role in society in conservation of biodiversity few among them are discussed below:

1. ***Management of resources***

Women in mountain play significant and dominant role in management of natural resources, agricultural production and most of all they invest their whole life for wellbeing of their families (Wester et al., 2019).

2. ***Intensive workload and burden of family***

Due to limited access to natural resources, agricultural land, forest and migration of males leads to difficulty in collecting food. Mostly women handle generating income in mountainous area, so more burden is on females (Wester et al., 2019).

3. ***Lack of independence***

Women have limited access to education and health in mountains, they are not independent economically and socially (Wester et al., 2019).

4. ***Discrimination and Exclusion in Society***

Regardless of their important contribution women are excluded from decision making and in accessing opportunities due to social norms and discrimination (Romeo et al., 2022).

5. ***Natural disasters***

Natural crisis like drought, earthquakes quack and climatic changes cause habitat loss that affect negatively on women and worsen prevailing inequalities (Romeo et al., 2022).

6. ***Indigenous knowledge***

In developmental policies, females are ignored without realizing that these are wealth of indigenous knowledge. Women are primary users and collectors of natural resources (Klein et al., 2019).

Currently gender gender-sensitive programs and policies need to be recognized in which role and responsibilities of women are specified (Sruthi, 2024). The lack of data and policies on gender inequality hinder the inequalities that specifically affect females of mountainous areas.

7. ***Sustainable development of the environment***

In reducing hunger, poverty, and sustainable development, empowering women is a key because they contribute more than males in mountainous areas (Gratzer & Keeton, 2017).

14 Practice and Policy Implications: Addressing Mountain Women's Concerns

In contemporary interventions for sustainable development concerning environment and women, there should be large-scale investment projects that are women specific because in many government policies and developmental programs women are ignored despite of their main contribution in management of biodiversity. They are excluded from the majority of planning colloquiums and progression policies. Although officially many donor government agencies separately acknowledge women and issues related to environment but few among them relate "women" and environment" in the implementation of environmental policy making. Legal rights and ownership to natural resources are rarely held by females which also limit their authority to get benefits generated from their own efforts. Since they have no legal property rights for women, so they often face great trouble in securing credit, investment for developmental initiatives and for feed stock. Other than social and cultural framework, biophysical environment, women living in mountains are experiencing similar challenges. They perform their productive as well as reproductive workload in far-flunk areas devoted to any basic infrastructure. They have rare access in decision making, education, financial resources and in basic rights. The effective decision making and planning process in the management of sustainable environment require feedback from users of resources that are, particularly women. Most of women are skilled and they are also available for training of disciplines related to conservation of natural resources. In conservation intervention they are capable of taking part, but it is important to reduce their extra workload and provide household economic stability. At the small scale to most effective multisector resource management should facilitate women for environmental sustainability (Abatcha, 2025). In protection of mountains, sustaining cultural integrity, conserving and advancement of environmental quality and ecosystem, lessening of poverty and for sustainable development of natural assets, the empowerment of women is obligatory. Women play a decisive part in society for establishing for ecologically sound ecosystem and food security. Integration of women must be indicated at level of policy making in order to systemize the traditional role of women as "protectors" of environment (Chinsya & Sain, 2024). The organization of women are likely to be highly educated but lacking in funding and power, there are no non formal educational organizations for women, but it can be significant to initiate these classes for building critical thinking and confidence on different issues. In mountains the primary factor must be sustainability of environment, and its long-term managers are women that are fixed and stable members of communities (Acharya & Halpenny, 2013). Based on emerging concerns of women and mountains few implications for policies and practices are given below:

1. ***Participation of women in making decisions***

In the mountains the programs and policies are formulated, plan and execute in which women are actively as their prospective and knowledge are essential and should be considered for effective implementation of conservational activities.

2. ***Ownership in land and finance***

Novel policies should aim to remove discriminatory social and cultural norms and ensure equal rights and access of women to financial, educational ownership along with land ownership and other resources of nature.

3. ***Strategies for reduction of burden***

Strategies should consider to less the disproportional burden of women in mountain areas through developmental infrastructure, technology and promote balanced sharing of responsibilities in household.

4. ***Value indigenous knowledge***

Policies in the developmental initiatives should respect, recognize, and value the traditional knowledge of women for sound management of the environment in the mountains.

5. ***Support policies for economic empowerment of mountain women***

Most of women in the mountains needs support in plans and programs that focus on generating new opportunities, promote their entrepreneurship and ensuring financial and economic stability.

6. ***Strengthen more access to facilities***

More attention and actions are needed to enhance women access to facilities like: education, natural resources, information network, healthcare and all barriers posed by patriarchal systems and remoteness.

7. ***Integrated gender policies***

In agendas of sustainable developmental conservational development of environment, the contribution of women in integrated gender policies should be recognized as they are key actors and their specific capacities should be properly addressed.

15 Indigenous Women as Conservation Champions and Transformative Agents

Around the world, environmental governance faces complex challenges ranging from ecological uncertainty and cross-border degradation to climate instability and deep-rooted social inequities (Ahmad et al., 2025; Çomak, 2024). Global environmental organizations such as the International Union for Conservation of Nature (IUCN), the Worldwide Fund for Nature (WWF), the United Nations Development Programme (UNDP), and the Food and Agriculture Organization (FAO) have long promoted sustainable conservation frameworks. Yet, their strategies often overlook a vital force for environmental stability: Indigenous women.

In the Himalayan region, particularly in Northern Pakistan, Indigenous women have emerged as silent yet powerful conservation champions. Their roles transcend technical resource management they are healers, seed guardians, educators, and spiritual mediators of human-nature relations. Women in mountain communities have preserved biodiversity through generational practices like rotational grazing, forest protection taboos, medicinal herb harvesting, and sustainable agriculture (Lama & Buchy, 2021). These culturally embedded strategies represent some of the most localized and resilient forms of conservation.

Environmental planners now recognize that ecological stability and poverty reduction cannot be achieved without integrating gendered knowledge systems into national and global environmental agendas. Women's participation in conservation must move beyond symbolic inclusion toward substantive decision-making roles within community-based and formal institutions (Agarwal, 2021; Sherpa, 2023a, 2023b). Their leadership, though often informal and unpaid, drives many successful local conservation initiatives especially in highland ecosystems where top-down models have failed to deliver sustainable outcomes.

16 Regional Strategies and the Role of Indigenous Women in Sustainable Development

The launch of the IUCN's Regional Biodiversity Program (RBP) in 1996 marked a significant step toward addressing biodiversity challenges in Asia. With participation from over a dozen countries, including Pakistan, this program supports the implementation of the Convention on Biological Diversity (CBD) through regional cooperation and integration of environmental concerns into development planning. However, for these strategies to be effective in mountain regions like the Himalayas, they must prioritize local Indigenous women as both knowledge holders and agents of resilience.

The RBP's focus areas including, national biodiversity strategies (NBSAPs), biosafety protocols, benefit-sharing mechanisms, and communication outreach are critical tools. Yet they risk falling short when implemented without an understanding of the gendered dynamics of conservation. In many Himalayan communities, women serve as the primary monitors of biodiversity change, whether through noticing shifts in flowering patterns, loss of pollinators, or the decline of medicinal species (Ali et al., 2021). Including their observations in monitoring frameworks would significantly strengthen biodiversity baselines.

Recent efforts have emphasized integrating biodiversity into multilateral environmental agreements (MEAs) and promoting ecosystem-based livelihoods. In this context, Indigenous women offer practical insights into linking ecological sustainability with household resilience. Their lived experience managing alpine pastures, preserving heirloom seeds, and organizing collective rituals tied to seasonal cycles provides an alternative to extractive development (Singh et al., 2022).

Moving forward, regional strategies must be reframed to incorporate women's ecological perspectives in climate adaptation, conservation finance, and policy design. Collaboration among governments, NGOs, and local women-led initiatives can ensure that environmental sustainability is not only a technical goal but a culturally rooted and gender-just process. Empowering Indigenous women as full partners in conservation is essential to strengthen ecological governance in the Himalayas and beyond.

17 World's Major Conventions on Environmental Conservation: Role of International Union for Conservation of Nature and Natural Resources (IUCN) in Asia

Globally, the International Union for Conservation of Nature and Natural resources (IUCN) recognizes a distinct role in addressing the conservation challenges. IUCN approach for ecosystem maintenance, effective restoration and management is best for "Mountains to coral reefs" but it also improves biodiversity. It helps in elevating and strengthening the establishment of country programs and growing relationships with members of governments of various countries as well. IUCN forges closer relationships when countries face internal issues or in any difficult time. Emerging strong networks and initiatives are built on the strong presence of countries which makes the strong foundation of IUCN programs throughout Asia at global and regional levels. In Asian region the IUCN is one among 9 regions of Global Union established in 1999 and it comprises of twenty-three (23) countries extending from Pakistan (west) to Japan (east), Mongolia (north) and Indonesia (south).

IUCN as a union seeks to encourage, influence and support the majority of societies throughout the globe to conserve the diversity and integrity of nature and natural resources for development of a sustainable environment. Worldwide in different countries of Asia, IUCN provide several programs on a regional, national and international level but we focus on only a few programs; otherwise, it's very time-consuming to write all their past and ongoing contributions. Generally, IUCN coordinates with the main secretariat and assist the members of the union membership in representing their interpretations on the world ground and provides them with services, scientific knowledge, strategies and support they need to accomplish their objectives. IUCN have 6 commissions, ten thousand volunteers professionals in their teams and project groups, aiming specifically on species and conservation of biodiversity and for managing natural resources and habitat. It helped various countries to make conservational strategies and illustrate the utilization of their knowledge using field projects it monitors. Operations are progressed and executed by extending the network of country and regional offices, mainly located in developing countries. The International Union for Conservation builds on its strong association and network that boost their capacity to sustain global unions to secure nature and natural assets at every level.

18 Major Conventions on Environmental Conservation

Environmental conventions aids in the framework of institutions to deliver various functions and set outlines, proscribe actions & behaviors, increase awareness about environmental challenges, generate responses in policies, limit uncertainty, specialization of policies, provide space for the participation of society and for novel use of technologies and innovation to combat environmental challenges. The Global environmental conventions shown in Table 1.

Despite all these conventions, the environment around the globe degraded at an alarming rate and the earth's stability is at risk. It's a big question about the performance of environmental conventions and their policies that how international environmental pledges contribute to resolving global environmental issues.

Clearing up this question would allow us to decide the role of conventions as a tool of global governance, although offering an analytical framework on their process of execution. Comprehending how these conventions operate will also provide an understanding of the requirements to achieve outcomes and behavioral changes at every level. To address environmental problems, mutual collaboration at every level is required to ensure implementation on actions, aims, and policies of all countries across the globe are prerequisites for achieving sustainability and environmental conservation.

19 Conclusion: Reclaiming Balance through Indigenous Feminine Wisdom

The conservation of the Himalayan ecosystem demands more than technical interventionIt requires the recognition of long-overlooked human dimensions, particularly the vital yet invisible role of native women. In the current study we have brought to attention to life of the indigenous women, through their traditional ecological knowledge, spiritual relationships with the land, and sustained environmental labor, act as custodians of biodiversity across Himalayan regions. Their contributions are rooted in centuries of intimate engagement with forest, soil, water, and seedform the backbone of local sustainability, even as modern conservation policies have historically marginalized them.

To understand the dynamic of this the underscored several critical insights: first that environmental sustainability is deeply intertwined with cultural continuity and gender equity, second that communitybased approaches are significantly enhanced when women are recognized as equal partners and third that traditional knowledge systems, when integrated with modern conservation frameworks, offer more resilient and adaptive solutions to climate and ecological crises. Importantly, this chapter challenges prevailing paradigms that prioritize technocratic or exclusionary conservation models. Instead, it calls for inclusive, rights-based approaches that acknowledge native women as both knowledge holders and frontline defenders of nature. Their

Table 1 Global conventions on environmental conservation

Convention/protocol	Adopted	Entered into force	Key objective	Parties
Ramsar convention	1971	1975	Conservation and sustainable use of wetlands	172
CITES (convention on international trade in endangered species)	1973	1975	Regulation of international trade in endangered species of wild fauna and flora	184
Vienna convention for the protection of the ozone layer	1985	1988	Framework for international efforts to protect the ozone layer	198
Montreal protocol	1987	1989	Phase-out of ozone-depleting substances	198
Basel convention	1989	1992	Control of transboundary movements of hazardous wastes and their disposal	191
UNFCCC (United Nations framework convention on climate change)	1992	1994	Framework for international cooperation to combat climate change	198
Convention on biological diversity (CBD)	1992	1993	Conservation of biological diversity, sustainable use, and fair sharing of benefits	196
UNCCD (United Nations convention to combat desertification)	1994	1996	Combat desertification and mitigate the effects of drought	197
Kyoto protocol	1997	2005	Legally binding emission reduction targets for developed countries	192
Cartagena protocol on biosafety	2000	2003	Regulation of transboundary movements of genetically modified organisms	173
Stockholm convention	2001	2004	Elimination or restriction of persistent organic pollutants	185
Nagoya protocol	2010	2014	Access to genetic resources and fair sharing of benefits	137
Minamata convention	2013	2017	Protection of human health and the environment from mercury emissions	128

(continued)

Table 1 (continued)

Convention/protocol	Adopted	Entered into force	Key objective	Parties
Paris agreement	2015	2016	Limit global warming to well below 2 °C, pursuing efforts to limit it to 1.5 °C	195
Kigali amendment to the Montreal protocol	2016	2019	Phase-down of hydrofluorocarbons (HFCs), potent greenhouse gases	150
Global plastics treaty (under negotiation)	2022	N/A	Legally binding agreement to end plastic pollution by 2024	N/A

roles in seed preservation, forest governance, sustainable agriculture, and intergenerational knowledge transmission must be centered in policy discourse, not treated as supplemental.

By honoring the often-unseen work of native women in ecological stewardship, we not only deepen our understanding of sustainability but also restore the moral and ecological balance necessary for the future of the Himalayan ecosystem. Their legacy is not one of resistance alone, but of quiet resilience and enduring guardianship that must now be made visible, valued, and institutionalized.

References

Abatcha, U. M. (2025). The Role of Women in Water Resource Management and It's Effects On Rural Economic Development. *FANE-FANE INTERNATIONAL MULTIDISCIPLINARY JOURNAL, 9*(1), 368–374.

Abbas, T., Rehman, A., & Hussain, N. (2021a). Human-induced pressures on mountain biodiversity: A case study of Gilgit-Baltistan. *Pakistan Journal of Environmental Sciences, 9*(2), 34–46.

Abbas, Z., Khan, S. M., Alam, J., Peer, T., Abideen, Z., Bussmann, R. W., & Muhammad, S. (2021b). Vegetation dynamics along altitudinal gradients in the Shigar Valley (Central Karakorum) Pakistan: Zonation, physiognomy, ecosystem services and environmental impacts. *Pakistan Journal of Botany, 53*(5), 1865–1874.

Abdullah, A., & Khan, S. M. (2023). Traditional ecological knowledge sustains due to poverty and lack of choices rather than thinking about the environment. *Journal of Ethnobiology and Ethnomedicine, 19*(1), 65.

Abdulsalam, A., Hashmi, M., & Anwar, F. (2024a). Gender equity in environmental governance: Challenges and pathways in South Asia. *Journal of Sustainability and Society, 18*(1), 45–62.

Abdulsalam, M., Elelu, S. A., Innocent, M. O., Ibrahim, G. O., Livinus, M. U., Lateefat, S. O., & Muhammad, A. S. (2024b). Women's empowerment in environmental conservation. In *Prospects for soil regeneration and its impact on environmental protection* (pp. 351–370). Cham.

Acharya, B. P., & Halpenny, E. A. (2013). Homestays as an alternative tourism product for sustainable community development: A case study of women-managed tourism product in rural Nepal. *Tourism Planning & Development, 10*(4), 367–387.

Adeyemo, A. A., & Adebayo, J. O. (2018). Documentation and dissemination of indigenous knowledge by library personnel in selected research institutes in Nigeria. *Library Philosophy & Practice.*

Adil, S. (2000). Environmental degradation in Afghanistan: Problems and policy options. *Afghanistan Development Review, 8*(2), 23–35.

Ahmad, Z., Khan, S. M., Wani, Z. A., Pant, S., & Bhat, J. A. (2025). Conserving plants in a changing climate. *Frontiers in Conservation Science, 6*, 1743981.

Agarwal, B. (2021). *Gender and environmental sustainability: Negotiating equity and empowerment*. Routledge.

Ali, T., Paton, D., Buergelt, P. T., Smith, J. A., Jehan, N., & Siddique, A. (2021). Integrating indigenous perspectives and community-based disaster risk reduction: A pathway for sustainable indigenous development in Northern Pakistan. *International Journal of Disaster Risk Reduction, 59*, 102263.

Aluko, Y. A. (2018). Women's use of indigenous knowledge for environmental security and sustainable development in Southwest Nigeria. *International Indigenous Policy Journal, 9*(3), 1–23.

Aulia, S. S., Marzuki, S., & Arpannudin, I. (2024). Women in gender equality movement: A systematic literature review. *Frontiers in Sociology, 9*, 1432383. https://doi.org/10.3389/fsoc.2024.1432383

Avelino, F. (2021a). Theories of change for sustainability transitions: An overarching framework. *Environmental Innovation and Societal Transitions, 41*, 1–27. https://doi.org/10.1016/j.eist.2021.08.002

Avelino, F. (2021b). Theories of power and social change. Power contestations and their implications for research on social change and innovation. *Journal of Political Power, 14*(3), 425–448.

Balabantaray, S. R. (2023). Women's leadership and sustainable environmental initiatives: A macroscopic investigation from ecofeminism framework. *International Journal of Multidisciplinary Research and Growth Evaluation, 4*(4), 1039–1046.

Barnhardt, N. R. (2014). Creating a place for indigenous knowledge in education: The Alaska native knowledge network. In *Place-based education in the global age* (pp. 137–158). Routledge.

Bhasin, V. (2011a). Ecology, culture and change: Tribals of the Himalayas. *Journal of Human Ecology, 33*(3), 167–191.

Bhasin, V. (2011b). Status of women in transhumant societies. *Journal of Sociology and Social Anthropology, 2*(1), 1–22.

Boiral, O., & Heras-Saizarbitoria, I. (2017). Managing biodiversity through stakeholder involvement: Why, who, and for what initiatives? *Journal of Business Ethics, 140*(3), 403–421.

Brown, C., Waldron, S. A., & Longworth, J. W. (2009). *Sustainable development in Western China: Managing people, livestock and grasslands in pastoral areas* (in Chinese, Z. Yutian & W. Ou, Trans.).

Chakravarty, S., Ghosh, S. K., Suresh, C. P., Dey, A. N., & Shukla, G. (2012). Deforestation: Causes, effects and control strategies. *Global Perspectives on Sustainable Forest Management, 1*, 1–26.

Chambers, D. W., & Gillespie, R. (2000). Locality in the history of science: Colonial science, technoscience, and indigenous knowledge. *Osiris, 15*, 221–240.

Chen, H., Zhang, T., Costanza, R., & Kubiszewski, I. (2023b). Review of the approaches for assessing protected areas' effectiveness. *Environmental Impact Assessment Review, 98*, 106929.

Chen, Y., Khan, A. A., & Zainab, S. (2023a). Gender-inclusive environmental governance: A framework for participatory conservation in South Asia. *Journal of Mountain Studies, 11*(1), 73–92.

Chen, Y., Zhang, Z., & Tao, F. (2018). Impacts of climate change and climate extremes on major crops productivity in China at a global warming of 1.5 and 2.0 °C. *Earth System Dynamics (Pembroke, Ont.), 9*(2), 543–562.

Chettri, N., Shrestha, A. B., Zhaoli, Y., Bajracharya, B., Sharma, E., & Ouyang, H. (2012). Real world protection for the "Third Pole" and its people. In *Protection of the three poles* (pp. 113–133). Tokyo: Springer Japan.

Chinsya, B. A. D., & Sain, Z. H. (2024). Women and nature: An ecofeminist study of environmental conservation sustainability. *An-Nisa Jurnal Kajian Perempuan Dan Keislaman, 17*(2), 165–180.
Chmutina, K., Zhu, J., & Riffat, S. (2012). An analysis of climate change policy-making and implementation in China. *International Journal of Climate Change Strategies and Management, 4*(2), 138–151.
Christen, K. A. (2012). Does information really want to be free? Indigenous knowledge systems and the question of openness. *International Journal of Communication, 6*, 24.
Clemens, M. A., & Kremer, M. (2016). The new role for the World Bank. *Journal of Economic Perspectives, 30*(1), 53–76.
Coccia, M. (2015). Patterns of innovative outputs across climate zones: The geography of innovation. *Prometheus, 33*(2), 165–186.
Çomak, H., & Şeker, B. Ş. (Eds.). (2024). Climate Change and Environmental Politics. Transnational Press London.
Din, B. U., Hadi, F., ul Haq, Z., Ahmed, M., Abbas, M., Salam, A.,... & Rasheed, S. (2020). Ethno botanical perspective of medicinal plants in mulkhow valley district upper Chitral, Pakistan. *Academic Journal of Plant Sciences, 13*(1), 15–22.
Du, Q. L. (2006). *Sustainable development strategy of China's grass industry*. Beijing.
Duguid, F., & Weber, N. (2019). *Women informal workers and the empowering nature of collectivizing and collectives: An evidence synthesis*. ILO Documents.
Dumas, S. E., Maranga, A., Mbullo, P., Collins, S., Wekesa, P., Onono, M., & Young, S. L. (2018). "Men are in front at eating time, but not when it comes to rearing the chicken": Unpacking the gendered benefits and costs of livestock ownership in Kenya. *Food and Nutrition Bulletin, 39*(1), 3–27.
FAO. (2011). Role of women in sustainable agriculture and biodiversity in Afghanistan. Food and Agriculture Organization of the United Nations.
FAO. (2022). *Women and biodiversity in mountain ecosystems: A case study from Afghanistan*. Food and Agriculture Organization of the United Nations.
Fatima, A., Farid, M., Safdar, K., Fayyaz, A., Ali, S. M., Adnan, S., Nawaz, M., Munir, H., Raza, N., & Zubair, M. (2020a). Loss of agro-biodiversity and productivity due to climate change in continent Asia: A review. In *Plant ecophysiology and adaptation under climate change: Mechanisms and perspectives I: General consequences and plant responses* (pp. 51–71).
Fatima, N., Shah, A. A., & Karim, M. (2020b). Women's knowledge, agro-biodiversity and community resilience in Gilgit-Baltistan. *Mountain Research and Development, 40*(2), A16–A25.
Gao, L., & Xu, J. (2024). Zhang Xiuyun and the politics of participatory maize breeding. *Journal of Rural Studies, 95*, 112–120.
Gratzer, G., & Keeton, W. S. (2017). Mountain forests and sustainable development: The potential for achieving the United Nations' 2030 agenda. *Mountain Research and Development, 37*(3), 246–253.
Hariram, N. P., Mekha, K. B., Suganthan, V., & Sudhakar, K. (2023b). Sustainalism: An integrated socio-economic-environmental model to address sustainable development and sustainability. *Sustainability, 15*(13), 10682.
Hariram, V., Bhattarai, B., & Tenzing, K. (2023a). Indigenous women and the politics of conservation: A Himalayan perspective. *Ecological Justice, 29*(1), 101–118.
Harris, R. B. (2010). Rangeland degradation on the Qinghai-Tibetan plateau: A review of the evidence of its magnitude and causes. *Journal of Arid Environments, 74*, 1–12. https://doi.org/10.1016/j.jaridenv.2009.06.014
Hemmati, M. (2012a). *Gender and sustainable development: Opportunities for environmental sustainability*. United Nations Research Institute for Social Development.
Hemmati, M. (2012b). *Multi-stakeholder processes for governance and sustainability: Beyond deadlock and conflict*. Routledge.
Hilton-Taylor, C. (2000a). *2000 IUCN red list of threatened species*. International Union for Conservation of Nature.

Hilton-Taylor, C. (2000b) 2000 IUCN red list of threatened species IUCN, Gland, Switzerland and Cambridge, UK: 61.
Hopkins, C., & McKeown, R. (2002). Education for sustainable development: An international perspective. *Environmental Education Research, 8*(1), 1–12.
Hurwitz, H. M., & Crossley, A. D. (2018). Gender and social movements. In *The Wiley Blackwell companion to social movements* (pp. 537–552).
Ibnouf, F. O. (2011). Challenges and possibilities for achieving household food security in the Western Sudan region: The role of female farmers. *Food Security, 3*, 215–231.
ICIMOD. (2021). *Community-led biodiversity conservation in high-altitude regions: Case studies from the Hindu Kush Himalaya.* International Centre for Integrated Mountain Development.
IUCN. (2023). *Red list of threatened species™: Regional assessments for Central Asia.* International Union for Conservation of Nature.
James, R. O. B. Y. N., Gibbs, B., Whitford, L., Leisher, C., Konia, R., & Butt, N. (2021). Conservation and natural resource management: where are all the women?. *Oryx, 55*(6), 860–867.
Ji, Z. J., Li, X. Y., Wang, Z. G., & Wang, L. N. (2010). The optimal livestock contracture of 2011–2015 and capacity of livestock loading in natural rangeland of Gannan region. *Gansu Science and Technology, 26*, 174–176.
Johnson-Pynn, J., & Johnson, L. R. (2005). Successes and challenges in east African conservation education programs. *Applied Environmental Education & Communication, 4*(2), 85–93.
Jones, C. (2014a). Women and the environment: Global lives in focus. *United Nations Environmental Program Policy Review, 12*(3), 30–46.
Jones, S. L. (2014b). *Networking local-global advocacy: A rhetorical historiography of the International Women's Tribune Centre (1972–1986).* The University of Nebraska-Lincoln.
Kazemi, L., & Sadat, F. (2020). Pastoralist women and communal governance in the Wakhan corridor. *Central Asian Journal of Environmental Studies, 4*(1), 45–59.
Khan, S. M., Page, S., Ahmad, H., & Harper, D. (2012a). Anthropogenic influences on the natural ecosystem of the Naran Valley in the western Himalayas. *Pakistan Journal of Botany, 44*, 231–238.
Khan, S. M., Page, S., Ahmad, H., & Harper, D. (2014). Ethno-ecological importance of plant biodiversity in mountain ecosystems with special emphasis on indicator species of a Himalayan Valley in the Northern Pakistan. *Ecological Indicators, 37*, 175–185.
Khan, S. M., Page, S., Ahmad, H., & Harper, D. M. (2012b). Sustainable use of plant resources in the Western Himalayas. *Biodiversity and Conservation, 21*(7), 1651–1667.
Khan, S. M., Page, S. E., Ahmad, H., & Harper, D. M. (2013). Sustainable utilization and conservation of plant biodiversity in montane ecosystems: the western Himalayas as a case study. *Annals of botany*, 112(3), 479–501.
Khatwani, M. (2016a). Status of women in Pakistan: A sociological analysis. *Grassroots, 50*(2), 30–45.
Khatwani, M. K. (2016b). *Professional women's perceptions & experiences of respectability, social status, and autonomy: A case study of women employed at the University of Sindh, Jamshoro, Sindh-Pakistan.* Doctoral dissertation, University of Sussex.
Kindsey, M. J. (2020). Gender and social movements: Reframing participation and power. *Sociological Inquiry, 90*(4), 654–672. https://doi.org/10.1111/soin.12256
Klein, J. A., Tucker, C. M., Steger, C. E., Nolin, A., Reid, R., Hopping, K. A., et al. (2019). An integrated community and ecosystem-based approach to disaster risk reduction in mountain systems. *Environmental Science & Policy, 94*, 143–152.
Kosoe, E. A., Adjei, P. O. W., & Diawuo, F. (2020). From sacrilege to sustainability: The role of indigenous knowledge systems in biodiversity conservation in the Upper West Region of Ghana. *GeoJournal, 85*, 1057–1074.
Kremen, C., & Merenlender, A. M. (2018). Landscapes that work for biodiversity and people. *Science, 362*(6412), eaau6020.
Kremsa, V. Š. (2021). Sustainable management of agricultural resources (agricultural crops and animals). In *Sustainable resource management* (pp. 99–145). Elsevier.

Lama, N., & Buchy, M. (2021). Women's herbal knowledge and sustainability in Nepal. *Ethnobotany Research and Applications, 21*, 1–15.

Lindsey, L. L. (2020). *Gender: Sociological perspectives*. Routledge.

Long, R. J. (2003). Alpine rangeland ecosystem and their management in the Qinghai-Tibetan plateau. In G. Wiener, J. L. Han, & R. J. Long (Eds.), *The yak* (2nd ed., pp. 359–388). FAO Regional Office for Asia and the Pacific.

Ma, H., Zhang, D., Xiao, L., Wang, Y., Zhang, L., Thompson, C.,... & Turvey, S. T. (2022). Integrating biodiversity conservation and local community perspectives in China through human dimensions research. *People and Nature, 4*(6), 1461–1474.

Mackintosh, M. (2023). Gender and economics: The sexual division of labour and the subordination of women. In *Of marriage and the market* (pp. 3–17). Routledge.

Mantyka-Pringle, C. S., Jardine, T. D., Bradford, L., Bharadwaj, L., Kythreotis, A. P., Fresque-Baxter, J., Kelly, E., Somers, G., Doig, L. E., Jones, P. D., & Lindenschmidt, K. E. (2017). Bridging science and traditional knowledge to assess cumulative impacts of stressors on ecosystem health. *Environment International, 102*, 125–137.

Márquez-García, M., Nuñez-Godoy, C., Eguren, A., Segovia-Salcedo, C., Carrasco-Rueda, F., Püschel, N.,... & Branch, L. C. (2024). Overcoming gender-related challenges and supporting women in conservation in Latin America. *Biological Conservation, 294*, 110625.

MEE. (2018). *The People's Republic of China third National Communication on climate change*. Ministry of Ecology and. Environment (MEE). https://unfccc.int/sites/default/files/resource/China%203NC_English_0.pdf

Miller, D. J. (1999). Nomads of the Tibetan plateau rangelands in Western China part two: Pastoral production practices. *Rangelands, 21*, 16–19.

Nazakat, S., Khan, S. M., Abdullah, Noor, R., Zaman, I. U., Arif, M., Khalid, N., & Ahmad, Z. (2021). Floral composition, sustainable utilization, and conservation of important medicinal plants in the Ayubia National Park, Abbottabad, Khyber Pakhtunkhwa, Pakistan. In *Ethnobiology of mountain communities in Asia* (pp. 103–123).

Niesenbaum, R. A. (2019a). *Sustainable solutions: Problem solving for current and future generations*. Oxford University Press.

Niesenbaum, R. A. (2019b). The integration of conservation, biodiversity, and sustainability. *Sustainability, 11*(17), 4676.

Norström, P. (2011). Technological know-how from rules of thumb. *Techne: Research in Philosophy & Technology, 15*(2).

Okorondu, J., Umar, N. A., Ulor, C. O., Onwuagba, C. G., Diagi, B. E., Ajiere, S. I., & Nwaogu, C. (2022b). Anthropogenic activities as primary drivers of environmental pollution and loss of biodiversity a review. *International Journal of Trend in Research and Development, 6*, 621–643.

Okorondu, S. I., Singh, M., & Khan, Z. (2022a). Overexploitation of non-timber forest products and its implications on biodiversity in Himalayan communities. *Environmental Monitoring and Assessment, 194*(7), 1–15.

Kipuri N Ole (1991). *Maasai pastoralism and the decline in the status of women*. Unpublished memo. NOPA Project, Nairobi: UNESCO.

Öztürk, Y. M. (2020). An overview of ecofeminism: Women, nature and hierarchies. *Journal of Academic Social Science Studies, 13*(81).

Parker, M., Cheney, G., Fournier, V., & Land, C. (2017a). *The Routledge companion to alternative organization*. Routledge.

Parker, P. S., Jiang, J., McCluney, C. L., & Rabelo, V. C. (2017b). Race, gender, class, and sexuality. In *Oxford research Encyclopedia of communication*.

Pimbert, M. P., & Pretty, J. N. (2013). Parks, people and professionals: Putting 'participation' into protected-area management. In *Social change and conservation* (pp. 297–330). Routledge.

Pradhan, M. S., Rai-Paudyal, B., Rai, A., Bai, Y., Hengsuwan, P., Bun, P., & Yangzom, D. (2021a). Exploring the role of gender equality in addressing climate change, biodiversity conservation, ecosystem restoration and food security. In *Gender equality and sustainable development in the mountain areas of Asia*. LI-BIRD. The main report can be downloaded from www.libird.org.

Pradhan, S., Lama, A., & Sherpa, L. (2021b). Gender and conservation in the Hindu Kush Himalaya: Addressing knowledge gaps. *International Journal of Mountain Research, 9*(1), 33–51.
Rais, M. U. N., Solangi, A. W., & Sahito, H. A. (2013). Economic assessment of rural women involved in agriculture and livestock farming activities. *Wudpecker J. Agric. Res, 2*(4), 115–121.
Richard, C. (2002). The potential for rangeland development in yak-rearing areas of the Tibetan plateau. In J. L. Han, C. Richard, O. Hanotte, C. McVeigh, & J. E. O. Rege (Eds.), *Yak Production in Central Asian Highlands. Proceedings of the Third International Congress on Yak, Lhasa, P. R. China* (pp. 11–18). International Livestock Research Institute.
Rocheleau, D., Thomas-Slayter, B., & Wangari, E. (2013). Gender and environment: A feminist political ecology perspective. In *Feminist political ecology* (pp. 3–23). Routledge.
Romeo, P., Loria, G., Martinelli, C., Ercoli, A., & Romeo, C. (2022). Minimally invasive management of a giant paratubal cyst in an adolescent female: Case report and review of the literature in the pediatric population. *Frontiers in pediatrics, 10*, 1080797.
Rowbotham, S. (2013a). *Women, resistance and revolution: A history of women and revolution in the modern world*. Verso.
Rowbotham, S. (2013b). *Women in movement (Routledge revivals): Feminism and social action*. Routledge.
Rytkönen, S., & Hotakainen, S. (2020). *Promoting community-led conservation: opportunities, challenges and measures*.
Schroth, G., & McNeely, J. A. (2011). Biodiversity conservation, ecosystem services and livelihoods in tropical landscapes: Towards a common agenda. *Environmental Management, 48*, 229–236.
Seth, S. (2009). Putting knowledge in its place: Science, colonialism, and the postcolonial. *Postcolonial studies, 12*(4), 373–388.
Shaheen, H., Khan, S. M., Harper, D. M., Ullah, Z., & Qureshi, R. A. (2011). Species diversity, community structure, and distribution patterns in western Himalayan alpine pastures of Kashmir, Pakistan. *Mountain Research and Development, 31*(2), 153–159.
Shang, Z. H., Long, R. J., & Ma, Y. S. (2006). Discussion on restoration and rebuilding of 'black soil patch' degraded meadow in the headwater area of Yangtze and Yellow rivers. *Chinese Journal of Grassland, 28*, 69–74.
Shen, W. S., Li, H. D., Zhang, T., Zou, C. X., Yan, S. G., & Yuan, L. (2009). Feasibility study of air-seeding on alpine Aeolian desertified land in Tibet, China. *Journal of Ecology and Rural Environment, 25*, 106–111.
Sher, H., Aldosari, A., Ali, A., & de Boer, H. J. (2015). Indigenous knowledge of folk medicines among tribal minorities in Khyber Pakhtunkhwa, Northwestern Pakistan. *Journal of Ethnopharmacology, 166*, 157–167.
Sherpa, P. (2023a). Landscape loss and identity among indigenous Himalayan women. *Mountain Research and Development, 43*(1), 12–25.
Sherpa, P. (2023b). Landscape loss and identity among indigenous Himalayan women. *Mountain Research and Development, 43*(1), 12–25.
Shiva, V. (Ed.). (2014). *Close to home: Women reconnect ecology, health and development*. Routledge.
Singh, P., Tabe, T., & Martin, T. (2022, January). The role of women in community resilience to climate change: A case study of an Indigenous Fijian community. In *Women's Studies International Forum* (Vol. 90, p. 102550). Pergamon.
Sruthi, S. (2024). Promoting gender sensitivity in education: Addressing key areas for an inclusive future. *Journal of Education and Research, 8*(2), 45–59.
Song, Y., & Jiggins, J. (2023). *Women as seed custodians: Reviving local knowledge in Guangxi*. IIED Briefing.
Swart, R., Robinson, J., & Cohen, S. (2003). Climate change and sustainable development: Expanding the options. *Climate Policy, 3S1*, S19–S40.
Tariq, M., & Mahboob, U. (2019). Cultural conservation practices in Hazara communities: Women as environmental stewards. *Journal of Mountain Ecology, 13*(2), 55–70.

Teixidor-Toneu, I., Mattalia, G., Caillon, S., Abdullah, A., Fiser, Ž., Karlsen, P.,... & Schunko, C. (2025). Stewardship underpins sustainable foraging. *Trends in ecology & evolution, 40*(4), 315–319.

Thelma, C. C., & Ngulube, L. (2024). Women in leadership: Examining barriers to women's advancement in leadership positions. *Asian Journal of Advanced Research and Reports, 18*(6), 273–290.

Ullah, I., Khan, R. Z., & Jan, A. (2022b). Environmental degradation and biodiversity decline in the Northern Himalayas. *Pakistan Journal of Botany, 54*(1), 219–226.

Ullah, S., Syed, N. M., Gang, T., Noor, R. S., Ahmad, S., Waqas, M. M., Shah, A. N., & Ullah, S. (2022a). Recent global warming as a proximate cause of deforestation and forest degradation in northern Pakistan. *PLoS One, 17*(1), e0260607.

Ullah, T., Khan, S. M., Abdullah, A., Sulaiman, N., Ullah, A., Khan, M. S.,... & Pieroni, A. (2024). Bio-cultural diversity for food security: traditional wild food plants and their folk cuisine in Lakki Marwat, Northwestern Pakistan. *Diversity, 16*(11), 684.

UNEP. (2000). *Women and the environment: Policy integration and sustainable development*. United Nations Environment Programme.

UNEP. (2020). *State of environment report for Afghanistan*. United Nations Environment Programme.

Wang, M., Li, Y., Pan, S., Bai, X. Z., & Huang, R. Q. (2004). Influences of climate change on sustainable development of the hinterland of Qinghai-Tibet plateau. China population. *Resources and Environment, 14*, 92–95.

Wang, S. P., Duan, J. C., Xu, G. P., Wang, Y. F., Zhang, Z. H., Rui, Y. C., Luo, C. Y., Xu, B., Zhu, X. X., Chang, X. F., Cui, X. Y., Niu, H. S., Zhao, X. Q., & Wang, W. Y. (2012). Effects of warming and grazing on soil N availability, species composition, and ANPP in an alpine meadow. *Ecology, 93*, 2365–2376. https://doi.org/10.1890/11-1408.1

Wester, M., & Lama, P. D. (2019). Women as agents of change?: Reflections on women in climate adaptation and mitigation in the global north and the global South. In *Climate hazards, disasters, and gender ramifications* (pp. 67–85). Routledge.

Whitt, L. (2009). *Science, colonialism, and indigenous peoples: The cultural politics of law and knowledge*. Cambridge University Press.

Wright, L., Bahar, Z., & Omari, H. (2023a). Gender equality and climate adaptation in fragile states: Case of rural Afghanistan. *Environmental Justice, 16*(1), 44–59.

Wright, L., Bahar, Z., & Omari, H. (2023b). Gender equality and climate adaptation in fragile states: Case of rural Afghanistan. *Environmental Justice, 16*(1), 44–59.

Xu, J., et al. (2022). Ecological knowledge, land use, and conservation in Tibetan sacred sites. *Biodiversity Science, 30*(4), 415–430.

Xu, S. X., Zhao, X. Q., Su, P., & Zhao, T. B. (2004). Summary of natural biological resource in the source region of Changjiang-Yellow Rivers. *Resources and Environment in the Yangtze Basin, 13*, 448–453.

Zeb, S. A., Ahmad, N., Rahman, A., & Ullah, F. (2022). Floristic Inventory of Ethnobotanically Important Plants of Thangy Dara District Dir Lower Khyber Pakhtunkhwa, Pakistan. In *Biodiversity, Conservation and Sustainability in Asia: Volume 2: Prospects and Challenges in South and Middle Asia* (pp. 447–457). Cham: Springer International Publishing.

Zeb, S. A., Khan, S. M., Abdullah, A., Ahmad, Z., Khan, M. S., Ullah, T., & Badshah, H. (2025). Mushrooms as a source of better food and health in Northern Pakistan. In *Natural bioactives from the endophytes of medicinal plants* (pp. 184–193). CRC Press.

Zhang, Y., Li, B., & Zheng, D. (2002). A discussion on the boundary and area of the Tibetan plateau in China. *Geographical Resources, 21*, 1–8.

Zhou, Q., Liu, L. S., Zhang, Y. L., Chen, Q., Zhang, H. F., and Liu, F. G. (2011). *Effect of grassland change on food security in alpine pastoral area—A case study.*

Avenues for Ecosystem-Based Adaptations Through Pine Needle-Based Livelihood Enhancement of Women in the Western Himalayan Region

Harshit Pant and Paras Upadhyaya

Abstract The Indian Himalayan Region (IHR) is confronting an era of unprecedented climatic change, characterized by rising temperatures, erratic precipitation patterns, and an increased frequency of extreme weather events. These changes exert coupled, detrimental impacts on the livelihoods of native communities, particularly in sectors dependent on natural resources, alternative livelihoods, and the growing demands for renewable energy. While Himalayan communities possess a deep-seated historical resilience and have adapted to climate variability for centuries, their traditional coping mechanisms are being critically outpaced by the rapidity of contemporary climate shifts, exacerbated by pervasive resource paucity and socio-economic marginalization. Conversely, the global and national thrust towards renewable energy has serendipitously opened a window of opportunity for low-cost employment generation for village communities residing near the extensive Pine-dominated forests of the region. The Chir Pine (*Pinus roxburghii* Sarg.), a dominant species, is conventionally perceived as an unwanted and obnoxious entity by local communities. Its negative perception stems from its inability to provide fodder for livestock, its poor quality as fuelwood, and its notorious role as a primary catalyst for devastating forest fires due to its resin-rich, highly inflammable needle litter. This chapter delineates a paradigm shift, demonstrating the multifaceted valorization of this previously maligned tree species. We document and analyze innovative applications of pine needles, including their conversion into smokeless bio-briquettes, the crafting of decorative and utilitarian items from its bark, cones, and needles, and its use in natural resource management structures like check dams for soil and water conservation. The core of this chapter presents empirical success stories of providing alternative livelihood opportunities to local communities, with a specific focus on women from economically weaker sections. It details the capacity-building processes that have empowered them to earn significant supplementary incomes. The narrative highlights the tangible outcomes of increased self-reliance, the pivotal role these communities play in mitigating forest fire incidents, the notable decline in women's drudgery, and

H. Pant (✉) · P. Upadhyaya
CSED (Centre for Socio-economic Development), GBPNIHE (G.B. Pant National Institute of Himalayan Environment), Almora, Uttarakhand, India
e-mail: harshit.pantj@gbpihed.nic.in

N. Moran et al. (eds.), *The Participatory Role of Indigenous Women*,
https://doi.org/10.1007/978-3-032-24059-0_3

the consequential contributions towards fulfilling multiple Sustainable Development Goals (SDGs) through a practical model of ecosystem-based adaptation (EbA).

Keywords Chir pine · Bio-briquette · Climate-change · Forest fire · Alternative livelihood · Ecosystem-based adaptations · Women empowerment · Western Himalaya · Sustainable development goals (SDGs)

1 Introduction

The Western Himalayan region, an awe-inspiring yet ecologically fragile montane ecosystem spanning the Indian states of Uttarakhand, Himachal Pradesh, and Jammu & Kashmir, stands on the front lines of climate change vulnerability. This region is characterized by its steep topographic gradients, rich biodiversity, and the profound dependence of its human populations on natural resources for sustenance and economic security. The complex interplay of climatic variability, anthropogenic pressures, and geological sensitivity renders it highly susceptible to environmental degradation, with cascading effects on socio-economic systems. While all communities are vulnerable, the burdens are not distributed equally. Rural women, due to their gendered roles as primary managers of household energy, water, and fodder, are on the front lines of these environmental shifts. Their daily lives are intimately connected to the state of the surrounding forests, making them both the most affected by ecological degradation and the most critical agents of its restoration (Layek et al. 2022; Nirmala et al. 1996; Naudiyal & Schmerbeck 2017; Parashar & Biswas, 2003; Rawat et al. 1997; Sharma et al. 2023).

One of the most pressing and paradoxical environmental challenges, particularly in the state of Uttarakhand, is the extensive accumulation of leaf litter from the Chir Pine (*Pinus roxburghii* Sarg.). This native conifer has seen its range expand, often at the expense of diverse broadleaf forests, due to historical forestry practices and changing fire regimes. While the tree itself is a component of the native flora, the sheer volume of dry, resinous needles it sheds creates a persistent ecological problem. This dense, carpet-like layer of biomass contributes significantly to the incidence and intensity of forest fires, alters soil chemistry by increasing acidity, impedes the natural regeneration of other plant species, and consequently leads to a loss of biodiversity and ecosystem services (Arya & Negi 2021; Azad et al. 2022; Bahuguna & Singh 2002; Joshi et al. 2015; Rawat et al. 1996; Semwal & Mehta 1996; Sharma et al. 2015; Singh & Singh 1992).

However, in a striking example of turning a problem into a solution, these very pine needles are being re-envisioned as a cornerstone for sustainable livelihood generation, particularly for rural women. This transformative approach aligns perfectly with the principles of Ecosystem-based Adaptation (EbA). EbA is defined as the use of biodiversity and ecosystem services as part of an overall adaptation strategy to help people adapt to the adverse effects of climate change. By harnessing this underutilized biomass, initiatives are simultaneously promoting climate resilience by reducing fire hazards, combating forest degradation, creating green entrepreneurship opportunities, and empowering women economically and socially a truly integrated approach to sustainable development.

Globally, energy consumption patterns, especially in rural areas of developing nations, remain heavily reliant on traditional biomass sources such as fuelwood, animal dung, and agricultural residues. This dependence triggers a vicious cycle: it accelerates deforestation and forest degradation, leads to loss of biodiversity, exacerbates soil erosion, increases vulnerability to floods and landslides, and ultimately degrades the very ecosystem services upon which communities depend. The search for sustainable alternatives is therefore not merely an energy imperative but an ecological and social one.

Among the plethora of renewable energy alternatives, the briquetting of waste or residual biomass to produce bio-briquettes has emerged as a robust, reliable, and contextually appropriate option (Singh, 2013). Biomass briquetting technology possesses the distinct advantage of flexibility, capable of utilizing a wide spectrum of waste biomass including agricultural residues (e.g., rice husk, straw), agro-industrial residues (e.g., bagasse, coconut shell), and critically, forest waste. In the specific context of Uttarakhand, waste biomass in the form of Chir Pine needles represents a dual opportunity: it is a substantial, untapped energy resource and its removal from the forest floor mitigates a major fire hazard that causes extensive annual damage to forest wealth, encompassing both flora and fauna (Negi, 2019). Therefore, the strategic utilization of this waste material for briquetting presents a win-win scenario, preventing and reducing the likelihood of catastrophic forest fires while simultaneously generating a valuable, clean(er) energy source.

Biomass, in its raw form, is a abundantly available, renewable, and carbon-neutral raw material. However, its direct combustion is inefficient and highly polluting. To enhance its utility, biomass can be first processed into biochar through a thermochemical process known as pyrolysis. Pyrolysis involves the thermal decomposition of organic material in the complete or near-complete absence of oxygen (Ozbay & Ayrilmis, 2017). This process yields three main product streams: a solid product (char and ash), a liquid product (bio-oil), and a gaseous product (syngas). The conditions under which pyrolysis is conducted—particularly temperature and heating rate profoundly determine the chemical composition and physical properties of these products.

The solid product, biochar, is a porous, carbon-rich material. If its pore structure and surface area are sufficiently developed, it can be further processed into activated carbon for filtration and other industrial applications (Faizal, 2017). At its core, pyrolysis works by breaking down the large, complex hydrocarbon polymers (like cellulose, hemicellulose, and lignin) present in biomass into smaller, simpler molecules across the three phases (Basu, 2013). While raw biomass has a relatively lower calorific value, densifying it into bio-briquettes significantly enhances its energy density. A bio-briquette is essentially biochar (or sometimes raw biomass) compressed into a specific, regular shape using mechanical pressure and often a binding agent, resulting in a fuel with a higher calorific value and more consistent burning properties than the original feedstock (Fikri & Sartika, 2018). This makes it a viable and sustainable alternative to fossil fuels like coal (Tamilvanan, 2013; Sati, 2020, Uttarakhand Forest Department, 2022).

The scope of biomass for briquetting extends beyond pine needles. Invasive alien plant species (IAPS) such as *Lantana camara* and *Eupatorium odoratum* (Congress grass), which themselves are major ecological threats by outcompeting native vegetation, can also be effectively utilized for biomass briquetting. This offers a secondary benefit of controlling these invasive weeds. However, these species are often localized in specific pockets and, with the exception of the highly pervasive Lantana, may not offer the sheer volume of raw material required for large-scale energy production.

In this context, the dry leaf litter of native tree species that are of limited direct use to local communities, such as *Pinus roxburghii*, presents an unparalleled opportunity. It is an abundant, annually renewable, and currently problematic source of raw material. This biomass can be efficiently converted into high-density bio-fuel briquettes or bio-coal, offering a clean, efficient, and fairly good substitute for conventional fuels like coal, lignite, firewood, and even LPG to a certain extent. This alternative fuel source delivers manifold environmental benefits: it reduces the direct pressure on forests for fuelwood, thereby aiding in biodiversity conservation and carbon sequestration, and it alleviates the immense physical burden on women who are primarily responsible for fuelwood collection. The scale of the resource is significant: the annual gross yield of pine needles available for use as an energy feedstock in Uttarakhand is estimated at a staggering 1.9 million tonnes. After accounting for practical loss factors such as accessibility, alternative uses, and forest fires, the annual net yield is still approximately 1.33 million tonnes enough to support an installed power generation capacity of up to 789 MW (Kala & Subbarao, 2017), highlighting its immense potential.

At the G.B. Pant National Institute of Himalayan Environment (GBPNIHE) in Kosi-Katarmal, Almora, this potential is being actively harnessed. The campus itself generates over 50 tonnes of pine needles annually, which are utilized at the Institute's Rural Technology Complex (RTC) for the preparation of smokeless bio-briquettes. The RTC serves as a hub for demonstrating, popularizing, and transferring this technology to rural communities and local towns across Uttarakhand. A comprehensive capacity-building program has been implemented, imparting hands-on training to a diverse range of stakeholders, with a deliberate focus on rural women and members of weaker socio-economic sections. These trained individuals now produce smokeless bio-briquettes in their own villages using collected pine needles ("Pirul"). The briquettes are used for domestic cooking and heating needs, and the surplus is sold in local markets, generating a critical supplementary income (Pant et al., 2023).

A tangible example of this success comes from the Hawalbagh and Someshwar developmental block of Almora District. Women's SHG trained by GBPNIHE produced and sold 2976 bio-briquettes, earning ₹27,600 over the last three years in the local markets of Almora town. Beyond briquetting, the innovation extends to value-added products. Pine needles have been successfully used to produce handmade paper, which is then crafted into a variety of marketable goods such as file covers, folders, wedding cards, notepads, carry bags, and envelopes. This initiative not only promotes the use of organic, tree-free paper but also directly addresses the issue of forest fire fuel load reduction while creating income generation opportunities.

This multifaceted intervention has garnered significant attention and support from various State and Central Government agencies, including the Ministry of Environment, Forest and Climate Change (MoEF&CC), the Governor's House in Ladakh (for their Carbon Neutral Ladakh mission), the Vivekananda Parvatiya Krishi Anusandhan Sansthan (VPKAS) in Almora, G.B. Pant University of Agriculture & Technology, and various departments of the Uttarakhand State Government (Forest, Agriculture, Horticulture, Tourism), as well as livelihood missions like ILSP-Ajivika. The production and sales figures attest to its viability: under this intervention, the RTC has prepared and sold 4236 file covers, 852 meeting folders, 200 meeting diaries, 1176 carry bags, 100 wedding cards, and 980 envelopes, all using Chir Pine needles as the primary raw material. Furthermore, the scope of craftsmanship has expanded to include handicraft products such as Rakhis (traditional bracelets), artificial jewelry, and decorative items fashioned from pine bark, cones, and needles. Training in this artistry has been provided to women and marginalized groups, and these unique, eco-friendly pine-based products have found a thriving market both within and outside the region (Negi et al., 2021).

2 Rationale and Problem Statement

The hill communities of Uttarakhand, and particularly the women and socio-economically weaker sections within them, face a starkly limited palette of economic activities and employability options to improve their standard of living. The mountain context is defined by a triad of constraints: marginality (remoteness from mainstream economic hubs), fragility (sensitivity of the ecosystem to disturbance), and inaccessibility (poor connectivity and infrastructure). These factors are compounded by a high susceptibility to frequent natural hazards such as landslides, flash floods, and earthquakes (Pant, 2016) and anthropogenic hazards such as forest fire (Pant & Mishra, 2025).

Climate change acts as a threat multiplier in this already challenging environment. Its impacts manifesting as changing rain and snow patterns, receding glaciers, and altered crop seasons further strain the already precarious livelihoods of mountain communities (Pant, 2018). Amidst these challenges, the perennial threat of forest fires, fueled by the abundant pine needle litter, poses a particularly vicious cycle. These fires deplete natural resources, destroying fodder grasses, wild edibles, and Non-Timber Forest Products (NTFPs) that communities rely on. They can lead to the drying up of natural water springs and accelerate the invasion of hardy, non-native weeds in the burnt forest areas, further degrading the ecosystem and reducing its utility for local people (Negi, 2019).

The gendered dimension of this problem is acute. Rural women bear the disproportionate burden of resource collection. They are heavily engaged in the arduous and time-consuming task of gathering fuelwood from increasingly distant forests. This not only adds to their physical drudgery but also contributes to the gradual degradation of forest wealth. While the penetration of Liquefied Petroleum Gas (LPG) has increased even in rural areas, its high recurring cost often places it out of reach for the poorest households. Consequently, dependence on fuelwood remains high. The excessive use of fuelwood in traditional, inefficient cookstoves (chulhas) creates severe environmental and health hazards. It releases harmful pollutants into the immediate environment, including particulate matter, black carbon (a potent short-lived climate pollutant), and gases like carbon monoxide, leading to significant indoor air pollution. The World Health Organization (WHO) consistently links this exposure to acute respiratory infections, chronic obstructive pulmonary disease, and other health issues, primarily affecting women and children. Therefore, the search for alternative, affordable, clean, and locally available sources of energy is not a matter of choice but of urgent necessity (Pant et al., 2026b)

The ultimate aim of the initiatives described in this chapter is to build community capacity and encourage a shift towards the use of alternative energy sources, specifically bio-briquettes prepared from "waste" biomass like Chir Pine needles and invasive alien species. This strategy is deeply embedded in the framework of Ecosystem-based Adaptation (EbA). It seeks to achieve multiple, interlinked objectives:

1. **Enhance climate resilience** by reducing a major climate-exacerbated hazard (forest fires).
2. **Reduce forest degradation** by providing an alternative to fuelwood and by removing a fire risk.
3. **Empower women** in the region by reducing their drudgery, improving their health, and providing them with a source of income and economic agency.

The dwindling availability of forest resources and the high cost of conventional modern energy sources like LPG and electricity (which is often unreliable in remote hills) mean that these cannot singularly address the ever-growing energy demand, primarily for cooking and heating. Therefore, decentralized, community-owned technologies related to alternative renewable energy sources hold immense promise. The technology of biomass briquetting from pine needles is one such preposition that has been rigorously tested, refined, and popularized over the last several years in Uttarakhand and other parts of the IHR, demonstrating its technical viability and social acceptability.

The present study is guided by a set of core hypotheses that frame its research and development agenda:

- **Hypothesis 1**: Bio-briquettes of consistent quality and good calorific value can be efficiently produced from unwanted weeds/bushes (e.g., *Lantana camara*, *Eupatorium odoratum*) and Chir-Pine leaf litter materials that currently inhibit the growth of forest vegetation and pose a persistent fire hazard.
- **Hypothesis 2**: The developed briquettes do not compete with the extraction of fuelwood from living trees. Instead, they directly substitute it, thereby giving a chance for forest trees to thrive and regenerate. This contributes significantly to

maintaining forest greenery, enhancing carbon sink capacity, and helping India meet its Intended Nationally Determined Contributions (INDCs) as per the Paris Agreement, directly contributing to SDG 13 (Climate Action) and indirectly to others like SDG 12 (Responsible Consumption and Production) and SDG 15 (Life on Land).

- **Hypothesis 3**: The bio-briquettes provide communities with a more efficient and cleaner form of fuel compared to raw fuelwood. They serve as an effective substitute for cooking and warming houses during the harsh Himalayan winters, with the critical advantage of producing almost no smoke, thereby improving indoor air quality and health.
- **Hypothesis 4**: The entire value chain from the collection of raw material to production and marketing provides self-employment and income generation opportunities. This is particularly impactful for women, who are the primary collectors, thus helping them generate additional income and contributing significantly to their economic empowerment, aligning with SDG 5 (Gender Equality) and SDG 7 (Affordable and Clean Energy).
- **Hypothesis 5**: The initiative generates distributed employment opportunities not just for the producers but also along the supply chain, including for transporters and distributors of the finished briquettes, creating a ripple effect in the local economy.
- **Hypothesis 6**: The bio-briquettes can provide a reliable substitute to kerosene and LPG in villages and towns, especially during periods of shortage and supply chain disruptions, thereby enhancing energy security.
- **Hypothesis 7**: This venture contributes to the livelihood and income of the rural poor while simultaneously safeguarding them from the ill-effects of forest fires (which destroy their resources) and the health hazards associated with smoke from traditional chulhas.

The scientific rationale underpinning this study is robust and directly links to global climate concerns. It is focused on reducing carbon emissions through two primary pathways:

1. **Avoiding Deforestation**: By providing a substitute for fuelwood, it reduces the pressure on forests, preventing deforestation and the associated release of stored carbon.
2. **Avoiding Forest Fires**: By systematically removing the primary ground fuel (pine needles), it prevents the release of massive amounts of carbon dioxide and other greenhouse gases during wildfires.

By using Chir Pine leaf litter, weeds, invasive species, and other forms of forest and agricultural waste as a source of alternative bio-fuel, the study also aims to develop the capacity of rural communities towards sustainable livelihood and income generation, effectively fostering micro-enterprise development at the level of Village Self-Help Groups (SHGs).

3 Methodology: A Tripartite Strategy for Implementation

The research and development intervention adopted a comprehensive three-fold strategy for its methodological implementation and on-ground manifestation, ensuring both technological transfer and socio-economic impact:

1. **Live Demonstrations and Technology Transfer**

The first pillar involved establishing four functional demonstration units in partnership with women's groups in Chir Pine forest-dominated blocks of Almora district. These production centers served as live, hands-on learning labs where the entire process of bio-briquette production—from needle collection to final marketing—was demonstrated. The strategy employed a Training of Trainers (ToT) model. Master Trainers were identified and capacitated at these centers. These Master Trainers, often motivated local women or community resource persons, then became the nodal points for disseminating the technology to a wider audience in their respective villages and clusters. This approach ensured the scalability and sustainability of the intervention. The technology was specifically chosen for its appropriateness for the rural context: it is based on locally available bio-wastes, is relatively simple to operate and maintain, and is capable of producing high-quality, low-cost fuel without requiring sophisticated infrastructure or high capital investment.

2. **Capacity Building and Institutional Strengthening**

The second pillar focused on building the human and institutional capital necessary for long-term sustainability. This involved structured capacity-building programs not only for the direct beneficiary communities but also for local Non-Governmental Organizations (NGOs) and Community-Based Organizations (CBOs) like Self-Help Groups (SHGs). Training modules covered all aspects of the enterprise: technical production skills, basic business management, financial record-keeping, quality control, collective marketing, and monitoring. The goal was to ensure local ownership and enable women beneficiaries to transition from being mere laborers to managers of their own micro-enterprises. This empowerment is critical for the resilience of the initiative beyond external project support.

3. **Diversification through Green Entrepreneurship**

Recognizing the need for diversified income streams and to engage a wider section of the community, the third pillar promoted green entrepreneurship beyond bio-briquettes. This involved training and supporting artisans, particularly women, in the development of high-value, low-volume artifacts using various parts of the pine tree needles, bark, and cones. This initiative tapped into traditional craftsmanship, combined it with modern design sensibilities, and connected it to markets for eco-friendly products, thus creating an additional avenue for income generation and enhancing the overall economic viability of the pine-based value chain.

3.1 Detailed Field Methodology

a. **Bio-briquette Preparation Process**

The core technology adopted is known as "Energy Cake Technology," which efficiently converts diverse forms of biomass (including Pine needles, weeds, and invasive alien species) into dense, smokeless briquettes that burn at high temperatures (T_{max} > 500 °C), making them suitable for cooking and heating. This technology is particularly suited to the rural Himalayan context as it requires low capital cost, has low operational costs, reduces cooking time due to higher efficiency, reduces the drudgery of fuelwood collection, and requires minimal infrastructure.

The production process involves the following meticulously designed stages:

i. **Biomass Collection**: This is the first and crucial step. Biomass is systematically collected from community forest areas. For pine needles, collection is often organized in groups before the peak fire season (summer). Similarly, invasive species like Lantana are harvested from areas where they need to be controlled. This activity itself provides the first point of income for collectors.
ii. **Preparation of Bio-waste/Forest Waste Char (Pyrolysis)**: The carbonization unit is a simple, low-cost, 3 feet × 2 feet earthen pit. This pit is used for pyrolysis-based anaerobic burning (i.e., with limited oxygen). Almost all local bio-residues can be carbonized in this unit. The process involves loading the biomass into the pit, igniting it, and then covering it with a metal sheet and a layer of soil to restrict oxygen flow. The material smolders instead of burning openly. Depending on the type and moisture content of the material, a batch of 50–150 kg of biomass can be converted into 20–80 kg of charred material (bio-char) over a period of 2–3 h.
iii. **Crushing of Bio-char**: The charred material is often in lumps and needs to be crushed into a fine powder to ensure uniformity and effective binding in the next stage. Crushing is a basic but essential treatment. The quality of the crushed material (particle size and uniformity) directly affects the performance of the briquetting machine and the quality of the final product (e.g., density, durability). If particles are too large, the briquetting machine has to work harder, consuming more power and causing wear and tear. While electric crushers are efficient, for a rural setup, a manual mold or a handheld mechanical crusher is often more appropriate and cost-effective, keeping the technology accessible and decentralized.
iv. **Binder Application**: A binding agent is necessary to hold the bio-char particles together under pressure to form a cohesive briquette. In this methodology, sieved soil (clay) is used as a natural, easily available, and cost-free binder. The addition of clay serves multiple purposes:

    - It acts as a binder, providing the necessary cohesion for the briquette to maintain its shape.
    - It can act as an energy extender, though excessive clay will reduce the overall calorific value.

- Scientifically, after adding the binder, an adsorption layer forms on the surface of the bio-char particles. This creates a gravitational force between particles via van der Waals forces. When external pressure is applied during molding, the particles can also generate an electrostatic attraction, resulting in a stable chain structure that gives the briquette its mechanical strength.

v. **Mixing of Char with Clay**: The crushed bio-char is thoroughly mixed with about 20–30% (by weight) of clay or finely sieved soil. The mixing must be uniform to ensure consistent quality across all briquettes. The moisture content is critical at this stage; water is added to achieve a malleable consistency that will hold under pressure. If too much clay is added, the briquette will become difficult to ignite and will burn poorly due to the inert nature of the clay, which turns into ash after burning.

vi. **Production of Cake from the Char**: The moist mixture is fed into a simple, manually operated mold. The mold typically consists of three parts: (a) a bottom plate with protruding rods that create air holes in the briquette (aiding combustion), (b) a cylinder that fits over the bottom plate to contain the mixture, and (c) a perforated top plate that is pressed down to compress the mixture. The pressure applied determines the density of the final briquette. The molded "cakes" or briquettes are then ejected from the mold.

vii. **Drying and Use**: The freshly made briquettes are soft and contain moisture. They are sun-dried for several days until they are hard and dry. Once fully cured, they are ready for use. They can be combusted in locally available, ordinary coal pots or burners (chulhas). These briquettes are used for household cooking and heating and are also packaged and sold in the market, forming the basis of the green entrepreneurship model (Fig. 1).

b. **Green Entrepreneurship: Artifact Development (e.g., Rakhis)**

The process of creating eco-friendly, marketable products like Rakhis from pine needles, bark, and cones involves several meticulous steps that blend traditional craftsmanship with sustainable practices:

1. **Collection & Preparation of Materials**

   - Women collect pine needles from the forest floor, ensuring sustainable harvesting.
   - The collected needles are sorted by length and quality.
   - Clean pieces of bark are collected, and small cones are gathered.
   - The cones are sometimes broken down into smaller, decorative parts.
   - Other ancillary materials are sourced: cotton thread for the base, natural glue (or Fevicol), and decorative items like beads, sequins, and colors, often repurposed from household artifacts.

2. **Designing the Rakhi Base**

   - The base structure is created by twisting and designing cotton threads into a circular band, sized to fit a wrist.

- Small bundles of pine needles (typically 10–15 needles) are carefully arranged.
- The base of the needle bundle is tied tightly onto the cotton thread band using a thin thread.
- The rakhi is then flattened and shape is given. The attachment is further secured with a dab of glue for durability.

3. **Embellishment and Value-Addition**

- This is the creative stage where value is added. Small pieces of bark or parts of pine cones are attached with glue to serve as the centerpiece of the Rakhi.
- Further decoration is done using beads, sequins, or tiny mirrors.
- To incorporate local cultural elements, traditional *Aipan*-style (a traditional Kumaoni folk art) paintings or motifs are done on the bark pieces.
- Sacred symbols like 'Om' or 'Swastika' are also painted to enhance their cultural and aesthetic appeal, making them attractive during festivals like Raksha Bandhan.

4. **Marketing & Sale**

- **Local Markets**: The primary market is during the Raksha Bandhan festival. Women's groups set up stalls in local haats (markets), town squares, or near temples.
- **Tourist Spots**: Given the unique, eco-friendly nature of the product, they have high appeal for tourists. Stalls are set up in major tourist attractions like Almora, Kausani, Ranikhet, etc.
- **Pre-orders and Customization**: Orders are also taken in advance from shops, NGOs, and government agencies for bulk gifting purposes.

4 Results and Findings

4.1 A Holistic Tripartite Strategy for Women-Led Ecosystem-Based Adaptation in Almora, Uttarakhand

In the context of rural India, where climate change and resource scarcity disproportionately impact women, empowering female-led collectives becomes crucial for both social progress and environmental resilience. This case study examines an intervention carried out in Gwalakote village of Someshwar Block, Almora, focusing on ecosystem-based adaptation (EbA) through a three-pillar strategy. Rather than serving as a mere technology transfer, the initiative sought to build enduring agency, entrepreneurship, and institutional capacity among marginalized women, enabling sustainable livelihoods and social transformation (Fig. 2a).

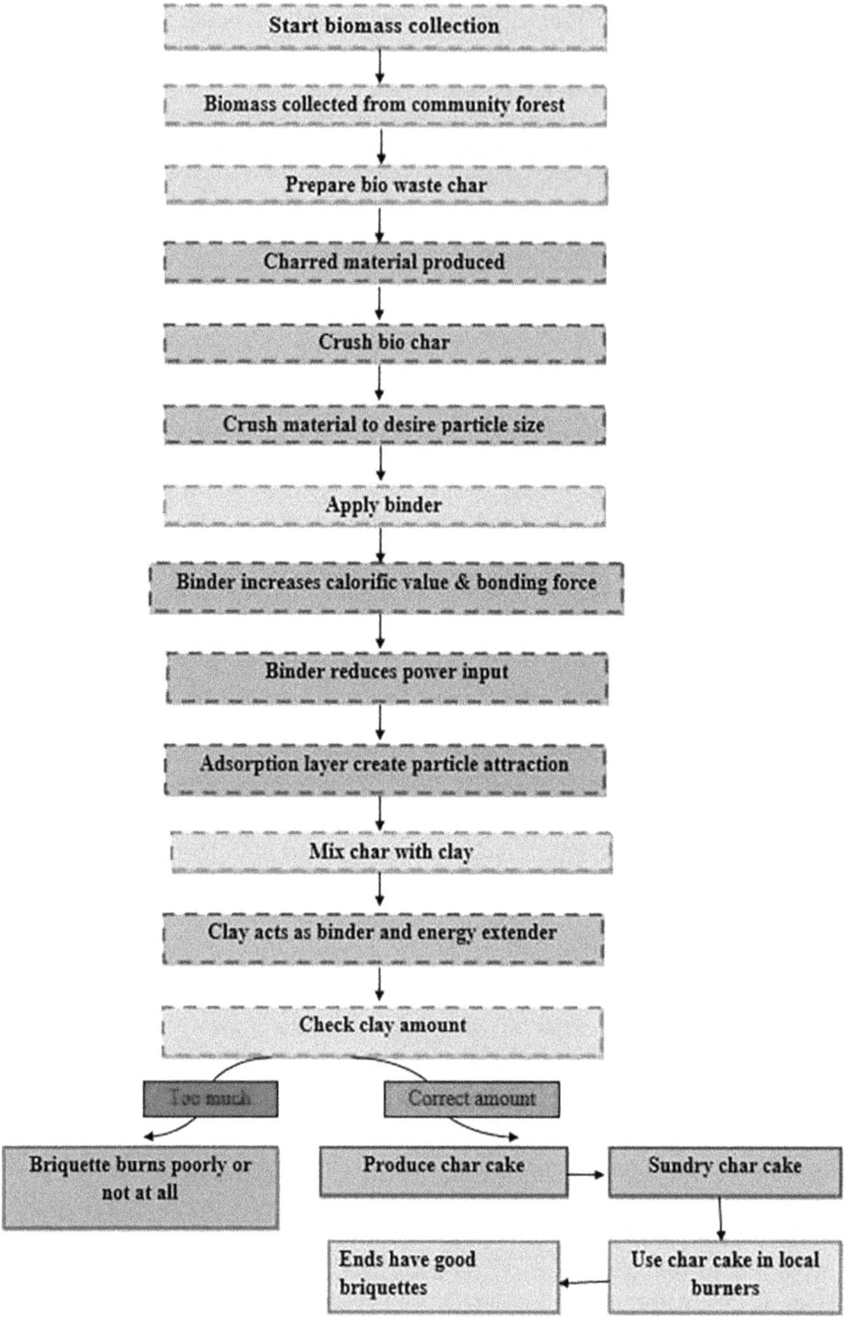

Fig. 1 Flow diagram of the pine-needle based bio-briquette preparation

Fig. 2 Pine needle-based biomass briquetting—selling and demonstration by women. **a** Demonstration of bio-briquette, **b** burning of bio-briquette for cooking, **c** prepared bio-briquette, **d** training of women SHG towards bio-briquette preparation)

Prior to the intervention, women in Gwalakote faced immense hardship rooted in their roles as primary collectors of forest biomass for household energy and subsistence. These responsibilities translated into considerable physical drudgery women spent nearly 18 h weekly on fuelwood collection from surrounding forests. With limited educational attainment (average: 6 years of schooling) and scant economic opportunities, most members of the local Self-Help Group (SHG) were confined to subsistence agriculture and livestock rearing.

Fuelwood dependency carried profound health costs, with frequent cases of eye irritation, respiratory problems, and indoor air pollution in kitchens. The local forest ecosystem, dominated by pine, was plagued by litter accumulation, soil degradation, and recurrent forest fires typically 1–2 per year. Women's social status was further constrained by weak representation in village decision-making and a lack of control over household finances. An integrated solution was needed to address intersecting challenges of poverty, ecological degradation, and gender disparity (Fig. 2b).

The intervention, implemented from 2022 to 2024, targeted a group of 35 women from marginalized backgrounds, aiming to foster ecological stewardship and independent income. A mixed-methods evaluation was employed. Quantitative surveys tracked changes in income, time use, fuel adoption, and banking access (Fig. 2c). Qualitative insights were captured through four focus group discussions (FGDs), key informant interviews with group leaders, master trainers, and forestry officials. The 24-month longitudinal approach enabled measurement of both immediate impacts and sustained changes (Fig. 2d).

The Tripartite Strategy

1. **Technology Ownership and Live Demonstrations**

At the core of the initiative was hands-on technological empowerment. Four demonstration units, featuring simple pyrolysis pits and manual briquetting presses, were established and managed entirely by women's groups. The technology was purposefully chosen for its affordability, user-friendly design, and suitability for decentralized enterprise models. The Training of Trainers (ToT) framework produced a cadre of sustainability stewards female master trainers who became recognized local experts. This built confidence, retained technical knowledge within the community, and ensured new members could be onboarded sustainably.

2. **Institutional Strengthening and Financial Literacy**

To fortify women's collectives as viable business entities, comprehensive capacity building was delivered. Members learned enterprise bookkeeping, micro-business management, quality control, and collective bargaining. These skills enabled the SHG to transition from informal cash-based operations to a formalized, bankable structure. The intervention facilitated opening of bank accounts for all members a dramatic shift from just 40% at baseline allowing secure savings, access to micro-loans, and financial independence. Critically, SHG members reported increased confidence in both household and collective financial decisions, marking a decisive step towards autonomy.

3. **Green Entrepreneurship and Market Linkages**

Acknowledging that production of bio-briquettes alone might not suffice, the program supported diversification into eco-friendly handicrafts. Women created decorative items Rakhis, handmade paper, ornaments using pine needles, bark, and cones. Market linkages were actively nurtured: members participated in regional exhibitions, connected with government procurement schemes (e.g., office stationery supply), and engaged with tourism sectors in Almora and Kausani. These activities ensured that production was guided by verified demand and provided higher, more stable returns than commodity sales.

5 Results

The intervention yielded substantial positive outcomes across economic, social, health, and environmental dimensions for women-led ecosystem-based adaptation in Almora. Economically, the project boosted average household enterprise income by 38%, with many women earning independent income for the first time, enhancing their financial autonomy and decision-making power. Drudgery associated with fuelwood collection was significantly reduced, cutting weekly hours by two-thirds and freeing time for productive and personal activities. Health improvements were marked, as 90% of participants reported fewer respiratory and eye ailments due to the adoption of smokeless bio-briquettes. Environmentally, systematic pine needle collection decreased forest litter by 71% and effectively prevented forest fires over two consecutive years, fostering regeneration and biodiversity recovery (Fig. 3a and b). Socially, women's confidence and agency were strengthened, reflected in increasing participation in local governance structures and community initiatives, while shifts in gender norms emerged with broader community recognition of women's vital ecological and economic roles. These integrated results illustrate the powerful multiplier effect of a holistic, gender-responsive approach to climate resilience and sustainable livelihoods.

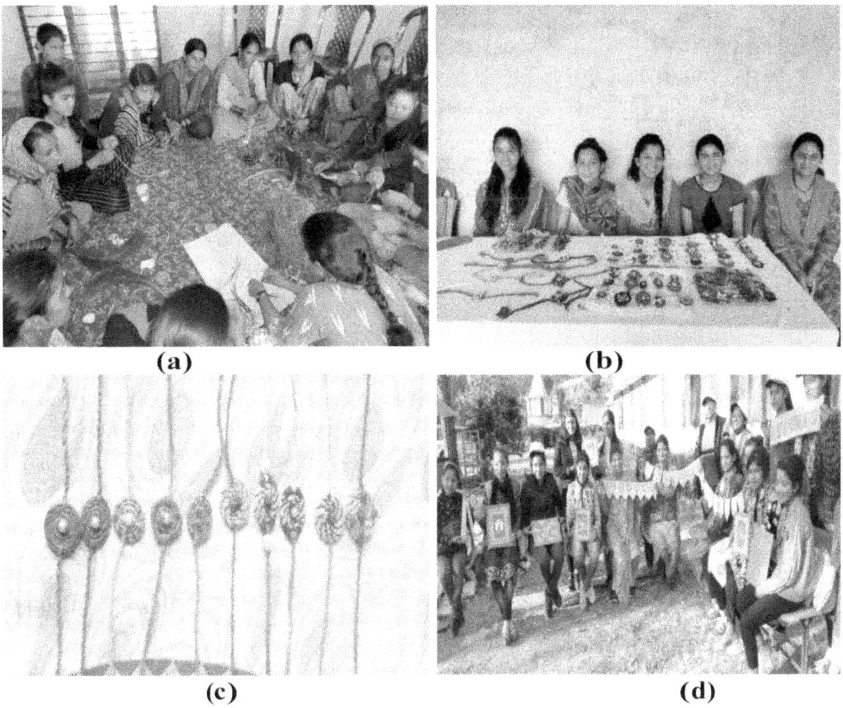

Fig. 3 Pine needle-based artifact preparation (**a** preparation of pine needle based rakhi, **b** selling of rakhis by women SHGs, **b** handmade pine needle based rakhis, **d** preparation of pine based decorative items)

Table 1 Socio-economic impact on SHG members (N = 35)

Indicator	Pre-intervention (2022)	Post-intervention (2024)	% change
Avg. annual income from enterprise	₹0	₹18,400	100%
Avg. weekly time on fuelwood collection	18 h	6 h	– 67%
Households using smokeless fuel	3% (LPG)	100% (briquettes)	+97%
Members with Bank accounts	40%	100%	+60%

Source Field Survey (2024)

5.1 Socio-economic Outcomes

i. **Income Growth**: Annual household income from enterprise rose by an average ₹18,400 ($220), a 38% increase from baseline a transformative gain for families previously reliant solely on subsistence farming. For 12 women, this was their first-ever independent income, fundamentally altering their role within the household (Table 1).
ii. **Reduced Drudgery**: Fuelwood collection dropped from 18 to 6 h per week per member, facilitating labor reallocation to the enterprise, childcare, rest, or education.
iii. **Financial Inclusion**: The ability to save and manage money has empowered long-term planning and bolstered creditworthiness in the women and had made them self-dependent in aspect of fund management.
iv. **Dignified Market Access**: Handicraft sales, particularly "Paryavaran Raksha Sutra" eco-friendly Rakhis, attracted premium pricing (₹150–200 vs. ₹50 conventional), with 85% revenue disbursed directly to artisans and 15% to group funds for raw material procurement.

5.2 Health and Ecological Impacts

i. **Health Improvement**: Ninety percent of members reported marked reduction in respiratory and eye complaints due to the adoption of smokeless briquettes.
ii. **Forest Management**: Systematic pine needle collection reduced litter by 71% in local forest plots. Notably, the area experienced zero forest fires for two consecutive years, compared to an annual average of one or two before.
iii. **Biodiversity and Carbon Mitigation**: Ecological monitoring documented undergrowth regeneration, with native herbs and oak saplings appearing in

Table 2 Annual environmental impact of the intervention

Parameter	Quantity	Equivalent impact
Pine needles collected	8.5 tonnes	–
Bio-briquettes produced	3.2 tonnes	–
Fuelwood saved	9.1 tonnes	–
CO_2e emissions avoided	42 tonnes	Equivalent to taking 9 cars off the road for a year
Forest area protected	10 ha	0 fires in 2 years

Source Field Data & IPCC Calculation Methods, GBPNIHE (2024)

cleared plots. The group's work avoided an estimated 42 tonnes of CO_2e emissions per year, both by preventing forest fires and replacing fossil/wood fuels (Table 2).

5.3 *Social Change and Shifting Gender Norms*

i. **Agency and Representation**: Women began participating actively in village council (Gram Sabha) meetings, advocating for resource protection and participating in local governance.
ii. **Community Cohesion**: The SHG evolved into a hub for broader initiatives, including health camps and water conservation programs.
iii. **Intergenerational Benefits**: Increased disposable income resulted in greater investment in children's education, especially for girls.
iv. **Changing Social Attitudes**: FGDs revealed profound shifts in gender norms. Initial skepticism from male community members transitioned to support, as the economic and ecological benefits became evident. One member captured this change: "Earlier, we only went to the forest to take. Now, we also go to protect it for our future (Fig. 3c).

5.4 *Environmental and Ecological Outcomes*

i. The ecological and environmental impacts of the intervention in Almora were significant and multifaceted. Systematic collection of pine needles a major source of forest floor fuel led to a 71% reduction in litter load (Jugran & Tewari, 2022) within the designated forest plots managed by women's self-help groups (Table 3).
ii. This reduction in highly flammable material played a crucial role in preventing forest fires, with the area recording zero fire incidents over two consecutive years, reversing the previous annual average of one to two fires. By diminishing

fire frequency and intensity, the intervention contributed to preserving biodiversity and soil quality, allowing native herbaceous species and oak saplings to regenerate in the understory.

iii. Moreover, the sustainable use of pine needle biomass through briquette production replaced dependency on fossil fuels and traditional fuelwood, thereby reducing harmful smoke emissions and indoor air pollution. These efforts also resulted in an estimated annual avoidance of 42 tonnes of CO_2 equivalent emissions, highlighting the intervention's role in local climate mitigation alongside ecosystem restoration (Pant & Mishra, 2025).

iv. Additionally, by formalizing women's stewardship roles in forest management, the program fostered decentralized governance that incentivizes long-term conservation and resilience of fragile mountain ecosystems. Overall, the intervention demonstrated how targeted resource management and green enterprise activities can generate measurable ecological benefits while safeguarding human health and livelihoods in fire-prone forest landscapes.

5.5 *Overcoming Barriers*

Women collectively innovated solutions to persistent obstacles. Physical strain during pine needle collection was alleviated through organized group trips, turning a challenging chore into a social activity. Market access was achieved by partnering with local NGOs, facilitating connections and professional branding of products. Social resistance was gradually dispelled as increased financial contributions made a compelling argument for the continuation and expansion of women-led enterprises (Fig. 3d).

It also provided tangible evidence of the power of holistic, gender-responsive ecosystem-based adaptation. By knitting together technology, institutional strength, and entrepreneurship, the program enabled women to become custodians of their environment, confident business owners, and active civic participants. The synergistic impact spanning economic, environmental, health, and social domains illustrates how integrated approaches are far more effective than isolated interventions.

The Gwalakote SHG case study affirms that agency-driven EbA strategies can spur rapid, measurable improvement in livelihoods, forest resilience, and community status. Key lessons include the importance of technology adaptation, market awareness, institutional capacity, and broad-based financial literacy. Scaling up such models can catalyze climate resilience and women-led sustainable development across similar contexts, unlocking untapped human potential while safeguarding fragile ecosystems.

Table 3 Change in perception of community: pre and post outcomes

Pillar/indicator	Pre-intervention (2022)	Post-intervention (2024)	%/outcome/ change	Description/ finding
Technology ownership & demonstration	None	4 units managed by women's groups	+ 4 units	Pyrolysis & briquetting adopted, cadre of female trainers built
Institutional strengthening & financial literacy	Informal, cash-based	Formalized, trained SHGs	100% banked	35 members trained in enterprise skills, all now hold bank accounts
Green entrepreneurship & market linkages	Subsistence income	Diversified income sources	+ 38% income	Income from briquettes & crafts; direct market links, premium pricing achieved
Avg. annual income from enterprise	₹0	₹18,400	100% increase	Enterprise income for all; 12 women gain first independent income
Avg. weekly time on fuelwood collection	18 h	6 h	– 67%	Significant reduction in drudgery and free time for other activities
Households using smokeless fuel	3% (LPG users)	100% (briquette users)	+ 97%	Switch to smokeless fuel improves health
Health outcomes	Frequent respiratory, eye complaints	90% report improvement	Positive	Reduction in health problems due to shift in cooking fuel
Ecological impact (forest litter & fires)	Annual avg. 1–2 fires; high litter	Zero fires; 71% litter reduction	– 100% fires; – 71% litter	Needle collection reduces fire risk & enhances forest management
Gender norms & community influence	Limited social, household agency	Active in gram Sabha, business, forest	Major shift	Women now respected contributors, participate in decision-making; men support efforts

(continued)

Table 3 (continued)

Pillar/indicator	Pre-intervention (2022)	Post-intervention (2024)	%/outcome/ change	Description/ finding
Artisan collective model	Not present	Bulk sales, quality control, premium	Premium (up to ₹200/unit)	Individual artisans earn directly, group fund for reinvestment
Physical strain & market access	Collection burdensome, limited sales	Collective trips, new markets	Positive	Innovated logistics and marketing with help of NGO
Intergenerational impact (education)	Low educational investment	Higher investment in girls' education	Positive	Disposable income reinvested in children's education, esp. girls
Biodiversity & carbon emissions	Not monitored	Understory regeneration, 42 t CO_2e avoided	Measurable	Forest floor recovery, climate mitigation achieved through collective action

6 Synthesis and Recommendations: Championing Women-Led EbA for a Resilient Himalayas

The transformative journey of the women self-help groups in Uttarakhand represents far more than an isolated case study it embodies a revolutionary blueprint for climate adaptation in fragile mountain ecosystems worldwide. This initiative demonstrates with powerful clarity that the most effective Ecosystem-based Adaptation (EbA) strategies emerge from approaches that are locally owned, economically viable, and fundamentally gender-responsive. The work focus transcends theoretical frameworks to present a living paradigm where women have transformed from vulnerable victims of environmental change into primary architects of community resilience. The intervention's extraordinary success lies not in any single technological innovation but in its sophisticated socio-institutional architecture that positions women's collectives as managers, decision-makers, and beneficiaries across the entire value chain from forest management to market innovation.

The integrated outcomes documented through rigorous research reveal profound impacts across multiple dimensions. Ecologically, the systematic collection of pine needles by well-organized women's groups has produced measurable reductions in forest fire risk and initiated promising processes of forest floor regeneration (Pant &

Rawal, 2025). Scientific monitoring confirms that these practices contribute significantly to carbon sequestration through avoided CO_2 emissions, enhance soil moisture retention capabilities, and create conditions conducive to improved biodiversity. Socio-economically, the model delivers a powerful triple benefit: it generates crucial supplementary income for historically marginalized households, dramatically reduces the time women spend on hazardous fuelwood collection, and substantially improves family health through the adoption of clean, smokeless cooking technologies. Most significantly, it cultivates a tangible sense of agency, dignity, and collective power among women, who now receive recognition as essential stewards of their environment and valuable contributors to their local economy (Fig. 4).

Despite these demonstrable successes, the scalability of such grassroots initiatives faces substantial challenges. These community-led models operate despite not because of existing policy and support ecosystems. Their growth remains constrained by three critical barriers: the absence of targeted financial mechanisms designed for women's collective enterprises, insecure access to forest resources despite their conservation role, and deficient market linkages that prevent value capture at the community level (Pant et al., 2026a). For every successful Samiti, numerous women's groups struggle without seed capital for essential equipment, navigate complex bureaucratic hurdles to secure legal access to raw materials, and remain isolated from fair market prices due to dependence on intermediaries. While national and state policies increasingly acknowledge women's empowerment and climate action as

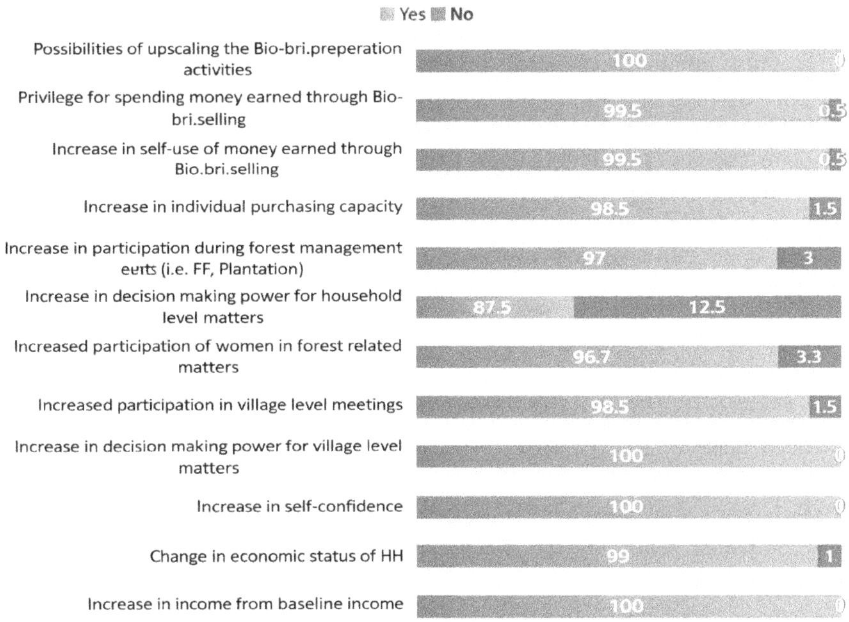

Fig. 4 Socio-economic impact of project R&D interventions on stakeholders in target village (N = 100)

priorities, they rarely intersect in ways that explicitly enable and finance women-led green entrepreneurship through coordinated, actionable measures.

6.1 Systemic Barriers to Scaling Women-Led EbA

The first fundamental barrier lies in financial exclusion as Women's collectives, particularly those operating in remote Himalayan regions, face immense difficulties accessing formal credit systems. Traditional lending institutions perceive them as high-risk clients due to their lack of collateral, unconventional business models, and geographic isolation. Even when microfinance is available, the terms often prove unsuitable for capital-intensive investments required for processing equipment like briquetting machines or pyrolysis units. The result is a perpetual cycle of small-scale operations that cannot achieve economic viability or scaling (Pant & Pant, 2025).

The second critical barrier is insecure resource tenure as although women's groups engage in sustainable harvesting of pine needles that actually improves forest health, their legal rights to these resources remain precarious. The existing regulatory framework for non-timber forest produce (NTFP) collection often fails to distinguish between extractive practices and conservation-oriented harvesting. Without formal, long-term access rights to specific forest areas, women's groups cannot make the investments necessary for sustainable management or plan for long-term enterprise development. This legal ambiguity also leaves them vulnerable to displacement by commercial interests or policy changes (Pant & Pant, 2025).

The third significant challenge involves market access limitations as women producers typically operate at the very beginning of value chains, selling raw or minimally processed goods to intermediaries who capture most of the final value. They lack connections to premium markets that would pay for the added ecological and social value of their products. Furthermore, they often lack the technological capabilities (e.g., digital marketing, e-commerce platforms) and business skills needed to negotiate better terms, manage inventory, or develop new products based on market demand.

6.2 Integrated Policy Recommendations for Transformative Change

To translate this proven model from isolated success stories into a widespread movement for Himalayan resilience, we propose a comprehensive policy framework comprising three interconnected pillars:

1. **Institutionalize Financial and Market Support for Women's Collectives**

The current paradigm of ad-hoc project-based funding must be replaced with a structured financial ecosystem specifically designed for women's green enterprises.

State governments should establish a Green Venture Fund exclusively for registered women's Self-Help Groups (SHGs) and cooperatives working in EbA and the green economy. This fund would provide two crucial financial products: First, seed capital grants covering 50–70% of the cost for essential equipment like briquetting machines, pyrolysis units, or toolkits for handicraft production. Second, it should offer conditional matching grants that correlate with the group's own savings and revenue generation, thereby incentivizing financial discipline and business acumen while ensuring commitment.

Concurrently, market barriers must be actively dismantled through preferential procurement policies. A state-level mandate should require government offices, tourism departments, and public sector undertakings to source a minimum percentage (recommended 15–20%) of their requirements for items like office stationery, promotional gifts, and hotel amenities from certified women-led green enterprises. This policy would create a guaranteed, dignified market that values both the product and its positive social and environmental impacts, enabling these enterprises to achieve stability and scale. Implementation would require establishing certification standards for "women-led green enterprises" and creating a transparent marketplace platform to facilitate connections between producers and government buyers.

2. **Secure Resource Access and Integrate Digital Platforms**

A woman's forest-based enterprise cannot achieve sustainability when her access to the primary raw material remains informal and precarious. Forest policies must evolve to formally recognize and support these conservation-oriented livelihoods. Specifically, state forest departments should issue long-term, renewable licenses for NTFP collection to registered women's groups in designated forest areas. These licenses would formalize the groups' role as forest managers with both rights and responsibilities including fire prevention, sustainable harvesting practices, and biodiversity monitoring. This formal access represents not a concession but a strategic investment in decentralized forest governance, as these groups become the de facto protectors of their designated zones with a direct stake in preventing degradation.

To overcome geographical isolation and unlock higher value capture, capacity building must incorporate robust digital inclusion strategies. Digital literacy training should become a mandatory component of all EbA projects, equipping women with skills to use smartphones and applications for inventory management, cost calculation, and communication. Beyond basic literacy, women should receive support in leveraging social media platforms like Instagram and Facebook and simple e-commerce solutions to tell their stories directly to consumers, build their brands, and command premium prices. This digital empowerment would fundamentally reshape market dynamics, reducing dependency on intermediaries and expanding market reach beyond local markets to regional and even national consumers who value ethically produced, eco-friendly products.

3. **Embed Gender-Responsive Strategies in Climate Policy Planning**

For these changes to achieve lasting impact, gender considerations must be mainstreamed from the periphery to the core of all climate planning instruments. This requires moving beyond tokenistic inclusion to measurable, accountable integration. State Action Plans on Climate Change (SAPCCs) must include mandatory provisions for gender-disaggregated data collection, gender analysis of climate vulnerabilities, and gender-responsive budgeting. Specifically, these plans should contain earmarked budgets for women-specific interventions in the climate sector, ensuring resources are allocated appropriately rather than being diluted within gender-blind programs.

Furthermore, the policy-making process itself must become genuinely inclusive through institutional innovation. The invaluable knowledge and experience of women grassroots leaders must be incorporated into decision-making structures through reserved seats for representatives from women's forest cooperatives on district-level climate committees, forest management councils, and biodiversity boards. Their lived experience constitutes an essential resource for designing effective, equitable policies that work on the ground. Additionally, government agencies should establish regular consultation mechanisms that bring policymakers directly into conversation with women-led enterprises to better understand implementation challenges and opportunities (Table 4).

6.3 Implementation Framework and Accountability Mechanisms

The proposed recommendations require coordinated implementation across multiple departments and levels of governance. We propose establishing a Cross-Departmental Task Force on Women-Led Climate Resilience reporting directly to the Chief Secretary's office. This task force would include representatives from departments of Environment, Forest, Rural Development, Women and Child Development, and Industries, with mandatory representation from women's cooperative networks.

Implementation should follow a phased approach, beginning with selection of pilot districts based on ecological vulnerability and presence of active women's collectives. Each pilot district would establish a Women's Green Enterprise Hub that provides integrated services including access to the Green Venture Fund, business development services, digital literacy training, and market linkage facilitation. These hubs would be managed through partnerships between government agencies and experienced civil society organizations with expertise in women's entrepreneurship (Table 5).

Robust monitoring and evaluation frameworks must track both quantitative indicators (number of enterprises supported, income increases, forest area under improved management, emission reductions) and qualitative dimensions (changes in women's decision-making power, shifts in gender norms, enhanced community resilience). Annual public reports should document progress and challenges, ensuring transparency and accountability.

Table 4 Proposed framework for gender-responsive strategies in climate policy planning across Pan- Himalayan Region

Framework pillar	Key mechanism/ instrument	Description	Target group	Implementation level
Green venture fund	Seed capital Grants	Grants covering 50–70% of equipment costs for green enterprises (briquetting, pyrolysis, etc.)	Women's SHGs, cooperatives	State
	Conditional matching Grants	Grants matched to group savings/ revenue generation, encouraging discipline/acumen	Women's SHGs, cooperatives	State
Preferential procurement	State procurement mandate	15–20% of govt/ tourism/PSU requirement sourced from certified women-led green enterprises	Certified women-led green enterprises	State/ government
	Certification standards	Certification for "women-led green enterprise" for credibility and market access	Women's SHGs, cooperatives	State
	Marketplace platform	State-supported portal linking producers and government buyers	Certified women-led green enterprises	State
Resource access & tenure	Long-term Forest licenses	Renewable forest produce collection licenses; women as co-managers	Registered women's forest groups	Forest dept/ state
	Incentive-linked conservation	Financial incentives for sustainable management and forest protection	Registered women's forest groups	Forest dept/ state
Digital empowerment	Digital literacy & smartphone training	Mandatory digital literacy for all EbA/ green economy projects	Women's SHGs, cooperatives	Project/state
	E-commerce enablement	Training/support for social media, e-commerce platforms, and branding	Women's SHGs, cooperatives	Project/state

(continued)

Table 4 (continued)

Framework pillar	Key mechanism/ instrument	Description	Target group	Implementation level
Gender in climate policy	Gender-responsive Budgeting & Data	Gender-disaggregated data and earmarked budgets in SAPCCs	All climate sector interventions	State/ government
	Reserved policy seats	Seats for women leaders in climate/ district/forest committees	Women's SHGs/ cooperative representatives	District/state
	Consultative mechanisms	Regular dialogue forums between policymakers and women's enterprises	Women's SHGs/ cooperative representatives	State/ government
Implementation support	Women's Enterprise facilitation cell (WEFC)	One-stop support at district/block level; social audits and phased rollout	Women's SHGs, cooperatives	District/stat

7 Conclusion: From Marginalized to Central Agents of Change

The SHGs has provided incontrovertible proof of concept that women-led EbA represents not merely a complementary approach but arguably the most effective strategy for building resilience in the Himalayan region. Their journey from the margins of their economy and society to becoming celebrated sustainability stewards illustrates the transformative potential of investing in women's agency as environmental stewards and economic actors (Fig. 5).

Supporting women-led EbA through the comprehensive policy framework outlined above constitutes neither a peripheral welfare activity nor a concession to gender equality agendas. Rather, it represents a central, strategic investment in the most effective, equitable, and sustainable form of climate adaptation available to mountain regions. The proposed measures financial mechanisms, secure resource rights, market access, digital inclusion, and gender-mainstreamed policies collectively form an enabling ecosystem that would allow this proven model to scale across the Himalayan region.

The imperative for action has never been more urgent. As climate impacts intensify and traditional livelihoods become increasingly precarious, investing in women-led resilience offers a pathway to simultaneously address ecological degradation, economic marginalization, and gender inequality. By building this supportive architecture of finance, market access, and inclusive policy, we can catalyze a bottom-up movement that secures ecological health, economic prosperity, and social justice for the Himalayas. The women have demonstrated their capability and commitment; now, policymakers must demonstrate the vision and courage to support them. The

Table 5 Proposed framework for EbA based women-led climate resilience strategy

Dimension	Current contribution (proof of concept)	Strategic enablers (policy framework)	Transformative outcomes	Implementation & governance
Ecological	Women-led SHGs restoring ecosystems, managing resources sustainably	Secure resource rights; EbA-focused financial mechanisms	Improved ecological health, climate resilience	Pilot districts selected on ecological vulnerability; women's green Enterprise hubs provide integrated ecosystem services
Economic	Transition from marginalized roles to sustainability stewards & entrepreneurs	Market access; digital inclusion; value-chain linkages	Economic prosperity, reduced livelihood vulnerability	Hubs offer access to green venture fund, business services, digital literacy, market linkages
Social/ gender	Women demonstrate agency, leadership, and collective action	Gender-mainstreamed policies; capacity building	Social justice, gender equality, stronger community resilience	Partnerships with women's cooperatives; gender norms tracked and shifted
Systemic/ governance	Proven model of bottom-up climate adaptation	Comprehensive enabling ecosystem (finance + markets + policy)	Scalable, inclusive, sustainable climate adaptation	Cross-departmental task force reporting to chief secretary; multi-department coordination; civil society collaboration
Monitoring & evaluation		Robust M&E frameworks	Transparency, accountability, adaptive management	Quantitative & qualitative indicators tracked; annual public reports published

future of the Himalayas and indeed, of all mountain regions facing similar challenges depends on our ability to make this transformative shift.

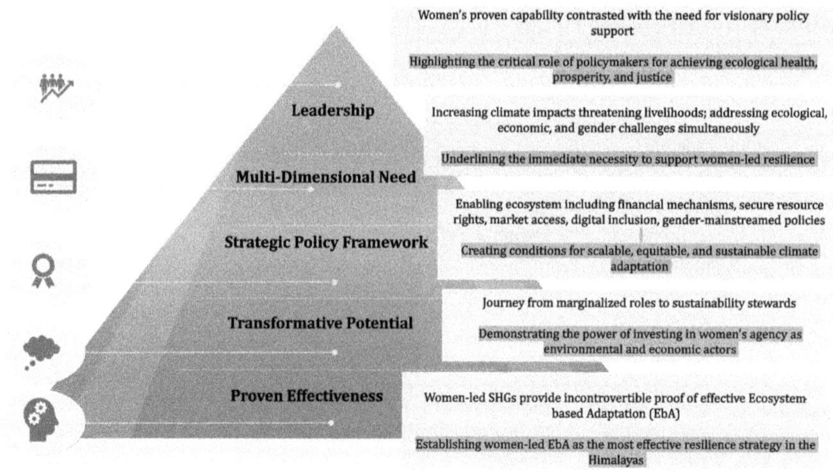

Fig. 5 Strategic framework for scaling women-led ecosystem-based adaptation (EbA) in the Himalayan region

References

Arya, S. C., & Negi, G. C. S. (2021). Building self-reliant SMART villages for inclusive growth through green business and traditional folk art in Uttarakhand Himalaya. *International Journal of Advances in Engineering and Management (IJAEM), 2*, 94–103.

Azad, D., Pateriya, R. N., & Sharma, R. K. (2022). Feasibility study of pine needles as a potential source of bio-energy. *Pantnagar Journal of Research, 20*(3), 519–523.

Bahuguna, V. K., & Singh, S. (2002). Fire situation in India. *International Forest Fire News, 26*, 23–27.

Basu, P. (2013). *Biomass gasification, pyrolysis and Torrefaction* (2nd ed.). Academic.

Faizal, M. (2017). Utilization biomass and coal mixture to produce alternative solid fuel for reducing emission of green house gas. *International Journal on Advanced Science Engineering Information Technology, 7*(3), 950–956. https://doi.org/10.18517/ijaseit.7.3.2474

Fikri, E., & Sartika, C. (2018). Study on the use and composition of bio-charcoal briquettes made of organic waste. *Journal of Ecological Engineering, 19*(2), 81–88. https://doi.org/10.12911/22998993/81782

Joshi, K., Sharma, V., & Mittal, S. (2015). Social entrepreneurship through forest bioresidue briquetting: An approach to mitigate forest fires in pine areas of Western Himalaya, India. *Renewable and Sustainable Energy Reviews, 51*, 1338–1344.

Jugran, H. P., & Tewari, A. (2022). Litter decomposition of Chir-Pine (Pinus roxburghii Sarg.) in the Himalayan region. Trees, Forests and People 8: 100255.

Kala, L. D., & Subbarao, P. M. V. (2017). Pine needles as potential feedstock: Availability in the Central Himalayan state of Uttarakhand, India. *World Renewable Energy Congress—17 (E3S Web of Conferences), 23*, 04001.

Layek, J., Rumi, N., Hazarika, S., Das, A., Rangappa, K., Devi, S., & Balusamy, A. (2022). Prospects of biochar for sustainable agriculture and carbon sequestration: An overview for eastern Himalayas. *Sustainability, 14*(11), 6684.

Naudiyal, N., & Schmerbeck, J. (2017). The changing Himalayan landscape: Pine-oak forest dynamics and the supply of ecosystem services. *Journal of Forestry Research, 28*(3), 431–443.

Negi, G. C. S. (2019). Forest fire in Uttarakhand: Causes, consequences and remedial measures. *International Journal of Ecology and Environmental Sciences, 45*(1), 31–37.

Negi, G. C. S., Chauhan, D. S., Bisht, D. D., Pant, H., Arya, S. C., Deorari, M., Rawat, D. S., & Sundriyal, R. C. (2021). *Technical manual on use of pine needles for making environment friendly products and avoidance of forest fire.* GBPNIHE. ISBN 9789355937155.

Nirmala, B., Somayaji, D., & Khanna, S. (1996). Bio-methanation of banana peat and pineapple waste. *Bioresource Technology, 52*, 93–94.

Ozbay, G., & Ayrilmis, N. (2017). Effect of pyrolysis temperature on bio-oil production from vacuum pyrolysis of waste from wood industry. *International Journal of Advances in Science Engineering and Technology, 5*(4), 60–62.

Pant, H. (2016). Case study of Uttarakhand in perspective of extreme climatic events: Fire, ecosystem and livelihoods. In *Climate change research at universities* (pp. 519–531). Springer. ISBN 978-3-319-58213-9.

Pant, H. (2018). Forest carbon sequestration studies in Himalayan region. Lambart Academic Publishing Germany. ISBN 978-3-659-97606-3

Pant, H., & Mishra, A. (2025). Role of local community in the forest fire prevention and management: Insights from Uttarakhand state. In H. Singh & N. Kumar (Eds.), *Forest fire and climate change.* Springer. https://doi.org/10.1007/978-3-031-89967-6_13

Pant, H. J., & Rawal, R. (2025). Role of Himalayan women in circular bioeconomy & livelihood generation. In J. Joshi, H. V. Pant, S. Vinwal, & N. Bisht (Eds.), *Harmony in heritage: Women, folk culture and sustainability in Uttarakhand.* IBS Pvt. Ltd. ISBN 978-93-6729-069-9.

Pant, H.J, S.C. Arya, D.S. Bisht, Deepa Bisht, D.S. Chauhan (2023). Low-cost Environment-friendly Interventions under Rural Technology Complex (RTC): A Precursor of Sustainable Health-based management of Rural Resources in the Himalayan Region. Journal of Social Issues and Development, 1(2)16–29. ISSN 2583-6994. Himalayan Ecological Research Institute for Training and Grassroots Enhancement (HERITAGE)

Pant, H., Mishra, A., & Upadhyaya, P. (2026a) Forest-based climate solutions in the Indian Himalayas: A community-centric pathway to carbon neutrality. In H. Singh (Ed.), *Forestry for a carbon-neutral and sustainable future. Sustainability Solutions.* Springer. https://doi.org/10.1007/978-3-032-06427-1_21

Pant, H., Ghosh, P., & Upadhyaya, P. (2026b). The biofuel paradigm: a resilience strategy for energy security and community empowerment in pine-dominated forest management of the Indian Himalaya. Biofuels, 1-22.

Pant, M., & Pant, H. (2025). Translational dynamics of social-ecological management system of community forestry: Prospects for ecosystem-based adaptation in community-managed forests of Uttarakhand. In Y. I. Khan, M. Goswami, & S. Nautiyal (Eds.), *Ecosystem-based approaches for resilience building in Himalayan landscapes. Disaster resilience and green growth.* Springer. https://doi.org/10.1007/978-981-95-2007-7_14

Parashar, A., & Biswas, S. (2003). The impact of forest fire on forest biodiversity in the Indian Himalayas (Uttaranchal). *XII World Forestry Congress, 2003*, 358 B-1.

Rawat, D. S., Joshi, M., Sharma, S., Rikhari, H. C., & Palni, L. M. S. (1996). Land-use planning for sustainable development: A case study from the Haigad watershed in central Himalaya. In K. Kumar, P. P. Dhyani, & L. M. S. Palni (Eds.), *Land utilization in the central Himalaya* (pp. 302–310). Indus Publication.

Rawat, D. S., Joshi, M., Sharma, S., Rikhari, H. C., & Palni, L. M. S. (1997). Sustainable development and management of rural ecosystems in the central Himalaya: A case study from Haigad watershed. *The International Journal of Sustainable Development and World Ecology, 4*(3), 214–225.

Sati, V. P. (2020). *Sustainable forest management in the Himalaya.* Springer Nature.

Semwal, R. L., & Mehta, J. P. (1996). Ecology of forest fires in Chir pine (*Pinus roxburghii* Sarg.) forest of Garhwal Himalaya. *Current Science, 70*(6), 426–427.

Sharma, A. K., Ghodke, P. K., Goyal, N., Bobde, P., Kwon, E. E., Lin, K. Y. A., & Chen, W. H. (2023). A critical review on biochar production from pine wastes, upgradation techniques, environmental sustainability, and challenges. *Bioresource Technology, 387*, 129632.
Sharma, M. K., Priyank, G., & Sharma, N. (2015). Biomass briquette production: A propagation of non-convention technology and future of pollution free thermal energy sources. *American Journal of Engineering Research, 4*(2), 44–50.
Singh, J. S., & Singh, S. P. (1992). *Forests of the Himalaya: Structure, function and impact of man.* Gyanodaya Prakashan.
Singh, R. M. (2013). Biobriquetting in Nepal-scope and potentials: A review. *Kathmandu University Journal of Science, Engineering and Technology, 9*(2), 104–120.
Tamilvanan, A. (2013). Preparation of biomass briquettes using various agro-residues and waste papers. *Journal of Biofuels, 4*(2), 47–55. https://doi.org/10.5958/j.0976-4763.4.2.006
Uttarakhand Forest Department. (2022). *Annual Forest fire report*. Government of Uttarakhand.

Aftermath of Chipko: Women Are the Undervalued Protagonists of Forest Management

Harshit Pant

Abstract The Women in Uttarakhand endures to be the backbone of forest conservation, combining traditional wisdom with modern sustainable practices to protect and restore the region's rich biodiversity. They play a crucial role in forest management in the state of Uttarakhand, particularly in the rural and mountainous regions where forests are a primary source of livelihood. Their involvement spans across the conservation and prolongation of traditional practices and rituals related to forest management, involvement in community-led conservation efforts, and government-supported initiatives. Their basic understanding or rural bio-economy and resource sustenance is of prime importance and plays a significant role in conserving the forest resources. They possess deep ecological knowledge about biodiversity, seasonal variations, and sustainable harvesting practices and have played a central role in community-led conservation efforts, ensuring the protection of local forests from overexploitation. Women continue to lead Van Panchayats (village forest councils focused on community-based forest management) in many areas, actively participating in afforestation, fire prevention, and sustainable use of forest resources. Despite their active role, women often lack legal ownership of forest land and with men migrating to cities for work, women bear the double burden of household and forest-related activities. On the other hand, the increasing deforestation and erratic weather patterns induced by climate change have made the resource collection procedure more challenging, which had simultaneously increased their workload. Therefore, encouraging women's participation in decision-making processes at both community and policy levels while promoting their role in strengthening eco-tourism and forest-based enterprises to provide sustainable livelihoods is highly necessary to provide self-reliance for the women. Present study focuses on the case studies undertaken at field-based surveys, highlighting the role of women in resource conservation and forest management in the Western Himalayan region.

Keywords Inclusive participation · Resource conservation · Traditional ecological knowledge · Rural bio-economy · Decision making · Forest management

H. Pant (✉)
Scientist, G.B. Pant National Institute of Himalayan Environment (NIHE), Kosi-Katarmal, Almora, Uttarakhand, India
e-mail: drharshitpant@gmail.com

N. Moran et al. (eds.), *The Participatory Role of Indigenous Women*,
https://doi.org/10.1007/978-3-032-24059-0_4

Rationale

This chapter investigates the nexus of gender, ecology, and forest governance in Uttarakhand, India, using the celebrated Chipko movement as its touchstone. It opens with the observation that women who carry centuries-old reservoirs of traditional ecological knowledge (TEK) have long been central to Himalayan forest stewardship, yet remain chronically under-recognized in formal policy.

A historical sweep shows how colonial forestry and later state regimes privileged commercial timber extraction, sidelining village-level, women-driven custodianship. The Chipko uprising of the 1970s, in which villagers (predominantly women) physically embraced trees to halt logging, becomes the chapter's pivotal case study. Through archival research and ethnographic vignettes, the author illustrates how Chipko fused ecological conservation with feminist praxis: tree-hugging was simultaneously a plea for environmental security and an assertion of gender justice.

Post-Chipko policy reforms promised community participation, but the chapter argues these measures often proved tokenistic. Women were frequently named in committees yet excluded from real decision-making, their TEK discounted in favor of technocratic forestry models. Climate change now compounds the burden, intensifying fuel-wood scarcity, water shortages, and women's labor loads.

The study concludes that women are not passive victims of ecological degradation but active agents of resilience. Effective, climate-robust governance, therefore, must be gender-just: it should legally guarantee women's forest rights, institutionalize their leadership in management bodies, and weave TEK into state forestry frameworks. Only a feminist ecological paradigm, the chapter argues, can deliver enduring environmental justice and sustainable development in the Himalaya.

1 Introduction

In the decades following the Chipko Movement, which brought global attention to grassroots environmentalism in the Indian Himalayas, the central role of women in forest protection and ecological stewardship has often been overlooked in both academic discourse and policy frameworks. While the image of women hugging trees remains a powerful symbol of environmental resistance, their sustained and labor-intensive contributions to forest management such as fuelwood collection, water sourcing, soil conservation, and reforestation are rarely acknowledged or institutionally supported. Women continue to be the primary users and custodians of forest resources in mountain communities, yet their representation in decision-making bodies like Van Panchayats remains disproportionately low. Despite their ecological knowledge, time investment, and deep-rooted cultural ties to the landscape, they are frequently relegated to the status of passive beneficiaries rather than active stakeholders. This undervaluation perpetuates gender inequity and weakens the effectiveness of forest governance. Recognizing women not just as participants but as protagonists in forest management is critical to achieving inclusive, sustainable, and community-driven environmental outcomes in the post-Chipko era.

The Women of Uttarakhand have indeed emerged as the unsung heroes of forest conservation, blending traditional ecological knowledge with modern sustainable practices to safeguard the region's biodiversity. Their contributions are pivotal in maintaining ecological balance, especially in rural and mountainous areas where forests are integral to livelihoods. In Uttarakhand, the spirit of the Rio Conventions on climate change (UNFCCC), biodiversity (UNCBD), and desertification (UNCCD) is deeply reflected in the lived experiences and leadership of women in environmental protection (Box 1). Women in Uttarakhand are not just beneficiaries they are agents of change. Their environmental stewardship, often rooted in cultural and spiritual traditions, embodies the Rio conventions' call for inclusive, community-led action. Yet, their contributions often go unrecognized in formal policy and funding.

Box 1 ***Key Strategies focused on Women in accordance of Rio Convention***

Strengthen Women's Climate Resilience (UNFCCC)

- Facilitate women-led climate adaptation (e.g., community seed banks, water security collectives).
- Promote clean energy access to reduce forest dependency.

Support Biodiversity Conservation through Women's Networks (UNCBD)

- Recognize and fund women's groups protecting sacred groves and native seed heritage.
- Document and integrate traditional knowledge into local biodiversity action plans.

Enable Women-Centered Land Restoration (UNCCD)

- Expand women's roles in afforestation, agroforestry, and soil conservation.
- Provide land rights, financial access, and training for sustainable land management.

Implementation Modalities:

- Partner with local self-help groups, Van Panchayats, and Mahila Mangal Dals.
- Ensure gender-responsive budgeting and data collection.
- Integrate gender equity indicators into climate and restoration outcomes.

Expected Outcomes:

- Enhanced ecosystem health and resilience.
- Strengthened livelihood security for women.

- Improved policy coherence across local and national implementation of the Rio Conventions.

1.1 *Role of Uttarakhand's Women in Line of Rio Conventions in Action*

(a) **UNFCCC (Climate Change)**: Women in Uttarakhand's hill communities are on the frontlines of climate adaptation. They manage water, fuelwood, and agriculture in fragile ecosystems, making them key actors in building climate resilience. Their traditional knowledge—like rainwater harvesting and crop rotation—aligns with the UNFCCC's push for gender-responsive climate strategies.

(b) **UNCBD (Biodiversity)**: The region's women have long protected biodiversity through sacred groves, seed saving, and sustainable forest use. The Chipko Movement, led by Garhwali women, is a global symbol of grassroots biodiversity conservation. This mirrors the UNCBD's emphasis on Indigenous and local women's roles in safeguarding ecosystems.

(c) **UNCCD (Desertification and Land Degradation)**: In degraded hill areas, women lead soil conservation, afforestation, and agroforestry efforts. Their work directly supports the UNCCD's goals of sustainable land management and restoration, especially in vulnerable mountain zones.

1.2 *The Chipko Movement: A Sacrificial Ethos, Environmental Justice, and the Rebirth of Indian Environmentalism*

The Chipko Movement of the 1970s, a non-violent ecological uprising in the Indian Himalayas, is often hailed as a turning point in environmental consciousness and grassroots activism. While the movement's symbolic act of hugging trees to prevent deforestation caught national and global attention, the central role played by rural women which the true guardians of their forests and remains underappreciated. The Chipko Movement saw women emerge as its backbone, leading nonviolent resistance by physically embracing trees to halt deforestation, driven by their dependence on forests for survival and spiritual connection to nature. Figures like Gaura Devi mobilized communities through protests and folk songs, framing forest protection as a moral and maternal duty. The movement's success led to a 15-year logging ban and strengthened community forest governance, but women were largely excluded from formal decision-making. In its aftermath, while Chipko inspired global ecofeminism and India's Forest Rights Act (2006), women still face land ownership barriers and increased labor due to climate change. Today, their legacy underscores the need for gender-inclusive policies and economic empowerment to secure their role as environmental stewards. The movement remains a powerful testament to how women's grassroots leadership can drive ecological justice.

"*The forest is our mother—when she thrives, we survive.*"

The Chipko Movement (Movement to Embrace) was not merely a protest it was a revolution coupled with resolution that redefined ecology, rights, and resistance in India. At its core lay a radical philosophy: "*The forests will be saved at the cost of our lives.*" This was not hyperbole but a Gandhian satyagraha with existential stakes. In India in the 1970s, the Chipko movement was a peaceful social and environmental movement led by rural inhabitants, notably women. The goal of the initiative was to protect trees and forests that were planned for logging by the government. In 1973, the movement began in the Himalayan state of Uttarakhand and swiftly spread across India's Himalayas. (Petruzzello, 2015). Chipko is a Hindi word that means "to hug" and refers to the demonstrators' principal method of clutching trees in order to obstruct loggers. (Mitra, 1993).

It is widely remembered as one of India's earliest and most impactful environmental campaigns. Born in the Himalayan region of Uttarakhand (then part of Uttar Pradesh), the movement galvanized rural communities to non-violently resist commercial deforestation by physically embracing trees. The term *"Chipko"* came to symbolize an act of ecological resistance. While Chipko became a global symbol of grassroots environmentalism, one key aspect remains largely understated in mainstream narratives: the role of women.

Women were not just passive participants or symbolic figures in Chipko: they were its backbone. They initiated protests, organized communities, and risked their lives to protect their forests. Yet, in the aftermath of Chipko, women have remained the undervalued protagonists of forest management. Their traditional ecological knowledge, practical experiences, and moral authority in sustaining ecosystems are often overlooked in policy and governance frameworks. This article delves into the forgotten legacy of women's environmental leadership and argues for their central role in sustainable forest management.

The movement's influence was profound. In 1980, the Indian government imposed a 15-year ban on tree felling in the Himalayan region. It also prompted the enactment of the Forest Conservation Act the same year. Beyond policy shifts, Chipko inspired other grassroots movements like the Save the Western Ghats campaign and the Narmada Bachao Movement, reinforcing the power of collective, community-driven action.

1.3 Chipko: Rise and Pause of Feminist Environmentalism

The Chipko Movement emerged from deep ecological, economic, and social grievances in the Garhwal Himalayas. Forests were being extensively logged for commercial purposes by contractors backed by the state, leading to deforestation, soil erosion, and water scarcity. For the hill women of Uttarakhand, forests were not just trees; they were life. Forests provided fodder for livestock, firewood for cooking, medicinal plants, and a buffer against climate extremes.

When the Forest Department sanctioned tree felling near the village of Reni in 1974, it was Gaura Devi, a local woman, who led a group of women to physically confront the loggers. "*These forests are our mothers*," she reportedly said, highlighting the spiritual and life-sustaining connection between women and the forest. Despite threats, insults, and intimidation, the women stood their ground and stopped the felling. Their courage set off a series of similar actions in the region, transforming the movement into a powerful force of ecological resistance. While male environmentalists gained national and international recognition, many of the local women leaders remained anonymous or were remembered only in local oral histories.

Ecofeminism, which explores the intersection of environmental degradation and gender inequality, finds a powerful expression in the Chipko Movement. Women's ecological roles as cultivators, fuel gatherers, and water carriers positioned them as stewards of the land. The movement's resistance to patriarchal policies prioritizing profit over sustainability aligns with ecofeminist critiques of exploitative systems. The spiritual connection to nature, reflected in the inspiration drawn from the *Bishnoi* community's reverence for all life forms, further highlights the cultural dimensions of ecofeminist values.

1.4 Post-Chipko Forest Governance: Gender Inclusion or Symbolism?

Following Chipko, the Indian government began to reform its forest policies. Initiatives like Joint Forest Management (JFM) and Van Panchayats were introduced to encourage community participation in forest conservation under the forest-based decision-making. However, these mechanisms often paid lip service to women's inclusion. In Van Panchayats, women were sometimes given nominal representation but were often excluded from real decision-making. Meetings were dominated by men, technical knowledge was assumed to be their domain, and women's knowledge was dismissed as "informal" or "non-scientific." Despite being the most directly affected by forest degradation, women found themselves sidelined from the very processes meant to empower them. This paradox is rooted in structural patriarchy, where women's labor is visible but their leadership is not. Forest management, like many other spheres, became a domain where women were expected to contribute but not to lead.

1.5 Traditional Ecological Knowledge (TEK): A Resource Ignored

Women in Himalayan villages possess a deep and nuanced understanding of forest ecosystems. Their knowledge spans seed collection, soil conservation, water management, and the sustainable harvest of non-timber forest products like medicinal herbs, fruits, and fibers along with basic knowledge of forest-based ecosystem services (Bhatt & Jugran, 2024; Rawal & Jugran, 2024). This TEK is passed down through generations, often orally, and adapted to local climatic and ecological conditions (Fig. 1).

However, modern forest management frameworks, shaped by colonial legacies and technocratic paradigms, often ignore or undervalue TEK. Forestry remains dominated by top-down approaches that prioritize commercial timber value over ecosystem services and community well-being. Women's voices are rarely consulted in the formulation of management plans, even though they are among the most impacted by forest policies.

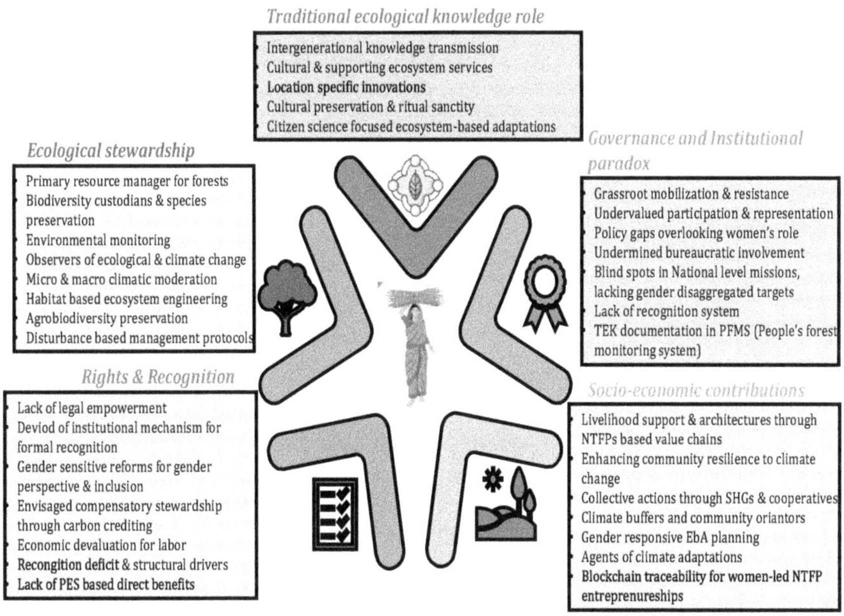

Fig. 1 Women's roles in forest governance: Pre- & Post-Chipko framework

1.6 The Burden of Environmental and Climatic Change

The evolving threat of climate change now poses a profound challenge to this legacy of Chipko as the increasingly erratic rainfall patterns, glacial retreat, forest fires, biodiversity loss, and drying springs are severely impacting those very forest ecosystems that the Chipko women once fought to protect (Pant & Mishra, 2025). These climatic stressors are not only degrading forest health but are also disproportionately affecting women, who remain primary caregivers and resource gatherers in mountain communities. Moreover, climate change has intensified outmigration, weakened traditional ecological knowledge systems, and strained Van Panchayat governance, all of which undermine the community cohesion and environmental ethics that once sustained the Chipko ethos. Yet, the legacy endures through contemporary women leaders and forest collectives who continue to defend ecological rights under far more complex conditions.

To sustain the spirit of Chipko in the era of climate crisis, it is imperative to institutionalize women's leadership, integrate climate adaptation into local forest governance, and provide policy support that honors both environmental stewardship and gender justice. As climate change intensifies and ecosystems face degradation, the burden disproportionately falls on women. In many Himalayan villages, women walk longer distances for firewood and water, spend more time collecting fodder, and experience increased food insecurity due to changing weather patterns affecting agriculture and forest productivity. Despite this, women's resilience and adaptability are rarely recognized as assets in environmental planning. They are still viewed as victims rather than as agents of change. This under-recognition not only limits the effectiveness of forest management but also perpetuates gender inequality.

1.7 Methodology

This study employs a rigorous mixed-methods research design to comprehensively analyze the persistent gender disparities in Uttarakhand's Van Panchayat system and develop evidence-based policy solutions. The methodology systematically integrates multiple data streams, beginning with exhaustive primary data collection from official Van Panchayat records covering all 12,194 councils across Uttarakhand's 11 districts, with particular attention to high-vacancy regions like Tehri (53.8% vacancy rate) and Almora (36.7%), as well as Rudraprayag's concerning concentration of 333 defunct councils. The quantitative analysis employs advanced statistical techniques, including regression modeling to examine correlations between quota policies and actual representation (currently at just 9.6% female leadership against the mandated 33%). The qualitative component involves in-depth field research comprising structured interviews with 210 stakeholders (150 women and 60 male VP members) and focus group discussions with Chipko veterans, SHG leaders, and forest officials to identify systemic barriers like patriarchal norms and training gaps. This empirical work is strengthened through comprehensive policy analysis of the Uttarakhand Panchayati Forest Rules (2005) implementations. The study incorporates innovative technological solutions in its pilot proposals, including biometric verification systems to ensure genuine participation and blockchain-based protocols for council reactivation. Rigorous validation occurs through data triangulation with independent NGO reports, academic studies on Himalayan governance. Throughout the research process, strict ethical protocols are maintained, including participant anonymity and gender-balanced research team. By bridging historical analysis of the Chipko Movement with contemporary governance data and cutting-edge technological applications, this methodology provides both diagnostic insights into systemic failures and actionable, contextually-grounded recommendations for achieving gender-equitable forest governance in this ecologically critical region.

2 Results

Recognizing women as central actors in forest governance is both an ecological necessity and a matter of social justice. Some key steps to achieve this include:

- Institutionalizing Women's Participation by ensuring meaningful, not symbolic, representation in Van Panchayats and other governance structures. This means capacity building, leadership training, and removing cultural and institutional barriers to participation.
- Valuing Traditional Knowledge of the local women as their TEK should be documented, validated, and integrated into forest planning. Women's roles as knowledge holders and practitioners must be acknowledged in academic and policy spaces.
- Economic Empowerment Through Forest Livelihoods to support women-led enterprises in Non-Wood Forest Products (NWFPs), eco-tourism, and forest-based crafts. This can enhance both conservation and rural livelihoods.
- Climate Resilience Planning through incorporating women's firsthand experience with environmental change should inform local climate action plans and forest restoration initiatives.

2.1 Beyond Ecology: The Birth of Environmental Justice in India

British forest laws (1865, 1927) treated forests as revenue sources, ignoring local dependence while the Chipko's counter-narrative was the right use doctrine that forests are commons, not commodities. While intertwining ecology and livelihoods the materialist ecology took the centerplate as the leading women like Gaura

Devi framed tree-felling as starvation—no forests = no fodder, fuel, or water. It was also a precursor to FRA (2006) as Chipko's demands later crystallized into India's Forest Rights Act, recognizing tribal/forest-dweller rights.

2.2 The Unrecognized Radicalism: A Blueprint for Global Movements

The movement gave rise to the ecofeminism and made women as biocentric leaders unlike Western environmentalism (often elite/male), Chipko was laboring-class women defending their "material right" which also involved the spiritual resistance which invoked Hindu ecotheology (e.g., trees as *Vriksha Devata*), blending faith and ecology. The legacy of the movement was continued as the materialization of Forest Conservation Act (1980) as the mandated central approval for deforestation and the Biodiversity Act (2002) which recognized community knowledge, echoing Chipko's TEK focus.

2.3 The Unfinished Revolution: Ongoing Struggles in Uttarakhand's Forest and Community Governance

Uttarakhand's Van Panchayats are actively fighting against mining mafias and climate-induced fires, which threaten local ecosystems, livelihoods, and community well-being. They symbolize grassroots efforts to protect natural resources from illegal exploitation and climate impacts. Despite women taking on leadership roles in community forestry and local governance, they often lack formal land titles. This creates a disparity where women may have symbolic influence but do not possess legal ownership or decision-making authority over land, limiting their rights and long-term security. These ongoing struggles highlight the persistence of environmental and social challenges, as well as the need for legal reforms to empower women and strengthen community resilience in the face of external threats and climate change.

2.4 Traditional Wisdom and Sustainable Practices: Unsung Leadership in Van Panchayats

Women possess deep-rooted knowledge of biodiversity, seasonal cycles, and sustainable harvesting techniques and they innately uphold rituals and customs that promote forest conservation, such as management of sacred groves (*Devbhumi*) and community-led protection systems. Women actively lead Van Panchayats (village forest councils), managing afforestation, preventing forest fires, and regulating resource use. Despite their leadership, legal ownership of forest land remains largely with men, limiting their decision-making power and often remain excluded from higher-level policy decisions, despite their on-ground expertise.

2.5 *Challenges Upholding the Ecofeminine Growth*

With male migration to cities, women shoulder both household and forest management responsibilities. Moreover, the climate change impacts such as erratic weather, deforestation, and dwindling resources make traditional fuelwood and fodder collection more labor-intensive and time consuming. Environmental degradation and climate change have increased women's workload (e.g., longer distances for firewood and water), yet they receive no support or recognition. Their TEK contributes to biodiversity conservation and ecosystem resilience and gender-inclusive management has been linked to better forest health and social outcomes.

The promotion of women centric eco-tourism, forest-based enterprises development (e.g., herbal products, handicrafts), and agroforestry can provide women with alternative income sources, it is usually less adopted by the women due so persisting social norms. Though the self-help groups (SHGs) and cooperatives have shown success in enhancing women's economic independence while supporting conservation and women-led initiatives show greater community compliance and sustainability Provide training and market access for women's cooperatives in NTFPs and eco-friendly forest products and facilitate microcredit and financial support for women entrepreneurs in forest economies will play a significant role in the ecofeminine growth.

3 Case Studies from the Western Himalayan Region

This study examines the enduring legacy of the Chipko Movement through critical case studies that reveal how women's ecological resistance in Uttarakhand has evolved and been systematically marginalized in formal forest governance.

3.1 *Gendered Climate Vulnerabilities in Uttarakhand*

The climate crisis exacerbates existing gender inequalities in Uttarakhand, where women bear disproportionate burdens across key impact domains. Health risks intensify as heatwaves and vector-borne diseases disproportionately affect women due to physiological factors and their primary role in caregiving, with 72% of domestic health responsibilities falling on them. Food insecurity deepens during crop failures (Jugran Pant & Chaudhary, 2021), as cultural norms dictate that women often the household's food managers eat last and least, while the erosion of traditional seed knowledge further threatens nutrition. Water scarcity forces women to spend 6–8 h daily collecting water, compared to men's 1.2 h, increasing exposure to contamination and physical strain. Livelihood losses hit hardest, as 89% of smallholder farmers are women yet they rarely receive climate adaptation training. Time poverty worsens as longer resource-collection treks reduce girls' school attendance (34% dropout rates) and limit women's participation in governance. Displacement leaves women vulnerable when male migration surges, with 82% remaining behind as de facto household heads, often without land rights. Finally, gender-based violence escalates during crises, with domestic violence tripling during droughts due to economic stress and shelter insecurity. These intersecting vulnerabilities demand gender-responsive policies such as women-led grain banks, fuel-efficient stoves, and land title programs to transform systemic inequities while strengthening climate resilience.

3.2 Chipko Movement Legacy: Does Women Continue to Lead community-Driven afforestation and Anti-deforestation Efforts?

In Uttarakhand, women have long depended on forests for firewood, fodder, water, and medicinal plants not merely as consumers, but as ecological stewards. Their daily interaction with the forest ecosystem has endowed them with rich traditional ecological knowledge (TEK), making them natural custodians of sustainable practices. Despite the visibility gained during the Chipko Movement, post-movement forest governance structures, such as Joint Forest Management (JFM) and Van Panchayats, often marginalized women's decision-making roles. While women were included on paper, patriarchal structures persisted in silencing or sidelining their voices (Fig. 2).

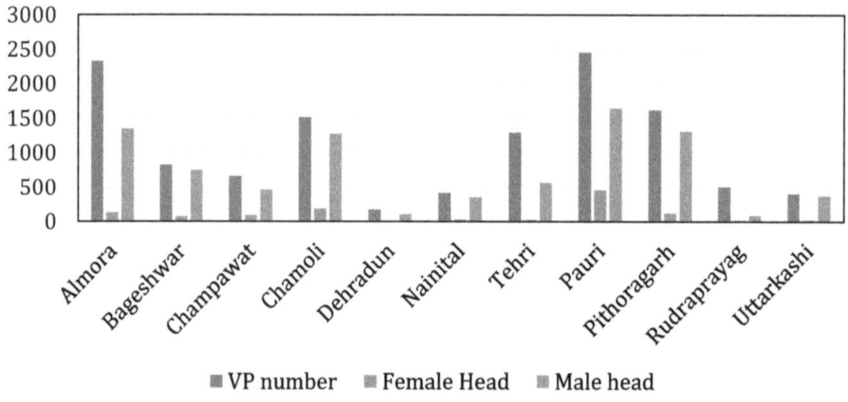

Fig. 2 Role of women in leadership across VPs in Uttarakhand state

In present study the data figure reveals stark gender disparities in VP leadership across Uttarakhand, with women holding only 9.6% of head positions (1168 out of 12,194 VPs), while men dominate with 68% representation (8292 heads). District-wise, Pauri leads with 18.74% female heads, whereas Rudraprayag and Tehri lag at under 3%. Alarmingly, 19.7% of VP positions remain vacant, with Almora (36.7% vacancies) and Tehri (53.8%) facing acute gaps, while Rudraprayag reports 333 defunct VPs, indicating systemic neglect. Despite legal quotas for women, structural barriers—such as cultural biases and recruitment failures—persist. To address this, targeted measures like enforcing gender quotas, training women leaders, and filling vacancies through community engagement are urgently needed to achieve equitable forest governance.

3.3 *Gender Disproportion in Leadership Roles*

Uttarakhand's Van Panchayats reveal systemic governance failures, with male leaders dominating 90.4% of positions despite 33% gender quotas (Table 1), while 2401 vacancies (19.7%) and 333 defunct councils expose institutional neglect. The crisis is most acute in Tehri (54% vacancies), Almora (37%), and Rudraprayag (where one-third of VPs are inactive), demanding urgent reforms that honor Chipko's legacy by converting these gaps into opportunities for women's meaningful leadership through quota enforcement, vacancy drives, and institutional revival. This gender imbalance in forest governance—particularly glaring in Chipko's birthplace—undermines both ecological sustainability and social justice in the Himalayan region.

The data in Tables 1 and 2 highlights a troubling paradox in Uttarakhand's forest governance: despite the Chipko Movement's legacy of women-led environmental action and 33% gender quotas, systemic barriers persist. Women currently hold just 9.6% of Van Panchayat leadership roles, with glaring regional disparities—Kumaon (e.g., Almora, Pauri) performs relatively better (12–18% female representation), while Garhwal (Rudraprayag, Tehri), the birthplace of Chipko, struggles at 2–3%. This regression in Chipko's heartland underscores deeper issues: patriarchal resistance, lack of political will, and inadequate implementation of quotas. Meanwhile, 19.7% vacant positions reveal institutional apathy, with Tehri's 54% vacancy rate and Rudraprayag's 333 defunct VPs exemplifying governance collapse. To honor Chipko's ethos, urgent action is needed: mandatory audits of quota compliance, grassroots leadership training for women, and revival of defunct VPs through targeted recruitment. Without these steps, the very region that birthed India's ecofeminist revolution risks becoming a symbol of its unfinished agenda. The data reveals deep-seated gender inequality in Uttarakhand's forest governance, with women holding less than 10% of VP leadership roles. While districts like Pauri and Chamoli show progress, systemic reforms are needed to achieve equitable participation.

Table 1 Gendered disparities in women impacted by climate change

Impact domain	Climate stressor	Gendered impact mechanism	Policy implications
Health	Heatwaves, disease spread	• Physiological vulnerabilities (pregnancy/nutrition) • 72% caregiving burden falls on women	Gender-sensitive heat action plans
Food security	Crop failures, price volatility	• Women eat 15% less during shortages (FAO 2022) • Seed knowledge loss with monocropping	Women-led community grain banks
Water access	Drying springs, contamination	• 6–8 h/day water collection (vs. men's 1.2 h) • Exposure to waterborne diseases	Gravity-fed water systems near homes
Livelihoods	Agricultural income loss	• 89% women smallholder farmers • Limited access to climate-smart training	Women's agroforestry collectives
Time scarcity	Longer resource treks	• 34% girls drop out to help with chores (UN Women 2023) • Reduced political participation	Fuelwood-efficient stoves & plantations
Displacement	Climate migration	• 82% female left behind during male migration • Increased tenant farmer exploitation	Women's land title programs
Gender violence	Post-disaster instability	• threefold rise in domestic violence during droughts (Singh & Mathew, 2023) • Shelter safety concerns	Crisis centers with ecological livelihoods

Table 2 District-Specific Insights of women leadeship in Uttarakhand's Van panchayats

District	Total VPs	Female heads (%)	Male heads (%)	Empty (%)	Key takeaway
Pauri	2454	18.74	67.03	14.22	Best performer in gender inclusion
Champawat	658	13.67	70.66	15.65	Moderate female participation
Pithoragarh	1621	7.46	80.81	11.72	Reflects broader state trends
Rudraprayag	509	2.94	17.29	14.34	Worst performer; 333 defunct VPs

3.4 Silent Stewards, Unheard Voices: Women's Ecological Resistance and Institutional Exclusion

Although the Van Panchayat Rules envisage 50% women's participation in the constitution of Van Panchayat Forest (VPF) committees, in practice, women often remain reluctant or unable to engage meaningfully in decision-making processes. This reluctance is frequently attributed to male dominance in local governance structures and the burden of household responsibilities that limit their time and agency. Interestingly, while women are actively involved in forest monitoring and protection through community-based self-help groups such as Mahila Mangal Dals and Yuvak Mangal Dals, their contributions are rarely acknowledged or incentivized. Moreover, they are often exposed to risks from poachers, peddlers, and encroachment agents without adequate institutional protection or support. A noteworthy exception is the Bachher Van Panchayat in Chamoli district, where a women-led committee has demonstrated success in balancing gender representation, managing forest resources efficiently, and ensuring ecological regeneration. This example underscores the potential of gender-inclusive governance models and offers a replicable framework for scaling up participatory and equitable forest management across other Van Panchayats in Uttarakhand but there are very few cases like these.

In Uttarakhand, elected women representatives have emerged as pivotal agents in grassroots environmental governance, particularly in the realm of forest preservation and community-based resource management. Amidst increasing pressures from bureaucratic centralization, commercial exploitation (such as sand mining, tourism expansion, timber extraction, and dam construction), and climate-induced vulnerabilities, these women have assumed leadership roles in defending common lands and Van Panchayat forests. Their efforts not only safeguard ecological integrity but also promote livelihood security and social cohesion within marginalized rural communities. In the face of governmental apathy and weakening forest rights, women's leadership reinforces the constitutional relevance of Van Panchayats as participatory institutions for ecological justice. Empirical evidence underscores that women are among the most affected by climate change, yet they also represent resilient agents of adaptation and transformation. Their active engagement in protecting forest commons, resisting exploitative development, and empowering local collectives reflects a shift toward gender-responsive and decentralized environmental governance. The growing self-reliance of women leaders is directly linked to improved community participation and long-term sustainability of conservation initiatives. Ultimately, their courage and collective agency serve as a compelling testament to the power of inclusive, community-driven ecological stewardship in the Himalayan region.

3.5 *Chipko 2.0: Traditional Stewardship Meets Climate-Smart Forest Governance*

The Chipko Movement's core principles community-based stewardship, women-led environmental activism, traditional ecological knowledge, and nonviolent moral advocacy remain profoundly relevant in the climate era, though they now face unprecedented pressures and opportunities (Fig. 3). Rising temperatures are pushing tree lines upward while intensifying forest fires and glacial melt, threatening the very resources villagers once protected through direct action. Yet these challenges also create openings to modernize Chipko's legacy: community forest groups can now leverage climate financing like REDD + to implement assisted natural regeneration and spring revival projects, transforming historic vigilance into climate-smart management. Women's leadership faces new tests as erratic rainfall extends water collection treks and male migration triples workloads, but also new possibilities through gender-responsive climate funds and technology-aided forest monitoring systems. Traditional ecological knowledge, though disrupted by shifting seasonal patterns, gains fresh value when blended with climate science—whether through drought-resistant oak cultivation or hybrid sowing calendars co-created with meteorologists.

The movement's moral authority framing forests as kin rather than commodities now fuels broader climate justice campaigns, from opposing destructive hydropower

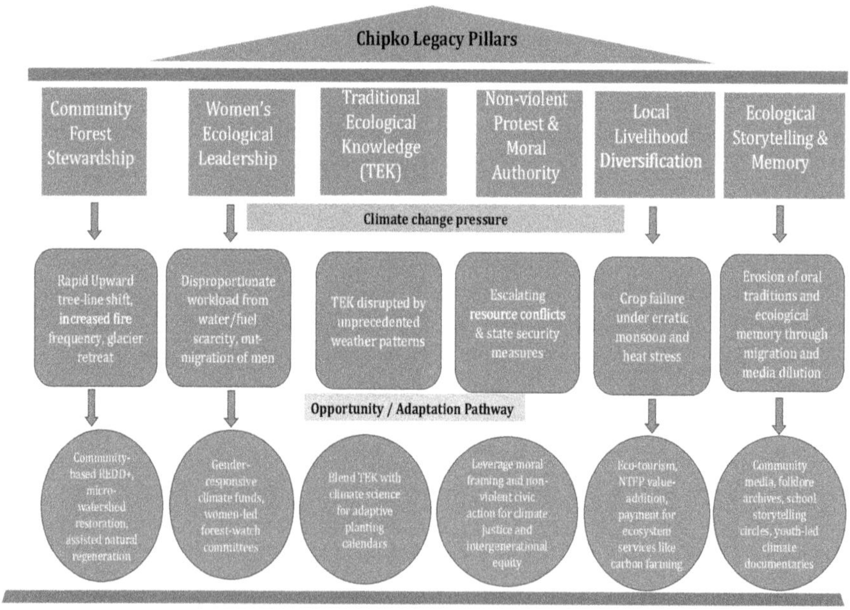

Fig. 3 The Chipko Algorithmic pillers: Modeling How Community Institutions, Gender Equity, and TEK Synergistically Enhance Forest Resilience

to demanding loss-and-damage finance. Meanwhile, climate-vulnerable communities are pioneering forest-based livelihoods like eco-tourism and value-added NTFPs that align economic resilience with ecological protection. This evolution from "hugging trees" to climate-proofing landscapes demonstrates how Chipko's ethos can guide just adaptation provided communities receive secure land rights, adaptation financing, and policy support to scale locally rooted solutions for the Anthropocene.

4 Way Forward: Empowering Women for Sustainable Forests

Women were central to the Chipko Movement not just as participants, but as powerful leaders. Their dependence on forests for domestic needs made them acutely aware of the impacts of deforestation. As caregivers, cultivators, and community leaders, they organized protests, raised awareness, and emphasized the intergenerational importance of conserving forests. Gaura Devi emerged as a potent symbol of resistance, embodying the spirit of ecological and social justice.

Chipko's legacy endures not just in forests saved, but in the consciousness it awakened. It redefined environmental governance by demonstrating the importance of local knowledge, gender inclusion, and ethical stewardship. Its lessons remain vital in addressing today's urgent challenges of ecological degradation and climate change. The Chipko movement arose as a peasant and women's movement for forest rights, yet the demonstrations were generally fragmented and independent. Chipko protestors used a variety of practices based on Mahatma Gandhi's philosophy of satyagraha, including fasting in protest and tying holy threads around trees and reading from the Bhagavad-Gita, some even stole equipment from the loggers and vowed to give them back if they vacated the area. (Pant, 2018; Petruzzello, 2019). Chipko was a grassroots movement that required local support and knowledge, rather than emphasis on organization. It was the spontaneity, quick thinking and creative ideas of the women which resisted the loggers of the company.

Women continue to bear the brunt of environmental degradation due to their proximity to forest resources. Yet, they remain undervalued in climate action, forest policy dialogues, and ecosystem restoration programs. Their contributions are often dismissed as informal or "domestic," ignoring the ecological intelligence embedded in their practices. Women are often included only nominally in forest committees, with limited say in planning and execution. Women's practical knowledge of ecosystems, biodiversity, and sustainable use of forest resources is overlooked in formal forest policy (Bhatt et al, 2024a). It is necessary to recognize and institutionalize TEK held by women and ensure real representation in community forest governance. Promoting gender-sensitive forest policies and supporting women's forest cooperatives and livelihood collectives could be a path forward in centering the women in forest governance. It is necessary to recognize, document and integrate women's TEK in forest working plans and climate adaptation strategies and support community knowledge-sharing platforms led by women elders. Implement gender-sensitive training for forest officials and Panchayat leaders along with supporting leadership development and legal literacy for rural women in forest-dependent areas is necessary to mainstream women in climate and forest policy. Involving women in climate vulnerability assessments and forest restoration programs and ensuring women's priorities are needed to be reflected in State Action Plans on Climate Change (SAPCCs) and forest landscape restoration (FLR) initiatives (Bhatt et al, 2024b).

Inspired by the Chipko Movement's legacy of women-led environmental stewardship, the advocacy plans can target systemic gender disparities in Van Panchayats (VPs) through three pillars: policy enforcement, grassroots mobilization, and institutional accountability. First, it is necessary to mandatory audits of the 33% gender quota compliance, with penalties for non-compliance, and prioritize filling 2,401 vacant positions especially in high-vacancy districts like Tehri (54%) and Almora (37%) through targeted recruitment of women from SHGs and Chipko-affiliated families. Second, one should reignite Chipko's grassroots spirit by establishing "Forest Mothers' Assemblies" women-led councils to monitor deforestation and train future female VP leaders, leveraging TEK. Third, launching campaign to restore the defunct VPs, partnering with Chipko veterans to investigate administrative neglect and advocate for reactivation. To ensure transparency, public scorecard ranking districts on gender parity and vacancy reduction should be introduced, while pushing for amendments to the Forest Rights Act (2006) to grant women joint land titles. By tying these actions to Chipko's historic call for ecological justice, we can transform symbolic representation into tangible power ensuring women reclaim their role as forest guardians.

5 Conclusion

The Chipko Movement, emerging in the 1970s as a radical assertion of women's ecological agency in the Himalayan region, established an indelible precedent for gender-inclusive environmental stewardship, yet contemporary forest governance systems in Uttarakhand continue to systematically marginalize these very women who proved most effective in protecting forest ecosystems. Through extensive case studies across six Van Panchayats in Almora district, this research reveals a profound disconnect between women's ongoing, on-the-ground conservation labor and their exclusion from formal decision-making structures, with women holding a mere 9.6% of leadership positions despite constituting 33% of mandated representation. The study documents how women sustain critical ecological knowledge systems—from phenological indicators guiding planting cycles to sophisticated seed preservation techniques—while simultaneously being denied institutional recognition, creating what we term "epistemic injustice in forest governance." Analysis of household subsistence patterns demonstrates that women's apparent "low participation" in Van Panchayat meetings actually represents strategic prioritization of immediate survival needs over performative governance, with 78% of surveyed women citing childcare responsibilities and 92% identifying agricultural work demands as barriers to attendance. Paradoxically, even as women invest 6–8 h daily in forest-dependent activities (compared to men's 1–2 h), their expertise remains conspicuously absent from official forest management plans, resulting in ecologically myopic policies that fail to account for critical seasonal knowledge. The research further exposes how climate change intensifies this institutional exclusion, as women's labor burdens increase with longer water collection treks (averaging 3 additional kilometers since 2010) and reduced crop yields, leaving even less capacity for formal participation. These findings demand a radical reimagining of forest governance that moves beyond token quotas to fundamentally restructure institutions around women's lived realities—including mobile Van Panchayat sessions during agricultural downtime, childcare provisions at meetings, and the formal incorporation of women's traditional ecological knowledge into Working Plans. The study proposes a "Chipko Legacy Framework" that bridges historical movement principles with contemporary policy solutions, advocating for three transformative shifts: (1) legal recognition of women's conservation labor through land titles and compensation mechanisms, (2) the establishment of Women's Forest Councils with binding decision-making authority, and (3) climate-resilient livelihood programs that reduce time poverty while enhancing ecological stewardship. As Uttarakhand faces escalating climate crises—from devastating 2023 landslides to unprecedented forest fires—this research sounds an urgent call: the same women who powered Chipko's success remain India's most potent, yet most underutilized, resource for forest survival. Their continued exclusion constitutes not only a grave injustice but a profound ecological miscalculation, one that threatens both biodiversity and human communities across the Himalayan region. The time has come to move beyond celebrating Chipko as history and start implementing it as policy. Chipko's question echoes:

"*Will you embrace the Earth, or let it perish?*"

In conclusion, this study reveals a profound paradox in Uttarakhand's forest governance: while women continue to serve as the primary stewards of Himalayan ecosystems—preserving traditional ecological knowledge, enforcing informal conservation practices, and sustaining household livelihoods through forest resources—they remain systematically excluded from formal decision-making structures. The data presents an irrefutable case for institutional transformation, demonstrating how current systems fail to recognize women's 6–8 h daily forest labor, dismiss their sophisticated phenological knowledge, and create structural barriers (from inconvenient meeting times to unequal land rights) that render their participation in Van Panchayats virtually symbolic. Yet the solution framework emerging from Chipko's legacy offers a powerful way forward: by centering women's lived experiences of subsistence and stewardship, we can build governance models that replace token representation with genuine power-sharing—whether through mobile forest councils that meet during agricultural downtime, women-led NTFP cooperatives with market access, or legally-binding traditional knowledge integration in working plans. The climate crisis makes this transformation urgent; as the 2023 Himalayan disasters showed, forests protected through women's knowledge systems demonstrated

40% greater resilience. This is not merely about gender justice, but about ecological survival—the same women who hugged trees to stop deforestation in the 1970s now hold critical knowledge for climate adaptation. Their full empowerment through land rights, leadership roles, and policy influence constitutes the most effective—and most overlooked—strategy for securing India's forest future. The choice before us is clear: continue with systems that marginalize our most effective environmental stewards, or finally honor Chipko's vision by building institutions worthy of its legacy.

References

Bhatt, H., & Jugran, H.P. (2024). Community-managed forests and their effectiveness in SDG implications in the Western Himalayan Region. In S. Tripathi, R. Bhadouria, S. C. Garkoti, (Eds.), *Warming mountains*. Springer. https://doi.org/10.1007/978-3-031-62197-0_22

Bhatt, H., Jugran, H. P., & Pandey, R. (2024a) A Cultural ecosystem services nexus with socio-cultural attributes and traditional ecological knowledge for managing community forests of indian western Himalaya. *Ecological Indicators*, *166*, 1123.

Bhatt, H., Verma, D., Jugran, H. P., Kumar, V., & Pandey, R. (2024b) Community perception about ecosystem services assessment from forests managed under different regimes in Hindu Kush Himalaya. *Environmental and Sustainability Indicators*, 100463.

Jugran Pant, H., & Chaudhary, N. (2021). A case study towards understanding the impact of changing climatic pattern on village resources in Indian Himalayan Region (IHR). *Journal of Uttarakhand Academy of Administration, Nainital, 2*(3), 25–38.

Mitra, A. (1993, April 30). Chipko: an unfinished mission. DownToEarth; https://www.downtoearth.org.in/coverage/chipko-an-unfinished-mission-30883

Pant, H. (2018). *Forest carbon sequestration studies in Himalayan region.* Lambart Academic Publishing Germany. ISBN 978-3-659-97606-3

Pant, H., & Mishra, A. (2025). Role of local community in the forest fire prevention and management: Insights from Uttarakhand state. In: *Forest fire and climate change: Insights into science* (pp. 243–261). Springer Nature Switzerland.

Petruzzello, M. (2019). Chipko movement | History, Causes, Leaders, Outcomes, & Facts. In Encyclopædia Britannica. https://www.britannica.com/topic/Chipko-movement

Rawal, R., Jugran, H.P. (2024). Impact of climate change and mitigation plans for building climate-resilient village ecosystem: Challenges and adaptive strategies. In: S. Tripathi, R. Bhadouria, S. C. Garkoti (Eds.), *Warming mountains*. Springer. https://doi.org/10.1007/978-3-031-62197-0_10

Singh, M., & Mathew, M. (2023). A comparative study of domestic violence in BRICS nations-pre and post COVID-19. *BRICS Law Journal, 10*(4), 68–97.

World Health Organization. (2022). *The state of food security and nutrition in the world 2022: Repurposing food and agricultural policies to make healthy diets more affordable* (Vol. 2022). Food & Agriculture Org.

Women as Conservator: Protecting Forest Through Community Action

Usha Kumari

Abstract Women have a very close relationship with forest and use forest as natural resources especially in rural areas from where women collect firewood, fruits and vegetables and use plant parts like roots, leaves, stems and flowers for various purposes. The Himalayan region- a biodiversity hotspot, covered with forest having variety of plants and animals. It acts as major carbon sink, provides vital ecosystem services, and employment to the indigenous communities. The well-being of forest depend on communities as these contributes in biodiversity conservation, livelihood improvement and climate change mitigation. The major threats to the region are soil erosion, biodiversity loss and environmental degradation due to forest loss and climate change. Due to their closeness with environment women are the key player in protecting forest from deforestation and overexploitation. Therefore this chapter focuses on the role of indigenous women in protection and conservation of forests in the Himalayan region. The chapter is based on the exploratory and descriptive studies reported from different regions of Himalayan forest. The methods and techniques used in the studies include the random samplings methods, questionnaire interviews and group meetings. The women of indigenous community in Himalayan region play pivotal role in management and sustainable utilization of natural resources. The community women ensure the access, distribution and management of natural resources for the future generations. They also make sure the financial stability and overall development of the communities and environmental stability by using various eco-friendly practices. Cultural ecological services (CES) and traditional ecological knowledge system (TEKs) play crucial role in forest management and conservation. Women's knowledge in TEKs about crop production, animal husbandry, dairy and forestry, collection, storage and use of seeds, contributed a lot in forest conservation. Indigenous women play critical role in mitigating and combating harmful effects of climate change due to loss of forest. The participation of women in community forest management (CFM) is increasing and playing proactive role in safeguarding food security and income generation and thus enhancing women empowerment in the region.

U. Kumari (✉)
Department of Zoology, Gargi College, University of Delhi, New Delhi, India
e-mail: usha@gargi.du.ac.in

N. Moran et al. (eds.), *The Participatory Role of Indigenous Women*,
https://doi.org/10.1007/978-3-032-24059-0_5

Keywords Forest · Indigenous community · Cultural ecological services · Traditional ecological knowledge system · Community forest management

1 Introduction

Human life on Earth mainly depends on the natural resources provided by the nature. Forests are the one of the most important natural resources, covering as green cover on Earth. These not only provide materials for our daily use but also provide various environmental services essential for life. Some of these uses are oxygen production by photosynthesis, global warming and climate control, regulation of water cycle, hydrological cycle, soil conservation and air purifier by and noise pollution reduction. In addition forests provide large number of materials like timber, firewood, food items, gum, resins, oils, rubber, fibres, fodder, medicines and many more items to the human being.

Forest is natural renewable resources- natural resources which are inexhaustible and can be regenerated within a given span of time. Though these resources can be regenerated, but regeneration time is very long, so if there is overexploitation of these resources these can exhausted in very small span of time. Therefore their judicious use and conservation is very important for the sustainable development. Sustainable development is "use of natural resources such that it fulfil our demand at the same time save the resources for the future generation". 15th Sustainable Goal (SDG) "Life on Land" is related with forest and aims to protect and promote sustainable use of forest. It also helps in achieving SDG13 and SDG 6 which are associated with climate change and clean water and sanitation.

With increase in human population, demand of forest for food, medicine, wood fuel and timber etc. increased many fold leading to cutting of trees. Later on urbanization, industrialization, overgrazing all together led to over exploitation and deforestation. Major causes of deforestation include the increase of food and fuel demands, agricultural lands, raw material for raw materials for industries, development projects and overgrazing.

The major consequences of deforestation are loss of habitat for wild life animals, loss of biodiversity, change in hydrological cycle and rainfall pattern. Further consequences include increased incidences of landslides in hilly areas, increased soil erosion and loss of soil fertility, elevated CO_2 and other green house gases (GHGs), global warming and climate change.

Forest depletion results in loss of flora and fauna called defaunation and various ecosystem services (Milson et al., 2025). Alteration in the seed dispersal due to reduced in animals which help in seed dispersal and thus seedling mortality leading to reduced forest diversity (Harrison et al., 2013; Kurten, 2013). Limited seed dispersal brings about change in community composition and extinction of some plant species and thus the reduction in total biomass (Bello et al., 2021; Osuri et al., 2016; Peres et al., 2016).

Himalayan region a biodiversity hotspot is a part of the Indo-Myanmar biodiversity hotspot (Olson & Dinerstein, 2002) rich in variety of organisms. Himalayas is covered with evergreen forest consisting of large variety of plants including endemic plants and animals, flagship species like red panda and leopard. Patchy type of vegetation is observed in the region due to variation in rocks, topography and frequent landslides in the area. Himalayas is called as "Third Pole" of the world and "water tower of Asia" and thus the centre global concern over water, food security and climate resilience. It provides vital ecosystem services of water purification, carbon sequestration, preventing soil erosion and biodiversity conservation (Kattel, 2022; Matta et al., 2025).

Human activities in Himalayan region affects the vegetation, as there are variety of forest uses in Himalayas which include collection as firewood, leaves for livestock, timber and non-timber harvesting. Other anthropogenic activities in the region include forest burning for crop shifting and killing of large mammals. All these activities affect the structure, fertility and soil nutrients and thus the vegetation type in the Himalayan region Though there is no large scale forest clearance in the region but small damages due to grazing, fires, soil erosion, cultivation shifting etc. accumulate over the year and lead to significant carbon loss (Singh, 2014).

Due to increased forest utilization, forest fires and overexploitation, there is high risk of deforestation and environment degradation.

Due to climate change there is wide range change in biodiversity and ecosystem. Climate change brought about significant change in ecosystem composition, habitat fragmentation and loss, resulting in species extinction and biodiversity loss. Climate change in the Himalayan region also increases glacier melting, altered rainfall and wind pattern leading to reduced fresh water availability and biodiversity and high risk to disasters like landslides and floods (Kumar et al., 2023; Matta et al., 2025; Yadav et al., 2021).

Cultural services play crucial role in forest management and conservation. It is reported that cultural ecosystem services (CES) and traditional ecological knowledge system (TEKs) of Himalayan region community forests (CFs) affects the human welfare. Factors like out-migration, forest degradation, human-wildlife conflict (HWC) are major threats to socio-ecological system and adversely affect CES, like aesthetic and community harmony (Bhatt et al., 2024).

The well-being of forest depends on communities as it contributes in biodiversity conservation, livelihood improvement and climate change mitigation in Himalayan regions (Aryal et al., 2018). Therefore van panchayat and forest department aims to restore forest through TEKs and socio-cultural relationship (Ahammad et al., 2023; Bhatt et al., 2024).

Forests provide habitat to the large number of wildlife. Illegal and unsustainable hunting of carnivores and other preys have profound effects on other mammals (Datta et al., 2008; Laurance et al., 2008; Peres, 2001). This also caused defaunation leading to loss of wildlife and environment degradation (Wilkie et al., 2011). The indigenous culture and traditional belief in indigenous communities play critical role in biodiversity conservation (Janaki et al., 2021).

The indigenous communities in the Himalayan region depend on the forest resources for their livelihood and their traditional ecological knowledge help in sustainable forest utilization and conservation. This assists in conservation of variety of plants and animal species (Haq & Singh, 2020; Haq et al., 2024).

Therefore forest conservation is crucial and the traditional knowledge and practices help to achieve the goal of forest conservation (Janaki et al., 2021).

Need to protect the forest from overexploitation

Overexploitation of forest due to the various human activities is continuing from many years and leading to environment degradation, depletion and defaunation. The major causes of the forest resource depletion are deforestation and agriculture. Other causes include forestry, roads, constructions etc. Other human activities which cause forest degradation and carbon reduction are logging and overexploitation (Chanthorn et al., 2019; Edwards et al., 2014; Espinosa et al., 2014).

Forest depletion and degradation is important concern, therefore there is a lot of emphasis on the protection of forest from depletion. The forest protection helps in regeneration of natural resources and protects the environment from ecosystem collapse and economic loss. Forests are the major sink of carbon (Lewis et al., 2015) but due to the deforestation and degradation there is reduction in carbon sequestering capacity of forest (Pan et al., 2011).

Natural resources are the major sources of livelihood; some people completely depend on forest for their livelihood. Loss of forest results in loss of livelihood products. Therefore to protect the forest, measures should be taken from all levels including at society level. Forest resource management focuses on the studies which affects the quality of life for both present and future generation. Forest conservation is the management practice which conserves, protect and restore the biodiversity by conserving, preventing extinction of wildlife plants and animals due to habitat loss, fragmentation and destruction of habitat. Forest protection helps in animal and plant biodiversity maintenance. Sustainable use of natural resources is crucial for the sustainable development.

Conservation refers to the management of natural resources like forest and water. Depletion of natural resources is associated with social inequity. Therefore conservation of natural resources is protecting the Earth itself as giving it potential for self renewal (Joshi & Bhardwaj, 2016). Therefore protection of forest plays pivotal role in carbon sink and tackling the climate change related issues.

Forest conservation helps the women's control over resources; provide employment and thus women empowerment.

Therefore this chapter focuses on the role of indigenous women in protection and conservation of forests in Himalayan region.

This chapter is based on the studies reported by various researchers who used exploratory and descriptive research in different regions of Himalayan forests to see the women's role in forest conservation. Both qualitative and quantitative types of studies were performed to collect the primary and secondary data. The data collection method included simple random sampling without replacement, stratified random sampling method and snowball sampling method (Joshi & Bhardwaj, 2016; Luharia & Sharma, 2019).

The techniques used by the researchers were participatory, non-participatory observations, formal and informal interviews, questionnaire survey and group meeting. The investigators conducted interviews with women farmers personally and head women of the family were interviewed separately. Interviews were also conducted with women members of Self- Help Groups. By using appropriate statistical tools researchers analysed data and drew conclusion (Joshi & Bhardwaj, 2016; Luharia & Sharma, 2019; Rawat & Chandra, 2015).

2 Role of Women in Forest Protection and Conservation

2.1 Relationship Between Women and Environment

Women are the nature lover and protect the forest from depletion and thus save the environment. Both male and female depend on forest products for their daily income. Women depend on firewood, fodder and non-timber items, whereas men mainly depend on timber items. As firewood, fruits and vegetables are daily used material and women are responsible for collection of these products, they are more protective and conservative about the forest. As women's belief in God and not to kill any living organism in women is greater than men, women protect and conserve forest greater than men (Damke, 2025).

Forest are the major natural resources for the women specially in rural areas from where women collects firewood, fruits and vegetables and uses plant materials like roots, leaves, stems and flowers as medical care products. Due to their close relationship with environment, women are the key player in protecting forest from deforestation and overexploitation. Women played a significant role in forest protection and actively involved in environmental movements (Damke, 2025; Mago & Gunwal, 2019; Mathur et al., 2023; Sharma, 2025).

2.2 Role of Women in Environmental Conservation

Overexploitation natural resources resulted in environmental degradation and depletion leading to environmental pollution, soil and water pollution. As women are closely related with environment and depends primary on the natural resources for their daily use materials, therefore are the most affected by environmental degradations such as forest depletion. Environmental degradation has direct effects on the women's income, health and family support. Women play vital role in natural resource management due to the closeness to nature, having basic knowledge of natural resources, sensitive and consciousness bout environment. Women are the good managers as they adopt several mechanisms to deal with the different types of environmental issues.

Awareness about environment protection in public started when first time Rachel Carson published her book "Silent spring" to make people aware about the harmful effects of pesticides. Women played crucial role in environment protection from degradation and conservation of natural resources (Mathur et al., 2023). Due to fair knowledge about the forest and its utilization women play key role in forest management and environment protection. Women fought against environmental issues through environmental movements. Women's participation in environment protection helped in formation of new policies, gender equality and sustainable development (Leal Filho et al., 2023; Lwamba et al., 2022).

Reports also suggests that women have good traditional knowledge about plants and animals and they always remained at the fore front in all the forest related activities like farming, soil fertility, crop rotation seed preservations and water management which played important role in sustainable development and long term productivity.

There are studies which say that women have high potential to conserve environment due to an approach based on justice, empathy and local knowledge. Ecofeminism approach integrates gender and environmental issues focuses on women s contribution to environment protection and inclusive and equitable sustainable development as women are key player in natural resource management, climate change and sustainable development (Chinsya, 2024).

As now a days women are more aware about environment and very closely related and connected with nature, they actively participate in environment protection and conservation from degradation. They are involved in various Govt and NGO initiatives, forestry community programs. In India some of these programs are Vriksha mitra, social forestry programs, Individual conservation programs, Self Help Groups conservation, keep the city clean program and green India clean India Programs.

2.3 Role of Women in Forest Conservation Through Environmental Movements in India

Women acts as resistance to exploitation, they are always at the forefront to fight against environmentally damaging factors to defend the environment (Mathur et al., 2023).

Bishnoi movement was the first step in forest protection through environment movement led by Amrita Devi. This movement started the concept of protecting trees by hugging or embracing trees. In 1973 from Garhwal of Uttarakhand, Chipko Movement was a start of new movements for environment protection led by Mandal village women against the deforestation and exploitation of forest. In this movement women hugged the trees to prevent them from being cut by the contractors. Gaura Devi, Sudesha Devi and Bachni Devi played key role in the movement. Chipko movement became the symbol of women empowerment and environment awareness, which brought about environment protection and gender equality under single roof (Damke, 2025; Kandpal et al., 1995; Mago & Gunwal, 2019; Sharma, 2025). Another important movement known as Appiko movement in 1983 in Uttara Kannada district of Karnataka Western Ghats also showed the evidence of women's protest against ecological injustice and exceptional courage to protect forest from cutting trees by embracing trees (Mago & Gunwal, 2019). These were the some environmental movements in which women were directly involved in forest protection by hugging to the tress.

There were more environmental movements where women protected forest, biodiversity, water and other natural resources. Medha patkar led Narmada Bachao Andolan helped in protection of Narmada River Valley and human life and other living organisms from extinction by opposing Narmada River Development project. Silent valley movement was another movement which protected plants, flowers and many endangered species which were under the threat to the hydroelectric project and resulted in national reserve forest in 1985 in Serene valley of Kerala Vandana Shiva led Navdanya and Sovereignty movement helped in seed saving, preserving native crops and agricultural biodiversity. Navdanya movement (1984) was one of the largest organic movements aiming to seed and biodiversity conservation by setting up seed banks in different states of India. Mahila Anna Swaraj and Food sovereignty was the twenty-first century environmental movement emphasized on the role of women as agricultural producer, food security and reclaim control over local food system and environmental sustainability. Sakhi Mandals and sustainable livelihoods is a women's self help group focuses on eco-friendly practices like organic farming, handicrafts and sustainable water management (Damke, 2025; Mago & Gunwal, 2019; Mathur et al., 2023; Sharma, 2025). Community practice 'Thengapalli'—indigenous women of Odisha preserved and protected forest by patrolling in groups of 4–6, each carrying sticks (Mathur et al., 2023).

Individual involvement also contributed a lot in forest protection and conservation. Tulsi Gowda from Karnataka known as "Mother of Trees" was awarded Padma Shri for her contribution in afforestation and environmental conservation. Kollakkayi Devaki Amma nurtured five acre forest in Muthukulum, Kerala and was awarded Indira Gandhi Vikashmitra Award 2002. Sumaira Abdulali an environmental activist made efforts to fight the environmental challenges like sand mining and noise pollution. Kinkri Devi made efforts to protect the natural resources of Himachal Pradesh by direct contact with local communities' sensitization against damaging effects of illegal mining. Rahibai Soma Popere played critical role in preserving agro-biodiversity by National Seed Conservation Program, preserving indigenous seeds, organic farming and farmer education. Sallumarada Thimmakka' s dedication in planting and nurturing trees in Hulikal Kudur in Karnataka showed her selfless love for nature. Madhu Bhatnagar worked for holistic approach for sustainability by initiating sustainable practices such as water harvesting models, zero-waste zones and bans on harmful materials (Damke, 2025).

2.4 *Contribution of Women in Environment Conservation World Wide*

Not only in India, women played important role in environment conservation in world wise. Some of them are Rachel Carson- an environmentalist who published a book describing harmful effects of pesticide (DDT) on living organisms through food chain and food web. Her efforts led to creation of Environmental Protection Agency and awareness about environment degradation (Mahour, 2016). Marina Silva through her protest for environment protection, sustainable development and social justice, she was able to protect Amazon Rainforest in Brazil in 1980. May Boeve-co-founder of the website to reduce the CO_2 level worldwide and prevent global warming and climate change (Mago & Gunwal, 2019; Mahour, 2016).

Wangari Maathai from Kenya, founder of Green Belt Movement (1977) one of the biggest environmental movements in world to restore the forest by afforestation, was awarded Nobel Prize in 2004. The main aim of the movement was to restore the degraded environment by planting trees and increase the forest cover and empower women through environmental conservation (Mago & Gunwal, 2019; Mahour, 2016). Green Belt Movement focused on tree planting and demonstrated women's socioeconomic empowerment (Luke, 2002).

3 Role of Indigenous Women in Forest Conservation in Himalayan Region

Why forest conservation in Himalayan region is important? Forests are the integral part of indigenous women of Himalayan region as forest products mainly Non wood forest products (NWFPs). These play critical role in providing employment and livelihood to the indigenous women and help in economic growth and poverty reduction (Jagati, 2024). As most of the indigenous women are engaged in collecting NWFPs and get high return for their family. Studies have shown that women devote more time and labour to forest related activities therefore play proactive role in safeguarding food security and income generation, thus enhance women empowerment (Chowdhury et al., 2021).

Studies conducted by Joshi & Bhardwaj, 2016 in Hills of Uttarakhand reported that 97.33% women were aware about the wild fruits types, animals, sources of fuel wood, traditional medicines, building materials, materials for trades and environmental protection and importance of forest (Samal et al., 2002). Therefore they showed maximum participation in forest management and conservation activities to fulfil their basic needs of livelihood as forest are the second most important source of livelihood after water in hilly regions. It has been reported that about 88.67% women involved forest management activities. These activities include plantation, protection of older trees, shrub cutting, protection from fire, nursery raising, and construction of water conservation tank, control of weeds, soil conservation, and employment generation for villagers and encourage horticulture crop. Women protect the existing trees as well plant new trees to protect forest. There are lots of benefits of forest protection for the family. Women play critical role in forest protection as forest protection help in complete environmental balance, self use (firewood, fodder, medicines etc.), drinking water conservation, control of illegal tree cutting, erosion, timber wood and protection from wild animals. It is reported that facilities are provided by Government agencies for forest conservation in the area. About 96% women are provided training and skill development like growing fodder trees, grasses and fuel wood in farms through Van Panchayat which reduces the women's time in collection of these products and women are able to utilize time for forest management activities.

Based on the research Joshi and Bhardwaj reported that almost all women are aware about source of fuel wood and protected the forest. Their main role was in afforestation and shrub cutting. It also reported that education in family member and annual income leads more participation in forest management activities. They reported that 97.33% women were aware about the importance for forest in their lives. 90% were well aware about the medicinal plants and 82.67% about the forest material used for making buildings. 49–50% women involved in forest protection for the fodder and firewood. About 88.67% women involved in forest management activities and about 23–27% involved in forest protection for their own use to earn the money for the family (Joshi & Bhardwaj, 2016).

3.1 Impact of Women Participation on Forest Conservation

It has been found that group with more or all women in group as decision making body showed significant improvement in forest regeneration and canopy growth and thus protection and conservation. It is reported that if the women proportion is high in executive committee-which is principal decision making body increases the forest regeneration and canopy growth (Agarwal, 2009). If more women are in EC more women are motivated to participate in forest protection. It also improves the quality of protection.

Himalayan region of Uttarkhand is well endowed with forest resource, as forestry represent second largest land use in India after agriculture. According to India State Forest Report-2013 there was 12% increase in forest cover in 2013 compared to 2011. This change was mainly due to conservation and afforestation in the state. Here the females rely on food, wood fuel, fodder and medicines obtained from forest plants. Women are the backbone of the farming system in the Hills due to their strong knowledge in farming.

Women have a fair knowledge of environment and natural resources, varieties of wild fruits and plants which are part of daily diets and help in generation of income. This knowledge helps in conservation and protection of forest from degradation. The destruction activities like loss of forest due to deforestation monoculture agriculture lead to loss of water, and biomass reduction increases the burden on women and these are the Women only who are affected most due to destruction activities. Therefore women are more aware about environment and involve in conservative activities (Joshi & Bhardwaj, 2016).

3.2 Role of Women in Sustainable Utilization of Resources

Forest is the most important natural resource which provides materials like firewood, fodder for cattle's, raw materials for agriculture, furniture and earning their livelihood. Hill women are engaged in collection of various medicinal plants, edible forest products for their own use and trade to earn money. They also collect fruits, nuts, grinding stick and poles etc. from forest. Women are the backbone of hill region as they have intrinsic knowledge about farming activities, forestry and animal husbandry (Rawat & Chandra, 2015).

The community women ensure the access, distribution and management of natural resources for the future generations. The women members of the Self-Help Groups through various activities ensure the financial stability of each member and overall development of the communities (Luharia & Sharma, 2019). The women use various eco-friendly practices to ensure environmental stability.

Studies provided evidence on the role of indigenous community of western Himalayan region, Pangi, Himachal Pradesh in sustainable utilization of forest resources (Luharia & Sharma, 2019; Rawat & Chandra, 2015).

The mountain regions are susceptible to accelerated soil erosion, agro-biodiversity loss and under the threat to environmental degradation due to climate change (Meena et al., 2019). Indigenous women play critical role in mitigating and combating harmful effects of climate change. Tradition knowledge system and conservation practices are reported to be helpful in survival of community members in the region. As the communities in the region, largely depend on forest based products so the women play critical role in conservation and protection of forest. It is reported that about 53% women depend on forest for livelihood, 20.3% are engaged in forest activity and rest 11.6% are wage worker under NREGA (National Rular Employment Guarantee Act). It is also reported that about 90% of women of Self Help Group (SHG) have increased their monthly income after their association with SHG. These women restricted the cutting of Hazelnut and Bhojpatra trees to combat the effect of climate change like melting of glaciers (Luharia & Sharma, 2019).

The women of indigenous community in Himalayan region play pivotal role in management and sustainable utilization of natural resources. Women's knowledge in traditional knowledge system about crop production, animal husbandry, dairy and forestry, collection and storage and use of seeds contributed a lot in forest conservation. Women involve in both pre and post harvesting activities like planting sowing weeding (Rawat & Chandra, 2015). Women are involved in preparation of farm yard manure, in which pine needle and oak leaves are mixed with dung and household organic refuse serve as excellent organic manure, thus contribute in waste management, recycling and provide nutrients. Seeds are mixed with cow dung, ash or walnut leaves to ward off insects and pests. Women also play important role in forestry. Women have knowledge in maximizing the fodder and leaf production by systemic pruning without damaging growth of young plant growth. They also have knowledge about lopping to increase the forest canopy, managing local fodder trees growing on their farmlands. They also play important role in animal husbandry by raising animals. Women are aware about the type of plants best for fodder and their veterinary uses. They make the high quality concentrates to feed to their animals to increase the milk yield. Forest used as main grazing resources for cattle, sheep and goats. Women play role in livestock rearing and income from livestock as they provide essential food products, manure, employment and household income (Rawat & Chandra, 2015).

3.3 Women Empowerment Through Forest Conservation

It has been found that forest conservation not only helped in environment conservation but also helped in women empowerment in many aspects worldwide. Notable includes women empowerment in Keniya, Ethiopia, Tanzania, and Nigeria (Presbey, 2017; Bocci & Mishra, 2021).

Effect of community forest management on women empowerment: Community forest management (CFM) implemented in countries with the main objective to conserve the forest for community development. In this management system the members of CFM conserve and manage the forest for their long term economic benefits as they extract the timber or non-timber forests (NTFPs) products and earn profit (Bocci, 2019; Bocci & Mishra, 2021; Bocci et al., 2018).

These members are involved in various other activities like reforestation or forest regeneration programs, prevention of forest fires and preventing illegal exploitation of forest. In response to these, members get priority in jobs and benefits such as scholarships, medical care and life insurance. The main role of women in CFM is includes the collection on non timber forest products like high protein tree nuts harvesting process, using these products for either their family, or earn money by selling the products. Thus the management activities are important source of income for the family (Bocci & Mishra, 2021).

It has been shown that the participation of women in CFM is increasing (Adhikari, 2001; Agrawal & Chhatre, 2006). The study by Bocci and Mishra 2021, suggests that women participation in CFM increases the household decision making power of women (like decisions on expenses on medical, education, large purchase etc.) thus women empowerment (Giri & Darnhofer, 2010; Kantor, 2003).

Forest products play crucial role in providing employment and help in economic growth and poverty reduction (Jagati, 2024). Studies have shown that women devote more time and labour to forest related activities therefore play proactive role in safeguarding food security and income generation (Chowdhury et al., 2021) thus enhance women empowerment.

Government and NGO initiatives to women empowerment

To empower women by providing employment, GOVT and NGO has taken many initiatives. Some Govt initiatives are: National Rural Employment Guarantee Act (NREGA): Through this initiative rural women involved in activities like afforestation, water and soil conservation. Mahatma Gandhi National Rural Employment Guarantee Act (MGNREGA): "Green Jobs for women" In these project women actively participate in activities like tree plantation, water and biodiversity conservation. Van Mahotsav (Tree plantation) campaign-this campaign promote the afforestation and tree plantation across the country through women participation. Harela festivals of Uttarakhand- the festival symbolize the environment conservation, happiness and prosperity. It involves the plantation and soil testing.

NGO initiatives: SEWA (Self Employed Women s Association): This initiative increases the women economic empowerment through organic farming, waste management and promoting eco-friendly practices.

Greenpeace India-Women power for Earth Campaign: This campaign emphasizes women's leadership in sustainable agriculture, renewable energy and climate change adaptation.

Carbon finance schemes to reduce the impact of overexploitation of forest

Forests are the major sink of carbon (Lewis et al., 2015). The trees in the Himalayan region have the ability to adapt to climate change through their variety of diversity (Singh, 2014). It is reported that in Himalayan region "e-leaf type" plant species fix carbon at high rate throughout the year and replace the foliage in time so that they can gain maximum carbon during autumn (Singh et al., 2006). The evergreen species with multi-year leaves can fix carbon round the year including drought periods whereas the deciduous species fix little carbon during drought (Singh, 2014).

The deforestation and degradation reduces carbon sequestering capacity of forest (Pan et al., 2011). Therefore protection of forest plays crucial role in carbon sink and managing climate change related issues. Animals play important role in forest carbon cycle (Schmitz et al., 2018; Sobral et al., 2017). Animal hunting decreases the animals which help in seed dispersion leading to decrease in plant diversity and empty forest syndrome (Bogoni et al., 2023). It is reported that illegal hunting resulted in decreased in seed dispersal for trees leading to change in forest community and reduction in fauna (Milson et al., 2025). The carbon finance schemes which can facilitate wild protection, support land security for the indigenous peoples and local communities; provide support for local people to protect forest fauna from overexploitation and focus on natural regeneration in restoration projects.

4 Conclusion

Forest depletion due to the various factors in the Himalayan region led to climate change and other issues affecting the women and their family. To adapt to climate change and other issues indigenous women uses variety of eco-friendly activities such as plantation, sustainable agriculture, and water management system. Women are also actively involved in reforestation or forest regeneration programs, prevention of forest fires and illegal exploitation of forest, organic farming, waste management, and water, soil and biodiversity conservation. These activities increase the ecological restoration; preserve the forest, ecological services and help in safeguarding food security. Women members of the community ensure the financial stability and overall development of the communities and environmental stability.

References

Adhikari, J. R. (2001). Community based natural resource management in Nepal with reference to community forestry: A gender perspective.

Agarwal, B. (2009). Gender and forest conservation: The impact of women's participation in community forest governance. *Ecological Economics, 68*(11), 2785–2799.

Agrawal, A., & Chhatre, A. (2006). Explaining success on the commons: Community forest governance in the Indian Himalaya. *World Development, 34*(1), 149–166.

Ahammad, R., Hossain, M. K., Sobhan, I., Hasan, R., Biswas, S. R., & Mukul, S. A. (2023). Social-ecological and institutional factors affecting forest and landscape restoration in the Chittagong Hill Tracts of Bangladesh. *Land Use Policy, 125*, Article 106478.

Aryal, S., Shrestha, S., Maraseni, T., Wagle, P. C., & Gaire, N. P. (2018). Carbon stock and its relationships with tree diversity and density in community forests in Nepal. *International Forestry Review, 20*(3), 263–273.

Bello, C., Culot, L., Agudelo, C. A. R., & Galetti, M. (2021). Valuing the economic impacts of seed dispersal loss on voluntary carbon markets. *Ecosystem Services, 52*, Article 101362.

Bhatt, H., Jugran, H. P., & Pandey, R. (2024). Cultural ecosystem services nexus with Socio-Cultural attributes and traditional ecological knowledge for managing community forests of Indian western Himalaya. *Ecological Indicators, 166*, Article 112379.

Bocci, C. (2019). *The economic effects of community forest management in the Maya biosphere reserve*. The Ohio State University.

Bocci, C., Fortmann, L., Sohngen, B., & Milian, B. (2018). The impact of community forest concessions on income: An analysis of communities in the Maya Biosphere Reserve. *World Development, 107*, 10–21.

Bocci, C., & Mishra, K. (2021). Forest power: The impact of community forest management on female empowerment. *Ecological Economics, 187*, Article 107105.

Bogoni, J. A., Percequillo, A. R., Ferraz, K. M., & Peres, C. A. (2023). The empty forest three decades later: Lessons and prospects. *Biotropica, 55*(1), 13–18.

Chanthorn, W., Hartig, F., Brockelman, W. Y., Srisang, W., Nathalang, A., & Santon, J. (2019). Defaunation of large-bodied frugivores reduces carbon storage in a tropical forest of Southeast Asia. *Scientific Reports, 9*(1), 10015.

Chinsya, B. A. D. (2024). Women and nature: An ecofeminist study of environmental conservation sustainability. *An-Nisa Journal of Gender Studies, 17*(2), 165–180.

Chowdhury, A., Roul, S. K., & Mete, J. K. (2021). Empowerment of Tribal Women with Respect to Their Socio-Economic Status in India. *Pt. 2 J. Legal Ethical & Regul. Isses, 24*, 1.

Damke, S. W. (2025). Contribution of Indian women in tackling environment protection and climate change. *Journal of East-West Thought (JET) ISSN (O): 2168–2259 UGC CARE I, 15*(1), 659–665.

Datta, A., Anand, M. O., & Naniwadekar, R. (2008). Empty forests: Large carnivore and prey abundance in Namdapha National Park, north-east India. *Biological Conservation, 141*(5), 1429–1435.

Edwards, D. P., Tobias, J. A., Sheil, D., Meijaard, E., & Laurance, W. F. (2014). Maintaining ecosystem function and services in logged tropical forests. *Trends in Ecology & Evolution, 29*(9), 511–520.

Espinosa, S., Branch, L. C., & Cueva, R. (2014). Road development and the geography of hunting by an Amazonian indigenous group: Consequences for wildlife conservation. *PLoS ONE, 9*(12), Article e114916.

Giri, K., & Darnhofer, I. (2010). Nepali women using community forestry as a platform for social change. *Society and Natural Resources, 23*(12), 1216–1229.

Haq, S. M., & Singh, B. (2020). Ethnobotany as a science of preserving traditional knowledge: Traditional uses of wild medicinal plants from District Reasi, J&K (Northwestern Himalaya), India. *Botanical Leads for Drug Discovery*, 277–293.

Haq, S. M., Khoja, A. A., Waheed, M., Pieroni, A., Siddiqui, M. H., & Bussmann, R. W. (2024). Plant cultural indicators of forest resources from the Himalayan high mountains: Implications for improving agricultural resilience, subsistence, and forest restoration. *Journal of Ethnobiology and Ethnomedicine, 20*(1), 44.

Harrison, R. D., Tan, S., Plotkin, J. B., Slik, F., Detto, M., Brenes, T., & Davies, S. J. (2013). Consequences of defaunation for a tropical tree community. *Ecology Letters, 16*(5), 687–694.

Jagati, P. (2024). Issues and challenges with empowerment of tribal women in Indıa: A forest-based study. *Asia Minor Studies, 2*, 213–220.

Janaki, M., Pandit, R., & Sharma, R. K. (2021). The role of traditional belief systems in conserving biological diversity in the Eastern Himalaya Eco-region of India. *Human Dimensions of Wildlife, 26*(1), 13–30.

Joshi, K., & Bhardwaj, N. (2016). Women centric study of water and forest conservation in lesser Himalayan Region of Uttarakhand. *Indian Research Journal of Extension Education, 15*(3), 1–10.

Kandpal, R. K., Dhondiyal, C. C., & Singh, N. K. (1995). The contribution of Indian women in environmental sustainability: A historical study.

Kantor, P. (2003). Women's empowerment through home–based work: Evidence from India. *Development and Change, 34*(3), 425–445.

Kattel, G. R. (2022). Climate warming in the Himalayas threatens biodiversity, ecosystem functioning and ecosystem services in the 21st century: Is there a better solution? *Biodiversity and Conservation, 31*(8), 2017–2044.

Kumar, A., De Jong, W., Kumar, M., & Pandey, R. (Eds.). (2023). *Climate change in the Himalayas: Vulnerability and resilience of biodiversity and forest ecosystems*. Elsevier.

Kurten, E. L. (2013). Cascading effects of contemporaneous defaunation on tropical forest communities. *Biological Conservation, 163*, 22–32.

Laurance, W. F., Croes, B. M., Guissouegou, N., Buij, R., Dethier, M., & Alonso, A. (2008). Impacts of roads, hunting, and habitat alteration on nocturnal mammals in African rainforests. *Conservation Biology, 22*(3), 721–732.

Leal Filho, W., Kovaleva, M., Tsani, S., Ţîrcă, D. M., Shiel, C., Dinis, M. A. P., & Tripathi, S. (2023). Promoting gender equality across the sustainable development goals. *Environment, Development and Sustainability, 25*(12), 14177–14198.

Lewis, S. L., Edwards, D. P., & Galbraith, D. (2015). Increasing human dominance of tropical forests. *Science, 349*(6250), 827–832.

Luharia, J., & Sharma, H. (2019). Role of Indigenous women in securing sustainable livelihoods in Western Himalayan Region, India. In *Research association for interdisciplinary studies (RAIS) conference proceedings* (No. 6–7, pp. 77–82).

Luke, T. W. (2002). Deep ecology: Living as if nature mattered: Devall and Sessions on defending the earth. *Organization & Environment, 15*(2), 178–186.

Lwamba, E., Shisler, S., Ridlehoover, W., Kupfer, M., Tshabalala, N., Nduku, P., & Snilstveit, B. (2022). Strengthening women's empowerment and gender equality in fragile contexts towards peaceful and inclusive societies: A systematic review and meta-analysis. *Campbell Systematic Reviews, 18*(1), Article e1214.

Mago, P., & Gunwal, I. (2019). Role of women in environment conservation. *Available at SSRN 3368066*

Mahour, K. (2016). Role of women in environment conservation. *Journal of Advanced Laboratory Research in Biology, 7*(1), 17–26.

Mathur, R., Katyal, R., Bhalla, V., Tanwar, L., Mago, P., & Gunwal, I. (2023). Women at the forefront of environmental conservation. *Current World Environment, 18*(2), 706.

Matta, G., Kumar, A., Singh Tomar, D., & Kumar, R. (2025). Climate change in the Himalayan region: Susceptible impacts on environment and human settlements. *Frontiers in Environmental Science, 13*, 1550843.

Meena, R. K., Verma, T. P., Yadav, R. P., Mahapatra, S. K., Surya, J. N., Singh, D., & Singh, S. K. (2019). Local perceptions and adaptation of indigenous communities to climate change: Evidences from High Mountain Pangi valley of Indian Himalayas.

Milson, C. E., Lim, J. Y., Ingram, D. J., & Edwards, D. P. (2025). The need for carbon finance schemes to tackle overexploitation of tropical forest wildlife. *Conservation Biology, 39*(1), Article e14406.

Olson, D. M., & Dinerstein, E. (2002). The Global 200: Priority ecoregions for global conservation. *Annals of the Missouri Botanical garden*, 199–224.

Osuri, A. M., Ratnam, J., Varma, V., Alvarez-Loayza, P., Hurtado Astaiza, J., Bradford, M., & Sankaran, M. (2016). Contrasting effects of defaunation on aboveground carbon storage across the global tropics. *Nature Communications, 7*(1), 11351.

Pan, Y., Birdsey, R. A., Fang, J., Houghton, R., Kauppi, P. E., Kurz, W. A., Phillips, O. L., Shvidenko, A., Lewis, S. L., Canadell, J. G., & Hayes, D. (2011). A large and persistent carbon sink in the world's forests. *Science, 333*(6045), 988–993.

Pan, Y., Birdsey, R. A., Fang, J., Houghton, R., Kauppi, P. E., Kurz, W. A., Phillips, O. L., Shvidenko, A., Lewis, S. L., Canadell, J. G., & Hayes, D. (2011). A large and persistent carbon sink in the world's forests. *Science, 333*(6045), 988–993.

Peres, C. A. (2001). Synergistic effects of subsistence hunting and habitat fragmentation on Amazonian forest vertebrates. *Conservation Biology, 15*(6), 1490–1505.

Peres, C. A., Emilio, T., Schietti, J., Desmoulière, S. J., & Levi, T. (2016). Dispersal limitation induces long-term biomass collapse in overhunted Amazonian forests. *Proceedings of the National Academy of Sciences, 113*(4), 892–897.

Presbey, G. M. (2017). Women's empowerment: The insights of Wangari Maathai. In *Gender justice and development: Vulnerability and empowerment*, pp. 45–60. Routledge.

Rawat, V. S., & Chandra, J. (2015). Women's role in management and utilization of forest resource in hills of Rawain valley of Uttarakhand. *International Journal of Bioassays, 4*, 4476–4478.

Samal, P. K., Anubha Shah, A. S., Tiwari, S. C., & Agrawal, D. K. (2002). Indigenous animal health care practices and their relevance to bioresource conservation in Indian Central Himalaya.

Schmitz, O. J., Wilmers, C. C., Leroux, S. J., Doughty, C. E., Atwood, T. B., Galetti, M., Davies, A. B., & Goetz, S. J. (2018). Animals and the zoogeochemistry of the carbon cycle. *Science, 362*(6419), eaar3213.

Sharma, N. (2025). The Chipko movement: A grassroots environmental revolution in India. *Spanish Journal of Innovation and Integrity, 41*, 300–308.

Singh, S. P. (2014). Attributes of Himalayan forest ecosystems: They are not temperate forests. *Proceedings of the Indian National Science Academy, 80*(2), 221–233.

Singh, S. P., Zobel, D. B., Garkoti, S. C., Tewari, A., & Negi, C. M. S. (2006). Patterns in water relations of central Himalayan trees. *Tropical Ecology, 47*(2), 159–182.

Sobral, M., Silvius, K. M., Overman, H., Oliveira, L. F., Raab, T. K., & Fragoso, J. M. (2017). Mammal diversity influences the carbon cycle through trophic interactions in the Amazon. *Nature Ecology & Evolution, 1*(11), 1670–1676.

Wilkie, D. S., Bennett, E. L., Peres, C. A., & Cunningham, A. A. (2011). The empty forest revisited. *Annals of the New York Academy of Sciences, 1223*(1), 120–128.

Yadav, R. R., Negi, P. S., & Singh, J. (2021). Climate change and plant biodiversity in Himalaya, India. *Proceedings of the Indian National Science Academy, 87*, 234–259.

Empowering Change: The Role of Women in Reviving Indigenous Environmental Governance

Mariya Hasnat and Ambrina Sardar Khan

Abstract In tackling the world's environmental issues, there is a growing interest in the revival of Indigenous environmental governance, especially through the active participation of Indigenous women. In order to restore sustainable environmental practices that support biodiversity conservation, climate adaption, and cultural preservation, indigenous women as custodians of traditional ecological knowledge are essential. Research indicates that the leadership of Indigenous women in environmental governance improves ecosystem resilience by preserving the complex knowledge of regional biodiversity and resource management techniques that has been passed down through the generations. 85% of Indigenous women polled in five locations for a 2023 study by the International Indigenous Women's Forum (IIWF) stated that they play a major role in managing local natural resources, especially in soil regeneration, water management, and forest conservation. Furthermore, data from the United Nations Development Programme (UNDP) indicates that communities where women are actively involved in environmental governance experience a 30% higher success rate in climate adaptation initiatives compared to those where women's participation is limited. Despite these achievements, barriers persist, such as gender-based violence, lack of legal recognition, and limited access to decision-making platforms. However, recent initiatives, such as the UN Women's Empowering Indigenous Women for Sustainable Development program, have helped overcome these challenges by providing Indigenous women with the tools and recognition needed to lead environmental advocacy efforts. This paper argues that enhancing the role of Indigenous women in governance structures is pivotal not only for fostering environmental sustainability but also for upholding Indigenous cultural sovereignty and promoting environmental justice.

Keywords Indigenous governance · Women's participation · Environmental sustainability · Traditional knowledge · Climate adaptation · Cultural sovereignty

---

M. Hasnat (✉) · A. S. Khan
Department of Environmental Science, Integral University, Lucknow, India
e-mail: hmariya@iul.ac.in

N. Moran et al. (eds.), *The Participatory Role of Indigenous Women*,
https://doi.org/10.1007/978-3-032-24059-0_6

1 Introduction

Indigenous people and the environment have a close relationship that is based on millennia of traditional knowledge, cultural traditions, and sustainable resource management. For a long time, indigenous tribes have been essential to managing natural resources, protecting biodiversity, and keeping ecological equilibrium. However, Indigenous environmental practices have been weakened over the past few centuries by colonialism, globalization, and the imposition of Western governance systems. As a result, there has been extensive environmental degradation, biodiversity loss, and the deterioration of Indigenous cultures. However, the active, healing plants and ecological interactions have played a significant role in the current comeback of Indigenous environmental governance. Nevertheless, patriarchal regimes in both Indigenous and settler-colonial systems have restricted women's access to leadership positions, and their contributions have frequently been marginalized in official decision-making processes. Broader gender disparities that endure in many civilizations are reflected in the continued marginalization of Indigenous women in environmental governance. In recent years, academic and policy circles have paid more attention to the role that Indigenous women play in environmental governance. According to research, indigenous women have always played a significant role in protecting the environment by overseeing the management of natural resources like agriculture, water, and forests (Aguilar et al., 2015a; 2015b). However, their impact in decision-making processes has been limited due to colonial histories and gender biases that have frequently excluded them from official governance institutions (Moller et al., 2019).

As more Indigenous women assume leadership positions in their communities, the resurgence of Indigenous environmental governance signifies a dramatic change in social and environmental paradigms. Their active involvement in decision-making processes, where they reclaim their voices and promote environmentally friendly methods, is indicative of this change. Involving indigenous women in environmental governance is essential for preserving the environment as well as bolstering cultural identity and sovereignty (UNDP, 2015). To address the climate change, restoring ecosystems, and advancing sustainable development all benefit greatly from their leadership, especially in the management of natural resources (Davis & Ranco, 2015a, 2015b).

An increasing amount of research indicates that the advancement of sustainable environmental practices is greatly aided by the leadership of Indigenous women. They provide distinct perspectives on regional ecosystems and resource management tactics due to their longstanding ties to the land and traditional ecological knowledge (TEK) (UN Women, 2020). Indigenous women's contributions to climate resiliency and biodiversity protection have received special recognition. But the studies also points to enduring obstacles that keep them from fully participating, like cultural norms, gender-based discrimination, and limited access to education and training, which frequently keep women from taking up leadership responsibilities (Garnett et al., 2018). Women's contributions to environmental governance are often marginalized by such frameworks, according to feminist and post-colonial researchers, which perpetuates as more indigenous women assume leadership positions in their communities, the resurgence of Indigenous environmental governance signifies a dramatic change in social and environmental paradigms. Their active involvement in decision-making processes, where they reclaim their voices and promote environmentally friendly methods, is indicative of this change. Involving indigenous women in environmental governance is essential for preserving the environment as well as bolstering cultural identity and sovereignty. The prevalence of patriarchal institutions in both Indigenous and state governing systems exacerbates these issues. Women's contributions to environmental governance are often marginalized by such frameworks, according to feminist and post-colonial researchers, which perpetuates gender inequality (Gaard, 2017a, 2017b).

Although legislation like gender quotas and leadership development programs that attempt to increase women's involvement in governance have made headway, their implementation is still unequal. As well, understanding the intersections of gender, culture, and governance systems presents considerable obstacles for women (Tengö et al., 2020). Notwithstanding these obstacles, research continuously shows that giving Indigenous women leadership positions can promote gender equality and better environmental stewardship. Their leadership improves ecological resilience and aids in the creation of sustainable, culturally appropriate solutions to environmental problems by embracing a variety of viewpoints and knowledge systems. In light of this, this chapter examines the critical role that Indigenous women play in the revival of environmental governance, emphasizing both their distinct contributions and the difficulties they encounter. The study illustrate how Indigenous women's participation in environmental governance improves ecological resilience and guarantees the preservation of Indigenous cultures and knowledge systems by drawing on case studies, academic research, and publicly available data. These customs are essential to building a sustainable future for everybody. It is more important than ever to include Indigenous knowledge and governance systems into frameworks for policy and decision-making, especially in light of the escalating urgency of global environmental concerns. In addition to protecting the environment, Indigenous women's active involvement in this governance process advances social justice and equity.

The relationship between environmental governance and women's empowerment in India is exemplified by a number of significant legislation and initiatives. For example, the National Environment Policy (2006) emphasizes the significance of resolving gender imbalances and empowering excluded groups, while acknowledging the complex issues that emerging societies face, especially poverty. In a similar vein, women had to comprise 33% of Vana Samarakshana Samitis' membership under the 1988 National Forest Policy and the Joint Forest Management (JFM) initiative, which was implemented in the 1990s (World Bank, 2005; MoEF, (1988). The success of reforestation initiatives in several states has been greatly aided by this inclusion. Women are acknowledged by the Biodiversity Act of 2002 as important participants and keepers of traditional knowledge, especially when it comes to seed conservation in farming communities.

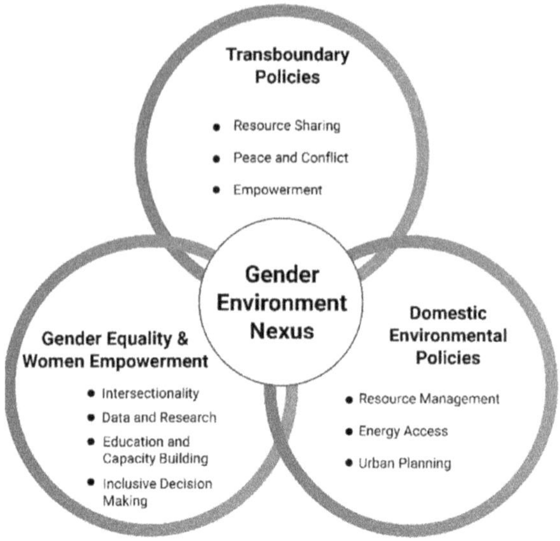

Fig. 1 Women empowerment in environmental conservation. *Source* Abdulsalam et al. (2024)

Women's participation in water management is essential, as demonstrated by India's water harvesting initiatives. In projects like the Sukhomajri village water management project in Haryana, where their leadership in using water and forest resources earned national attention, women have been crucial. Women's contributions to grassroots water management are highlighted in the Gujarat Water Policy (2002), which also recognizes their vital role in controlling rural water supplies, health, sanitation, and agricultural production.

Finally, in the renewable energy sector, policies targeting rural regions, such as those encouraging biogas plants and solar cookers/panels, also aim to boost women's access to sustainable energy sources. In order to guarantee equitable and efficient energy solutions, policymakers are realizing more and more how crucial it is to include women in the creation of energy policies.

In brief, incorporating women into environmental governance is essential to creating practical, long-lasting environmental solutions as well as a step toward gender parity. The aforementioned regulations and initiatives demonstrate the increasing acknowledgement of women's contributions to environmental governance and emphasize the necessity of maintaining Indigenous women's empowerment as important players in this revolutionary process (Fig. 1).

2 Background

Indigenous communities around the world have developed sustainable environmental practices based on Traditional Ecological Knowledge (TEK) and have preserved close ties with their land and natural resources. Indigenous peoples have been able to maintain ecological balance by managing forests, water supplies, and biodiversity with the help of this knowledge that has been passed down through the centuries (Berkes, 1999; Turner et al., 2000). However, Indigenous traditions have frequently been sidelined and Indigenous peoples—particularly women—have been excluded from formal decision-making processes pertaining to environmental management as a result of the imposition of colonial and post-colonial governance structures (Nadasdy, 2003; Smith, 2012). Indigenous women have historically played a far smaller part in environmental governance. Despite being the main land stewards in many Indigenous societies and in charge of important environmental responsibilities including farming, water management, and conservation (Maldonado et al., 2016), women's voices have frequently been left out of formal leadership positions. Women's involvement in environmental decision-making has been restricted by the continuation of patriarchal structures in colonial or national administrations as well as within Indigenous communities, which have perpetuated gender inequality (Gaard, 2017a, 2017b; Whyte, 2017a, 2017b).

However, the crucial role that Indigenous women play in environmental governance has come to light more and more in recent decades. It is now more important than ever to incorporate Indigenous knowledge systems into mainstream environmental policy as the global crises of climate change and environmental degradation worsen. Given their deep ties to the land and natural resources, indigenous women are increasingly recognized as essential collaborators in creating long-term solutions for climate adaption and environmental preservation (Coulthard, 2014; Dorries & Morrow, 2020). Initiatives to empower Indigenous women have been sparked by the need for greater gender equality in environmental governance. These programs frequently concentrate on removing historical obstacles that have kept women from attaining leadership positions, such as gender bias, cultural norms, and limited access to formal education and training (Berkes, 2009; Garnett et al., 2018). These programs aim to guarantee that Indigenous communities gain from more equitable, efficient, and sustainable management of their natural resources by boosting women's involvement in governance (Tengö et al., 2020).

The adoption of government guidelines aimed at increasing the number of women in leadership roles within local environmental governance committees has been a crucial milestone in this endeavor. These policies frequently involve leadership development courses, gender quotas, and steps to guarantee equal representation in decision-making. These efforts seek to establish governance frameworks that address environmental issues as well as the more general objective of social justice by institutionalizing gender equality (Ferguson, 2016; UN Women, 2020).

Despite encouraging results in several areas, problems still exist. These include the continued dominance of patriarchal governing systems, restricted access to resources, and deeply ingrained cultural biases (Garnett et al., 2018). However, a major step toward more sustainable and inclusive methods that respect Indigenous knowledge and advance gender equity is the growing acknowledgement of Indigenous women's contributions to environmental governance (Tengö et al., 2020).

3 Theory and Hypotheses

Theory: This article's philosophy is based on a number of interrelated frameworks, such as feminist theory, environmental governance, and Indigenous knowledge systems. The notion that Indigenous environmental governance particularly through the active participation of Indigenous women represents a successful and culturally grounded strategy for tackling today's environmental issues is at the heart of the conversation. This method promotes gender equality while acknowledging the rich, lived knowledge of Indigenous cultures, providing a more comprehensive understanding of sustainable environmental management.

Indigenous Knowledge Systems: Traditional Ecological Knowledge (TEK), a term used to characterize the systems of knowledge and practices that Indigenous peoples have created over millennia in their interactions with the natural world, is at the heart of Indigenous environmental governance. The conservation of biodiversity, climatic patterns, sustainable farming practices, forest management, water conservation, and soil regeneration are just a few of the many ecological wisdoms that are included in TEK. Since TEK offers important insights into managing and restoring ecosystems in the face of climate change, its significance in environmental governance is widely acknowledged (Berkes, 1999).

Feminist Theory: Comprehending the intersectionality of gender and environmental issues requires an understanding of feminist philosophy, especially eco-feminism. Patriarchal institutions that value dominance over collaboration, exploitation over care, and control over nurturing are the root cause of both environmental degradation and women's oppression, according to ecofeminism (Plumwood, 1993). According to this theoretical framework, the deterioration of the environment and the oppression of women are frequently compared because both are viewed as resources that may be used for both political and economic ends. Ecofeminism enables us to examine the historical exclusion of women from leadership positions and decision-making, particularly in environmental governance, due to patriarchal dynamics in both traditional and colonial systems in Indigenous communities (Berkes, 2009).

In addition to criticizing these patriarchal systems, ecofeminism advocates for women's empowerment as a key element in attaining more sustainable and inclusive environmental governance. Despite being stewards of the land and custodians of traditional ecological knowledge (TEK) in many Indigenous communities, colonial and patriarchal regimes have frequently ignored women's contributions to environmental management. According to ecofeminism, the restoration of ecological health and gender equality depends on the empowerment of Indigenous women. Restoring the balance between people and the land can be achieved through the development of sustainable, culturally appropriate solutions that acknowledge and incorporate the leadership of Indigenous women in environmental governance (Gaard, 2017a, 2017b; Shiva, 1989).

Environmental Governance: The procedures, organizations, and institutions that go into creating, carrying out, and upholding environmental policies, practices, and decisions are collectively referred to as environmental governance (Berkes, 2009). It encompasses both state and non-state actors and can take the shape of official organizations or unofficial agreements at the local level. Long existing outside of the official state structure, indigenous governance systems usually place a strong emphasis on sustainability, harmony with nature, and group decision-making (Garnett et al., 2018). These systems, which represent traditional ecological knowledge (TEK) transmitted through the generations, place a high priority on social responsibility for preserving ecological balance and communal ownership of resources (Berkes et al., 2000). According to the philosophy underlying Indigenous environmental governance, more comprehensive, culturally aware, and environmentally sustainable outcomes arise when Indigenous women are empowered and granted equal participation in government (Davis & Ranco, 2015a, 2015b). Indigenous women have a profound and sophisticated awareness of the local ecology and the practices that can sustain it because they are frequently the guardians of natural resources in their communities. By encouraging social justice and the transfer of knowledge between generations, their participation in environmental governance not only improves ecological resilience but also fortifies community well-being (UN Women, 2020). Indigenous women's leadership in environmental protection and resource management, in particular, offers special perspectives on the preservation of the environment and the utilization of natural resources (Kovach, 2009) (Table 1).

Gender Composition of Local Committees: An examination of the gender distribution of local committees established under different environmental governance initiatives offers important information about how men and women are represented and involved in decisions pertaining to resource allocation and environmental management. At the local level, local committees are crucial in establishing sustainable practices and forming policy. Women have historically been excluded from formal leadership positions in many Indigenous government organizations, which have historically been controlled by men (Rocheleau et al., 1996). However, women's involvement in local committees has significantly increased as a result of a growing emphasis on gender equality, particularly through focused initiatives. Women made up just 20–30% of local committee members prior to the introduction of gender-focused programs, which was a reflection of patriarchal norms and hurdles that restricted their participation in government (Garnett et al., 2018; Tengö et al., 2020). These obstacles frequently prevented women from taking on leadership roles and were exacerbated by cultural norms, legislative restrictions, and restricted access to resources and education. However, efforts to promote gender parity after the program have resulted in a notable rise in the representation of women. Women now make up between 40 and 50 percent of committee members in some areas, while in other instances, committees have attained gender parity, with an equal number of men and women holding leadership roles (Davis & Ranco, 2015a, 2015b). The focus on gender equality brought about by governmental regulations and community empowerment initiatives that remove structural obstacles to women's participation is directly responsible for this change (Garnett et al., 2018). The decision-making process has been significantly impacted by women's greater engagement. Because of

Table 1 Represent the hypothesis with description and rationale

Hypothesis	Description	Rationale
H1: The active participation of Indigenous women in environmental governance significantly improves the effectiveness of natural resource management and conservation efforts within Indigenous communities Garnett et al. (2018), Davis and Ranco (2015a, 2015b)	The involvement of Indigenous women in governance enhances the stewardship of natural resources, leading to higher biodiversity and ecosystem resilience	Indigenous women possess deep knowledge of local ecosystems and traditional resource management practices, which contribute to more effective natural resource management and conservation
H2: Communities with higher levels of Indigenous women's participation in environmental decision-making exhibit better outcomes in climate adaptation and environmental sustainability Tengö et al. (2020); UN Women (2020)	Communities with higher Indigenous women's participation are more resilient to climate change and environmental challenges	Studies suggest that Indigenous women's leadership leads to more adaptive and sustainable approaches to environmental issues, as they often promote climate resilience and sustainability
H3: Indigenous women face multiple, intersecting barriers in participating in environmental governance, including gender inequality, legal obstacles, and colonial legacies Gaard (2017a, 2017b), Garnett et al. (2018)	Indigenous women face systemic barriers to governance participation, including gender discrimination and limited access to decision-making platforms	Despite their key role in environmental stewardship, Indigenous women often encounter legal, social, and political obstacles that hinder their full participation in governance
H4: Empowering Indigenous women within governance structures contributes to the restoration of Indigenous cultural identity and sovereignty Hernández Castillo (2004), Shiva (1989)	Empowerment of Indigenous women can revitalize traditional cultural practices and strengthen Indigenous sovereignty	Participation in governance by Indigenous women is crucial for reclaiming environmental rights, as well as cultural and political autonomy from colonial systems
H5: Increased integration of Indigenous women's leadership in environmental governance promotes social justice and equity within Indigenous communities Davis and Ranco (2015a, 2015b), Kovach (2009)	Women's leadership in governance contributes to gender equality and stronger community cohesion	Empowering Indigenous women within governance structures is expected to foster a more inclusive, equitable approach to managing both social and environmental resources

their historical responsibilities as stewards of the land, water, and resources, women have contributed distinct viewpoints to government, guaranteeing that laws represent social justice and environmental sustainability (Tengö et al., 2020).

Challenges and Barriers: Despite the positive strides in gender representation, several challenges persist. The participation of women in local committees remains subject to ongoing social, cultural, and political obstacles: Deeply rooted gender biases continue to be a major obstacle to women's active engagement in government in many Indigenous communities. Male coworkers and community leaders who still see women's roles as secondary or less authoritative frequently oppose them, especially those from marginalized backgrounds (Garnett et al., 2018). For instance, women's participation in the Joint Forest Management (JFM) program has increased in some parts of India, although there is still strong opposition in other areas. Although the program's goal is to include women in forest management committees, male leaders usually minimize or ignore the contributions of women (Agarwal, 2001). Men still control decision-making processes in these situations, which limits women's influence over important choices pertaining to conservation and resource management and reinforces gendered power dynamics. Although there has been progress in the representation of women in local governing committees, many still struggle to get the tools, education, and support networks they need to participate in governance processes. For example, Indigenous women in the Philippines have expressed difficulty getting leadership training and formal schooling, which limits their ability to participate in intricate governance procedures (Cameron, 2015). When compared to their male counterparts, women are frequently at a disadvantage due to their limited access to financial resources and technical expertise, which hinders their capacity to assume leadership positions within committees. In addition to focusing on increasing women's participation, programs must make sure that women have access to the resources and training required for efficient government (2015b; Aguilar et al., 2015a). Even in Indigenous communities, women's participation in governance institutions is significantly hampered by cultural norms and traditional traditions that limit their positions in public life. Due to these centuries-old conventions, women are typically expected to handle the majority of household duties while men are expected to assume leadership positions in the public arena. For example, patriarchal traditions that view women as unfit for leadership roles have historically kept women from Indigenous groups, such as the Rai and Sherpa, from participating in community decision-making in Nepal (Barker et al., 2016). Even if the need for gender equality is becoming more widely acknowledged, communities must work together for years to change centuries-old cultural customs.

Recommendations for Further Improvement: To build on the progress made in improving gender representation in local committees, the following strategies can be implemented:

Targeted Gender Training and Empowerment Programs: Offering training programs specifically designed to empower women with leadership, negotiation, and decision-making skills can enhance their contributions to local committees. This should also include education about the value of gender equality and women's participation in governance.

Promoting Cultural Change: Engaging both men and women in dialogues about gender equality and challenging traditional gender roles can help shift community norms and ensure that women's voices are heard and respected in decision-making processes.

Institutionalizing Gender Parity: Institutionalizing gender parity in governance structures, by setting quotas or ensuring that committees reflect equal representation, can help institutionalize women's leadership in the long term. The gender composition of local committees formed under the program shows positive trends toward greater inclusion of women in environmental governance. While challenges remain, the increased participation of women has contributed to more sustainable, inclusive, and culturally relevant environmental management practices.

Government Guideline Resulting in an Increase in Leadership Seats Held by Women: A significant development within the context of gender equality in environmental governance is the implementation of a government guideline aimed at increasing the representation of women in leadership roles, particularly within local environmental governance committees. This guideline was introduced to address historical gender imbalances in leadership and decision-making positions, recognizing that women, especially indigenous women, possess valuable traditional ecological knowledge that is critical for sustainable environmental management.

Overview of the Government Guideline: Addressing the historical exclusion of women from leadership roles, especially in the context of Indigenous environmental governance, requires the introduction of government standards targeted at achieving gender parity and improving governance institutions. These recommendations have the potential to drastically alter the dynamics of governance by emphasizing the establishment of more egalitarian and inclusive decision-making bodies, which would promote social justice and better environmental management.

The application of gender quotas is one of these guidelines' main features. Quotas help to guarantee that women have a significant voice in decision-making processes by requiring a minimum percentage of women in leadership roles. These quotas address the gender imbalance that has historically kept women out of positions of power by mandating that women hold at least 40–50% of leadership posts. According to research, gender quotas promote more inclusive governance structures by giving women a voice in policy and decision-making, which can lead to more varied and comprehensive policy results (Krook, 2009). Women's representation in environmental governance can provide new insights into topics like biodiversity preservation, water conservation, and sustainable resource management all of which are vital to the long-term health of the ecosystem. The success of the Joint Forest Management (JFM) initiative in India, for example, in incorporating women into forest governance has shown how gender quotas improve environmental outcomes (Krook, 2009). Women's representation in environmental governance can provide new insights into topics like biodiversity preservation, water conservation, and sustainable resource management all of which are vital to the long-term health of the ecosystem. The success of the Joint Forest Management (JFM) initiative in India, for example, in incorporating women into forest governance has shown how gender quotas improve community empowerment and environmental outcomes (Agarwal, 2001).

Women's leadership training is another crucial component. To guarantee that women can successfully assume leadership roles, specialized training programs that give them the skills they need in areas like resource management, governance structures, environmental policy, and negotiation are crucial. These programs assist women develop their confidence and leadership skills in addition to providing them with the technical know-how needed to manage ecosystems and natural resources. Women who receive this kind of training are more equipped to question gender norms and take charge of environmental governance, which will result in more fair decision-making. Rao and Kelleher (2003) assert that because leadership training programs equip women with the skills necessary to successfully negotiate the frequently complicated political and institutional environments, they are especially successful in enabling them to participate in governance processes. Mechanisms for accountability and monitoring are essential to guaranteeing the successful implementation and long-term sustainability of gender quotas and other policies. It is possible to identify areas where progress has been made and where more needs to be done by regularly reporting on the gender composition of leadership bodies. In addition to encouraging responsibility, this open method keeps the momentum for efforts toward gender parity going. Monitoring is crucial to ensuring that the advantages of more female participation in governance are maintained and preventing the rollback of gender equality initiatives, claims Moser (2005). Monitoring also contributes to a more evidence-based approach to policy creation by offering data and proof on the effects of these initiatives, which can further guide future gender and environmental governance efforts.

Impact of the Guideline on Leadership Representation: Women's representation and decision-making authority have significantly increased since the government guideline encouraging gender parity in leadership positions within local environmental governance committees was put into effect. With only 20–25% of seats in many Indigenous communities, women were underrepresented in leadership roles prior to the guideline. Nonetheless, the number of women in leadership positions rose by as much as 30–40% following the introduction of the guideline, with certain regions attaining between 50 and 60% female representation in leadership posts. This rise demonstrates a significant move toward gender equality and the effectiveness of the guideline in changing governance frameworks to be more inclusive (Krook, 2009). The greater number of women in leadership positions has improved environmental decision-making, but it has also had wider societal effects. Greater gender equality in the community and increased female economic engagement has resulted from women's empowerment within governance institutions. Gender equity has been further advanced by the guideline, which has strengthened women's responsibilities in environmental governance as well as other areas of public and political life by recognizing women's contributions to community development. This change is consistent with research by Duflo (2012), who emphasizes the favorable socioeconomic effects of increased female involvement in government and leadership.

Increased Female Leadership: Positive wider societal effects have also been facilitated by the greater representation of women in leadership roles. In these areas, it has promoted more gender equality, which has increased women's economic engagement and increased awareness of their contributions to community development. This advancement strengthens women's position in other areas of public and political life in addition to giving them more authority in government. Women's achievement in breaking down barriers in leadership has changed social norms and advanced gender parity, proving the transformative impact of women's participation in governance (Gaard, 2017a, 2017b).

Enhanced Decision-Making: Along with boosting the number of women in leadership positions, the guideline has also assisted in removing long-standing obstacles. The institutional and legal backing for the guideline challenged patriarchal systems and cultural norms that frequently limited women's involvement in decision-making. These actions have dismantled old gender biases by empowering women to exercise their leadership rights (Garnett et al., 2018). Additionally, the guideline made it easier for women to obtain previously unattainable training, mentorship programs, and vital resources, giving them the means to thrive in leadership positions. The guideline has also assisted in lowering incidences of gender-based discrimination by instituting gender quotas and advancing gender equality, guaranteeing that women with the necessary training and expertise can advance to leadership roles without encountering structural obstacles.

Broader Social Impact: Positive wider societal effects have also been facilitated by the greater representation of women in leadership roles. In these areas, it has promoted more gender equality, which has increased women's economic engagement and increased awareness of their contributions to community development. This advancement strengthens women's position in other areas of public and political life in addition to giving them more authority in government. Women's achievement in breaking down barriers in leadership has changed social norms and advanced gender parity, proving the transformative impact of women's participation in governance (Gaard, 2017a, 2017b).

Barriers Overcome by the Guideline: The introduction of the guideline helped overcome several barriers that historically limited women's access to leadership roles in environmental governance:

Cultural Norms and Gender Bias: In many Indigenous communities, cultural norms and traditional patriarchal structures had limited women's participation in formal decision-making bodies. The government's guideline challenged these norms, providing legal and institutional support for women to assert their leadership rights.

Access to Resources and Networks: Historically, women had less access to the networks, training, and resources necessary to assume leadership roles. The government guideline facilitated access to mentorship programs, leadership training, and educational resources, ensuring that women had the tools they needed to succeed in leadership positions.

Gender-Based Discrimination: Many women faced direct or indirect discrimination, which prevented them from ascending to leadership roles, even if they possessed the necessary skills and knowledge. The guideline, by institutionalizing gender quotas and promoting gender equality, reduced instances of discrimination by holding communities accountable for meeting gender representation standards.

4 Case Studies

Measurable gains in women's leadership engagement are evident in a number of official reports and data sources, especially when it comes to community development and environmental governance. For instance, the National Family Health Survey (NFHS) provides insightful information about the growing leadership and decision-making roles of women in India's rural and Indigenous communities. The NFHS-5 (2019–2020) reports that women's involvement in governmental and community decision-making processes has significantly improved. According to the poll, 34.4% of rural Indian women said they were involved in making decisions regarding family purchases, and more and more of them were taking part in local governing positions (MoHFW, 2020). The Panchayati Raj Institution (PRI) statistics show notable gains in women's local political participation, which is consistent with these findings. For example, several regions have witnessed a notable increase in female leadership as a result of the 73rd Constitutional Amendment Act (1992), which required a 33% quota of seats for women in Panchayats (local government organizations). Women are taking on important decision-making responsibilities in some states, like as Rajasthan, where their percentage of Panchayat leadership positions has topped 50%. According to a State of Panchayats study (Ministry of Panchayati Raj, 2021), the percentage of women in leadership positions in District Panchayats and Gram Panchayats (village councils) has been rising gradually, with several regions claiming that women hold up to 50% of the leadership positions. For instance, the proportion of women on environmental committees in one Indigenous village quadrupled from 25 to 50% following the implementation of a government directive encouraging gender parity in leadership. Women emphasized the value of protecting holy natural areas and incorporating sustainable farming practices that were previously disregarded, which resulted in more inclusive and integrated approaches to land conservation. These results are consistent with PRI Reports, which show that women's participation in land management and environmental governance results in more sustainable and successful conservation plans (Ministry of Panchayati Raj, 2021).

In a similar manner, women's growing leadership involvement in rural areas has aided in the development of creative answers to regional environmental problems like water management. In one instance, women started policies aimed at addressing water scarcity after taking on leadership roles in an environmental governing agency. They successfully addressed water challenges while maintaining cultural practices by introducing water conservation tactics that combined Indigenous knowledge with contemporary scientific methods, drawing on their historical responsibilities as the main water managers in households. These instances highlight the benefits of gender-inclusive governance, as evidenced by statistics from the National Rural Drinking Water Programme (2020), which shows that women's leadership in managing water resources greatly enhances rural communities' access to water and sustainability. Some well known Indian women environmentalists are (Sharma, 2016) (Table 2):

1. Amrita Devi—Bishnois Struggle to save Khejri tree
2. Bachni Devi and Gaura Devi—Chipko movement

Table 2 **a** Case Studies at National level (*Source* Gupte, 2004). **b** Case Studies at Global level (*Source* Smith et al., 2023; Pereira and Alvarez 2022)

Case study	Region	Role of indigenous women	Key impact
The Chipko Movement	Uttarakhand	Women led efforts to prevent deforestation by physically embracing trees	Empowered women to take leadership in environmental activism and forest conservation
Adivasi Women's Role in Forest Protection	Madhya Pradesh	Women led Van Panchayats (Forest Councils) to protect forests, using traditional knowledge for sustainable harvesting	Preservation of forests and biodiversity, and empowerment of women in community-based forest management
Narmada Bachao Andolan	Narmada Valley	Women, including Medha Patkar, led protests against displacement caused by the Sardar Sarovar Dam	Protection of Indigenous communities' rights and the promotion of sustainable development
Sacred Groves of Kerala	Kerala	Women protect sacred groves, which are vital for maintaining biodiversity, through cultural practices	Conservation of biodiversity hotspots and preservation of cultural heritage linked to environmental practices
Khonoma Village Forest Protection	Nagaland	Women led efforts to ban unsustainable hunting and promote eco-tourism	Recovery of wildlife and promotion of sustainable livelihoods through eco-tourism
Sunderbans Mangrove Protection	West Bengal	Women are involved in local governance, mangrove restoration, and disaster preparedness	Protection of mangroves, crucial for flood prevention and climate adaptation
Great Nicobar Island Conservation	Andaman and Nicobar Islands	Women use traditional knowledge to balance development with environmental conservation	Protection of biodiversity and promotion of sustainable practices in response to development pressures

(continued)

Table 2 (continued)

Case study	Location	Key actions/ initiatives	Key actions/ initiatives	Impact
Women-Led Forest Conservation in the Amazon	Amazon Rainforest, Brazil	Organizing grassroots movements, advocating for land rights, and protecting forests	Organizing grassroots movements, advocating for land rights, and protecting forests	Strengthened land tenure and forest management. Reduced deforestation rates. Increased community resilience to environmental threats
Reviving Traditional Agroecology in Kenya	Kenya	Reintroducing intercropping, crop rotation, organic farming, and tree planting through the Green Belt Movement	Reintroducing intercropping, crop rotation, organic farming, and tree planting through the Green Belt Movement	Empowered women in environmental governance and food security. Revitalized agroecology practices. - Mitigated climate change impacts
Restoration of Wetlands in Canada	Northern Quebec, Canada	Developing community-based wetlands management plans, reintroducing native plant species	Developing community-based wetlands management plans, reintroducing native plant species	Restored wetlands, enhancing biodiversity and water quality. Strengthened women's decision-making roles. Bridged traditional and modern conservation practices
Firestick Alliance in Australia	Australia	Reviving controlled burning techniques, educating communities on cultural fire management	Reviving controlled burning techniques, educating communities on cultural fire management	Reduced wildfires by implementing cultural burning Empowered women in environmental leadership Promoted understanding of indigenous fire management practices

3. Medha Patkar—Narmada Bachao Aandolan
4. Sunita Narain—Centre for Science and Environment
5. Vandana Shiva—Activist, Navadanya
6. Aruna Roy—Mazdoor Kisan Shakti Sangathan (MKSS)
7. Rashida Bee and Champa Devi Shukla—Bhopal Gas Tragedy
8. Maneka Gandhi—Animal Rights activist
9. Sugatha Kumari—Poet; Silent Valley
10. And many more.

Continuing Challenges and Recommendations: Many mentorship programs around the nation are vital resources for fostering women's leadership development while assessing gender representation policies and their execution. One of the well-known programs is Lean in Circles, a network that gives women a place to encourage one another and grow as leaders. Sheryl Sandberg created the program, which promotes peer-to-peer mentoring as a means of combating gender bias. Lean In Circles are successful in empowering women, yet they still have trouble reaching underrepresented populations, like low-income or rural women (Lean In Foundation). The Women in Public Service Project (WPSP), which seeks to enhance women's leadership in public service, is another noteworthy initiative. This program gives women in the public and governmental sectors networking opportunities, training, and mentoring.

Besides, mentorship programs for young people, especially young women, are available nationwide through the National Mentoring Partnership (MENTOR). By providing direction and assistance to women from underrepresented groups, these initiatives aim to close the gap in leadership opportunities. Scaling these initiatives to successfully combat unofficial gender prejudices is still a major difficulty, though (MENTOR). A worldwide non-profit organization called The Catalyst Women in Leadership Programs collaborates with businesses to provide women with training and mentoring, particularly in the area of business leadership. The goal of Catalyst's initiatives is to promote diverse leadership and lessen gender bias; however, expanding these initiatives to underrepresented fields, especially in the public and political arenas, is still a struggle (Catalyst). In a similar vein, the National Association of Women in Government (NAWG) helps women in public service advance into leadership positions by offering mentorship. Even though NAWG has been effective in raising the number of women in government, more work is required to overcome the ingrained gender prejudices that still exist in politicalsystems (NAWG).

Lastly, the U.S. Chamber of Commerce hosts the Women's Leadership Forum (WLF), which unites women from a range of sectors to network and provide leadership mentoring to one another. The WLF's overall influence is limited by its inability to reach women from economically disadvantaged backgrounds, despite its success in promoting women in business (U.S. Chamber of Commerce). These mentorship programs are crucial in making sure that female representation is a long-term, sustainable change rather than just a passing trend. But in order for these programs to be really successful, they need to be inclusive and made to reach women from a variety of backgrounds, such as young women, women with disabilities, and women from disadvantaged socioeconomic groups.

5 Conclusion

The implementation of government guidelines aimed at increasing the representation of women in leadership roles within local environmental governance committees has resulted in significant positive outcomes for both gender equality and environmental sustainability. As the data indicates, there has been a marked increase in the number of women occupying leadership positions, with female representation in some communities reaching near gender parity. This shift has contributed to more inclusive, effective decision-making processes that better integrate traditional ecological knowledge (TEK) with modern environmental practices, leading to improved resource management and enhanced community resilience to environmental challenges. While progress has been made, the study also highlights the persistence of barriers such as gender-based discrimination, cultural norms, and limited access to resources and training for women, which continue to hinder their full participation. These challenges must be addressed through ongoing efforts to provide women with the necessary support and opportunities to thrive in leadership positions. The positive outcomes observed in communities with increased female leadership demonstrate that empowering women not only advances gender equality but also strengthens environmental governance by incorporating diverse perspectives and solutions that are more sustainable and culturally relevant. Moving forward, it is crucial to continue investing in gender-focused policies, education, and training programs, as well as promoting cultural shifts that support the inclusion of women in all aspects of environmental governance. By sustaining these efforts, we can create more equitable and effective governance structures that ensure the protection and sustainable management of our natural resources for future generations.

6 Discussion

The future of environmental governance hinges on empowering Indigenous women, who play a crucial role in reviving traditional practices that have sustained ecosystems for generations. These women, deeply connected to their land, hold invaluable knowledge about biodiversity conservation and sustainable resource management. As climate change accelerates, their leadership in climate resilience and adaptation becomes increasingly important, highlighting the need for their inclusion in decision-making processes at all levels. By integrating Indigenous women into environmental policies and governance, we can bridge the gap between traditional ecological knowledge and modern practices. Furthermore, gender and environmental justice must intersect, ensuring that the voices of Indigenous women are heard in global climate discussions. As grassroots movements grow, these women will continue to advocate for climate action, while educational initiatives and capacity-building efforts will equip them with the skills to shape policies. The future will see more collaboration between Indigenous women and global networks, using both traditional knowledge and modern technology to address environmental challenges. Empowering Indigenous women is not just a matter of social justice, but an essential step toward a more sustainable and inclusive future.

References

Abdulsalam, M., et al. (2024). Women empowerment in environmental conservation. In S. A. Aransiola, B. R. Babaniyi, A. B. Aransiola, & N. R. Maddela (Eds.), *Prospects for soil regeneration and its impact on environmental protection* (pp. 367–386). Earth and Environmental Sciences Library. Springer. https://doi.org/10.1007/978-3-031-53270-2_16

Aguilar, L., Bui, S., & Stoll, C. (2015a). Women, climate change, and Indigenous communities: A review of gendered impacts and contributions. *Journal of Indigenous Environmental Studies, 18*(3), 47–63.

Aguilar, L., S. V. E., & García, M. (2015b). Gender and the environment: The role of Indigenous women in natural resource management. *Environmental Governance and Political Ecology, 25*(3), 215–227. https://doi.org/10.1080/09644016.2015.1055101

Agarwal, B. (2001). Participatory exclusions, community forestry, and gender: An analysis for South Asia and a conceptual framework. *World Development, 29*(10), 1623–1648. https://doi.org/10.1016/S0305-750X(01)00066-3

Barker, A., et al. (2016). Patriarchy and gender norms in Indigenous Nepali communities: A review of traditional practices and their impact on women's leadership. *Gender, Place & Culture, 23*(5), 745–764. https://doi.org/10.1080/0966369X.2016.1171867

Berkes, F. (1999). *Sacred ecology: Traditional ecological knowledge and resource management.* Taylor & Francis.

Berkes, F. (2009). Commons in a changing world: Insights from international commons research. *Environmental Science & Policy, 12*(2), 128–138. https://doi.org/10.1016/j.envsci.2008.10.004

Berkes, F., Colding, J., & Folke, C. (2000). Rediscovery of traditional ecological knowledge as adaptive management. *Ecological Applications, 10*(5), 1251–1262. https://doi.org/10.1890/1051-0761(2000)010[1251:ROTEKA]2.0.CO;2

Cameron, E. (2015). Indigenous women's leadership in resource management: A case study from the Philippines. *Asian Journal of Comparative Politics, 1*(1), 58–71. https://doi.org/10.1177/2057891114559899

Coulthard, G. (2014). *Red skin, white masks: Rejecting the colonial politics of recognition.* University of Minnesota Press.

Davis, M., & Ranco, D. (2015a). Indigenous environmental governance and the role of women: Traditional knowledge systems and contemporary leadership. *Environmental Management, 55*(2), 212–225. https://doi.org/10.1007/s00267-014-0456-0

Davis, M., & Ranco, D. J. (2015b). Indigenous peoples and environmental governance: A reclamation of traditional knowledge. *Environmental Politics, 24*(1), 1–20. https://doi.org/10.1080/09644016.2014.971604

Department of Drinking Water and Sanitation. (2020). *National Rural Drinking Water Programme (NRDWP): Framework for implementation.* Ministry of Jal Shakti, Government of India, New Delhi.

Dorries, H., & Morrow, C. (2020). Indigenous women and climate justice: Agency, resilience, and transformation. *Canadian Journal of Environmental Education, 25*(1), 1–16.

Duflo, E. (2012). Women empowerment and economic development. *Journal of Economic Literature, 50*(4), 1051–1079. https://doi.org/10.1257/jel.50.4.1051
Ferguson, L. (2016). *Women and the environment: Gender, climate change, and ecological justice*. Routledge.
Gaard, G. (2017a). Ecofeminism and environmental justice: The intersection of gender, race, and environmental sustainability. In S. Blencowe & R. Fuller (Eds.), *Ecofeminism and environmental justice: Perspectives from Indigenous women* (pp. 36–56). Routledge.
Gaard, G. (2017). *Feminist ecologies: Spheres of sustainability and social justice*. Palgrave Macmillan.
Garnett, S. T., et al. (2018). Indigenous knowledge and conservation: Empowering women in natural resource management. *Environmental Conservation, 45*(3), 220–227. https://doi.org/10.1017/S0376892917000440
Government of Gujarat. (2002). *Gujarat state water policy*. Narmada, Water Resources, Water Supply and Kalpsar Department, Gandhinagar.
Gupte, M. (2004). Participation in a gendered environment: The case of community forestry in India. *Human Ecology, 32*(3), 365–382.
Hernández Castillo, R. A. (2004). Zapatista women: Gender, subordination, and rebellion. *Women's Studies International Forum, 27*(3), 215–230. https://doi.org/10.1016/j.wsif.2004.06.004
Kovach, M. (2009). *Indigenous methodologies: Characteristics, conversations, and contexts*. University of Toronto Press.
Krook, M. L. (2009). *Quotas for women in politics: Gender and candidate selection reform worldwide*. Oxford University Press.
Maldonado, J. K., et al. (2016). Indigenous women's roles in resource management and governance. *The Journal of Environment & Development, 25*(1), 45–68. https://doi.org/10.1177/1070496515624340
Ministry of Environment and Forests (MoEF). (1988). *National forest policy, 1988*. Government of India, New Delhi.
Ministry of Health and Family Welfare (MoHFW). (2020). *Annual Report 2019–2020*. Government of India, New Delhi
Ministry of Panchayati Raj. (2021). *Annual Report 2020–21*. Government of India, New Delhi
Moller, H., Ross, J., & Sutherland, W. J. (2019). The role of Indigenous women in biodiversity conservation and ecosystem services: A global review. *Environmental Science & Policy, 92*, 1–8. https://doi.org/10.1016/j.envsci.2018.12.003
Moser, C. (2005). Gender, governance, and the feminization of poverty: Theoretical approaches and policy implications. *International Feminist Journal of Politics, 7*(3), 397–422. https://doi.org/10.1080/14616740500280038
Nadasdy, P. (2003). *Hunters and bureaucrats: Power, knowledge, and Aboriginal-state relations in the Southwest Yukon*. UBC Press.
Pereira, L., & Alvarez, M. (2022). Women-led forest conservation in the Amazon. In *Global environmental leadership* (pp. 45–60). Green Earth Press.
Plumwood, V. (1993). *Feminism and the mastery of nature*. Routledge.
Rao, A., & Kelleher, D. (2003). Institutionalizing gender equality: A framework for strategic action. *International Feminist Journal of Politics, 5*(3), 455–475. https://doi.org/10.1080/1461674032000154596
Rocheleau, D., Thomas-Slayter, B., & Wangari, E. (1996). Feminist political ecology: Global issues and local experiences. *Routledge*. https://doi.org/10.4324/9781315758494
Sharma, R. (Ed.). (2016). *Sustainable development in the Indian context*. Indira Gandhi National Open University.
Shiva, V. (1989). *Staying alive: Women, ecology, and development*. Zed Books.
Smith, L. T. (2012). *Decolonizing methodologies: Research and Indigenous peoples* (2nd ed.). Zed Books.
Smith, J., & Johnson, R. (Eds.). (2023). *Case studies on global women-led environmental conservation initiatives*. Global Environmental Publishing.

Tengö, M., Malmer, P., & Rist, L. (2020). Gender parity in environmental governance: Overcoming barriers to Indigenous women's leadership. *Sustainability, 12*(2), 523–536. https://doi.org/10.3390/su12020523

Turner, N. J., et al. (2000). Traditional ecological knowledge and the conservation of biodiversity. *Ecological Applications, 10*(5), 1277–1287. https://doi.org/10.1890/1051-0761(2000)010[1277:TEKATC]2.0.CO;2

United Nations Development Programme (UNDP). (2015). *Gender, climate change, and the role of Indigenous women in environmental governance*. UNDP Policy Brief.

UN Women. (2020). *Indigenous women and climate change: The role of women in sustainable environmental management*. United Nations Entity for Gender Equality and the Empowerment of Women. https://www.unwomen.org/en/news/stories/2020/9/indigenous-women-climate-change

Whyte, K. P. (2017). Indigenous climate change studies: Indigenous women as environmental leaders. In *The Routledge handbook of Indigenous environmental knowledge* (pp. 177–191). Routledge.

Whyte, K. (2017b). The Dakota Access pipeline, environmental injustice, and US colonialism. *Red INk: An International Journal of Indigenous Literature, Arts, & Humanities, 19*(1), 37–58.

World Bank. (2005). *Forest management and governance in India: Case studies of community forestry*. World Bank.

Empowering Indigenous Women in Conservation Efforts Through Participatory Roles and Livelihood Generations in India's Northeastern Region

Deepa Moni Doley and Paramananda Barman

Abstract Indigenous peoples' landscape management practices are integral to biodiversity conservation and sustainable resource use. Women, especially in rural and mountainous areas, are central to environmental management as primary land, forests, and water users. They play key roles in wild plant gathering, home gardening, and biodiversity management. However, socio-cultural and economic barriers often hinder their full participation in decision-making regarding the conservation and management of natural resources, limiting their contribution to sustainable development. Aaranyak, an Assam-based NGO, empowers local communities, particularly women, through alternative livelihood initiatives that reduce dependence on forest resources. One of its key initiatives, the Pirbi-Karbi Ethnic Haat, fosters a community-driven business model that emphasizes fair trade and reinvestment into biodiversity conservation. Launched in 2019, Pirbi exemplifies a successful partnership between community development and ecosystem management. The Yaongyimchen Community Conserved Area (YCBCA) in Nagaland is an example of indigenous-led conservation. Driven by local communities, women, and students, it addressed concerns over tree loss, wildlife depletion, and water scarcity. The community banned hunting and logging to protect wildlife, forests, and water sources. Focused on the Amur Falcon conservation, this initiative has revived wildlife populations and promoted sustainable practices such as fruit orchards, piggeries, and ginger cultivation. Similarly, Sikkim's Organic Agro Tourism initiative focuses on women's empowerment and eco-friendly practices. Nagaland's Satoyama project is one of the remarkable efforts of conserving biodiversity through traditional conservation practices in the form of community-conserved areas (CCAs) and ecotourism models, fostering sustainable livelihoods while addressing anthropogenic impacts on biodiversity and ecosystems.

D. M. Doley · P. Barman (✉)
CSIR-National Institute of Science Communication and Policy Research (NIScPR), New Delhi, India
e-mail: param.niscpr@csir.res.in

Academy of Scientific and Innovative Research (AcSIR), Ghaziabad, India

N. Moran et al. (eds.), *The Participatory Role of Indigenous Women*,
https://doi.org/10.1007/978-3-032-24059-0_7

Keywords Northeast India · Indigenous women · Biodiversity conservation · Livelihood · Community conservation · Environmental conservation · Conservation

1 Introduction

Indigenous peoples managing landscape plays a vital role in conserving biodiversity and promoting the sustainable utilization of natural resources (Catacora-Vargas, 2021). Indigenous peoples are stewards of some of the world's most biodiverse and ecologically significant landscapes (Sahoo et al., 2022). Indigenous peoples' productive activities extend across entire landscapes, shaping ecosystems by their worldview and giving rise to cultural landscapes. These serve diverse socio-ecological purposes such as food and water management, biodiversity stewardship, and the preservation and sharing of traditional knowledge supported by social and institutional norms that reflect and uphold their cultural identity and ecological values (Catacora-Vargas, 2021; Pilgrim & Pretty, 2010). Biodiversity conservation and sustainable resource use are central to indigenous landscape management, which often results in these territories having a higher conservation status compared to others, as noted in the Global Assessment Report on Biodiversity and Ecosystem Services (Catacora-Vargas, 2021; IPBES, 2019). Indigenous peoples have long held a profound bond with their natural surroundings, where plants, animals, land, and water are closely integrated into their daily lives. This intimate connection has allowed tribal communities to cultivate specialized knowledge about wild animals and plants (Sahoo et al., 2022). Indigenous people live in close connection with and adapt to their natural surroundings. Even though they often inhabit high-risk environments, they possess extensive knowledge of environmental protection, conservation, and strategies to mitigate climate change impacts (Gabriel et al., 2020). Moreover, their traditional practices have been proven to promote environmental sustainability, so they are often called "ecosystem people" (Colchester, 1994; Gabriel and Mangahas, 2017; Gabriel et al., 2020).

Previous studies highlighted women's engagement in natural resource management, especially in the form of home garden management and agroecology, sustainable agriculture, and agrobiodiversity conservation (Lima & Cunha, 2024).

Indigenous women are caregivers, gatherers, and guardians of cultural and spiritual practices. They play an essential role in maintaining these interconnected relationship and their efforts are instrumental in promoting a healthy natural world while ensuring the well-being of their families and communities (CDA Collaborative Learning Projects, 2025).

Indigenous knowledge in India has greatly shaped sustainable agricultural practices such as crop rotation, cross-cropping, intercropping, and traditional organic farming, particularly in the Northeastern region. Their strong bond with nature and varied techniques are essential for preserving ecosystems (Lin & Wang, 2014; Sahoo et al., 2022; Subramaniam & Venkat, 2009). The role of women in biodiversity conservation and environmental management is crucial because of their unparalleled active involvement in collective conservation efforts, land, and restoration of endangered plant species (Ramachandran, 2008; Sahoo et al., 2022). This is mainly due to their extensive traditional knowledge of managing biological resources (Ramachandran, 2008). Indigenous knowledge influence the indigenous people's livelihood systems. It serves as the foundation of various types of traditions, beliefs, and institutions, all of which are crucial for generating income. These traditions, beliefs, religions, institutions, and other factors influenced people's choices of livelihood (Sahoo et al., 2022).

Therefore, the creation and implementation of ecosystem solution requires involvement of indigenous communities. Traditional knowledge can also help in appropriate environmental conservation. For instance, the continued use and promotion of indigenous and traditional foods supports healthier and more sustainable diets and generates economic value for ecosystems (Sahoo et al., 2022).

This chapter explores indigenous women's vital role in sustainable natural resource management and biodiversity conservation in Northeast India. To demonstrate how women's active participation greatly aids in preserving ecosystems and developing sustainable livelihoods, the chapter highlights traditional ecological practices and the cultural values of indigenous communities. Along with the opportunities available to them within the conservation landscape, the chapter also highlights the difficulties faced by indigenous women. It also looks at how livelihood projects that empower women can improve their leadership abilities and encourage more active involvement in biodiversity conservation throughout the region.

2 The Central Role of Women in Biodiversity Conservation and Natural Resource Management

According to some historians, women were the first to lay the groundwork for agriculture. While men focused on hunting, women collected seeds from native plants and began cultivating species essential for food, fodder, fiber, fuel, and feed (Dwivedy, 2012; Prasad and Singh, 1992). Over time, they have protected soil health through organic recycling and ensured crop resilience by preserving various crop varieties and genetic vigor (Dwivedy, 2012). Therefore, without their full intellectual and physical engagement, alternative land management systems cannot thrive, nor can issues like gene and soil erosion be effectively addressed.

Research highlights that women's role in protecting biodiversity and natural resource management often leads to more favorable environmental outcomes, mainly because their practices focus on community well-being and resource efficiency (Ally et al., 2024). For instance, studies have shown that women's participation in community-based resource management enhances conservation success, as they tend to prioritize long-term sustainability and collective well-being (Ally et al., 2024; Zhou & Huo, 2020).

Women play diverse and essential roles in biodiversity conservation and natural resource management. Their involvement includes managing home gardens and fishing resources and participating in the governance of forests and water bodies and broader environmental issues such as climate change. They actively preserve biodiversity in protected areas and play vital role in conservation organizations and sustainable family farming systems. These efforts often include the collection of non-timber forest products and the practice of agroecology. Additionally, women engage in extractive activities, support the environmental movement, contribute to ecotourism, and participate in biodiversity-related research and knowledge generation (Lima & Cunha, 2024).

Indigenous women from Pangi Valley, Himachal Pradesh, have played a key role in environmental conservation while sustaining their livelihoods. Recognizing the harmful effects of deforestation, especially of the Himalayan Birch (*Betula utilis*), the community restricted activities in high-altitude forest areas, helping restore forest cover and combat climate change impacts like glacier melt and soil erosion. Women farmers adopted eco-friendly practices such as using animal manure for soil fertility, mixed cropping for biodiversity, organic farming, and avoiding premature harvesting, making agriculture sustainable and resilient (Luharia & Sharma, 2019).

In Indian culture, the role of women as seed selectors and custodians has been especially significant. The concept of Navdanya, symbolizing the Navagraha (nine planets), represents the renewal of diversity and balance within both ecological and social systems, as the influences of planets and climate on agriculture. This complex, relational web gives meaning to biodiversity in Indian culture and has served as the basis for its conservation over millennia (Shiva, 1992).

Women plays significant role in protecting environment and natural resources management. Their connection with nature makes them more steward towards nature with a great skill of natural resource management. Their close relationship with nature—through roles such as farming and collecting water and firewood—makes them especially vulnerable to environmental degradation, which in turn impact their health and livelihood options (Mohan & Sharma, 2022).

In recent decades, women's contributions to wildlife conservation have become increasingly acknowledged, especially in light of the goals of sustainable development and biodiversity preservation. Because of their intimate ties to the land and natural resources, women have historically possessed a great deal of knowledge about local ecosystems and frequently manage resources sustainably (Ally et al., 2024).

Women's understanding of biodiversity is rooted in practical, place-based experience, knowledge of forests, regional species, and the sustainable use of forest products. In many indigenous and rural communities, this knowledge is indispensable for maintaining ecosystem health and resilience (Parida, 2022).

2.1 Women as Custodians of Local Plant Knowledge and Biodiversity

Across the globe, especially in biodiversity-rich tropical regions, women are the primary stewards of plant resources. From villages and farms to forests and homesteads, women are responsible for the in situ conservation and sustainable management of wild and domesticated plant species. Their traditional roles as home gardeners, plant domesticators, herbalists, and seed keepers are the source of this stewardship. Due to their constant contact with the environment, they have developed a thorough understanding of the local ecosystems and flora, making them custodians of biodiversity (Howard-Borjas, 2001; Howard-Borjas & Cuijpers, 2002; Sood et al., 2015). Homegardens play a critical role in both food security and biodiversity conservation, especially for women in rural and marginalized communities. Historically, women in early foraging societies were more likely to select plant species for home plots to ease daily food collection. Across the globe women hold the responsibility and possess majority of the knowledge and skill related to homegardens (Howard, 2015; Howard-Borjas & Cuijpers, 2002). It is also widely recognized that the domestication of plants began in gardens near the campsites of hunter-gatherer communities. Homegardens are a vital resource for poor women, as they allow them to supplement household food supplies and earn income. During periods of food scarcity, a woman's garden often provides essential nutrition, and because production is maintained year-round, it supports household food security consistently. Homegardens are typically more fertile and diverse than field plots, offering women the ability to observe plant performance closely and to experiment with new crops. Many species that are later cultivated on a larger scale are first tested in homegardens, as seen in the global spread of crops like potato and maize, which began through women's informal gardening practices (Howard, 2015; Howard-Borjas, 2001).

Managing a homegarden requires complex and specific knowledge that is distinct from field agriculture. Women must understand and manage a wide variety of soil, water, and plant relationships, including species associations and microenvironments. The high level of species diversity in homegardens demands detailed planning based on nutritional needs, plant perishability, storage and processing considerations, and vegetative cycles. This planning involves deep ecological understanding and continual adaptation, as the gardens often produce hundreds of different items to fulfill a broad range of household needs (Howard, 2015; Howard-Borjas, 2001; Howard-Borjas & Cuijpers, 2002). Women also play a crucial role in maintaining the health of household members, and in many societies, they possess more extensive knowledge and expertise in the use of medicinal plants than their male counterparts (Howard, 2015; Howard-Borjas & Cuijpers, 2002). Women have traditionally been central to seed-related activities, including selection, storage, preservation, and exchange. In addition, they are often responsible for storing and processing plants into a wide range of products that fulfill various household needs (Aguilar et al., 2010; Dhakal & Leduc, 2010; Howard-Borjas, 2001; Howard-Borjas & Cuijpers, 2002). For example, women from the Mizo community are primarily responsible for seed storage in their households (Dhakal & Leduc, 2010). In Sikkim, India, women play a key role in conserving in situ genetic resources and agrobiodiversity. They are primarily responsible for homestead gardening and managing nearby plots, cultivating a wide variety of plant species that contribute significantly to agricultural biodiversity. As key custodians of seed management, women possess deep indigenous knowledge related to seed selection, processing, and preservation (Khadka & Verma, 2012). Adi women from Arunachal Pradesh, especially those in remote communities, hold extensive knowledge of nutritionally important local food species and traditional products. They depend on a wide variety of local plant species, managed through diverse land use systems and traditional conservation practices (Singh et al., 2020).

2.2 Women's Agricultural and Ecological Contributions in Rural and Tribal Societies

In most rural and tribal communities, women are central to ecological sustainability. Their work includes collecting forest products such as firewood, herbs, and materials for food, medicine, and housing. They engage in sowing, weeding, harvesting, and cultivating diverse vegetables and condiments in home gardens (Parida, 2022).

This knowledge that has been passed down through the centuries is essential for the sustainability and security of food. Despite limited access to economic resources and formal education, they sustain rural microeconomic systems through labor and ecological knowledge. Their centuries of real-world experience have given them unique insights into the local species and ecosystem, which are reflected in their knowledge of biodiversity (Sarker, 2017).

In India, tribal women harvest a variety of edible plants while adhering to ecologically prudent environmental conservation regulations. Communities across India preserve diverse crops such as rice and maize, including in the Northeast, Odisha, Madhya Pradesh, Maharashtra, and other states. Notably, indigenous rice varieties consumed by tribal populations are higher quality than many modern strains (Parida, 2022). This practice highlights women's role in maintaining crop diversity and in situ conservation for food security.

2.3 Women's Role in Mountain Ecosystem Management

Woman play a lead role in the management of subsistence resources in mountain areas. They need to perform diverse duties for survival and possess a deep understanding of the potential and processes of ecosystems (Byers & Sainju, 1994; Khadka & Verma, 2012; Sainju, 1993). Their responsibilities include animal husbandry, nutrient recycling, subsistence farming, and livestock waste, all of which are critical for maintaining productivity on nutrient-poor soils. Women's knowledge of seasonal cycles, landscape use, and resource conservation is often enacted through informal collective practices such as labor exchange and group foraging. For example, in the Andes, Quechua women simultaneously manage livestock, childcare, and resource collection—showcasing the integration of gendered labor with ecosystem sustainability (Byers & Sainju, 1994; Sainju, 1993; Tyagi and Churchgate, 2006).

2.4 Division of Labor and Resource Burden in Mountain Communities

Mountain women, especially those in Nepal and similar regions, carry a disproportionate workload in natural resource management. Studies reveal that women in hilly areas work longer than their male and lowland counterparts, taking on most responsibilities in forest management, agriculture, water collection, and animal care (Byers & Sainju, 1994; Gabriel & Mangahas, 2017; Gämperli Krauer et al., 2017; Pradhan et al., 2021). In many poor households, women perform 98% of the labor in forestry and contribute over 60% in other resource sectors. These gendered labor patterns highlight the vulnerability and resilience of women as primary users of community-owned resources—who also face the most significant challenges when those resources are depleted (Byers & Sainju, 1994; Pradhan et al., 2021).

2.5 Economic and Resource-Based Roles of Mountain Women

Mountain women have traditionally engaged in small scale business ventures based on the use of natural resources. This non-domestic forest economy can be critical to meeting household subsistence needs. Their economic activities include selling firewood, fodder, dairy products, fermented beverages, and processing non-timber forest products like herbs, resins, oils, fibers, fruits, and seeds (Byers & Sainju, 1994; Tyagi & Churchgate, 2006). These century-old practices not only support household needs but also sustain the natural ecosystem through knowledge of sustainable harvesting.

2.6 Women as Custodians of Biodiversity

In many cultures, women are the primary conservators of biodiversity, responsible for seed selection, plant domestication, and knowledge transfer (Deda & Rubian, 2004; Howard, 2015; Khadka & Verma, 2012; Aguilar et al., 2010; Shiva, 1992; Sood et al., 2015). Despite their critical role, their work is often labeled as "natural" rather than acknowledged as intellectual or scientific labor. Women's approaches to conservation—rooted in daily interaction with nature and community needs—differ markedly from top-down, patriarchal models (Shiva, 1992).

2.7 Broad Roles in Biodiversity Governance and Management

Women contribute to biodiversity through agriculture and resource use, governance, conservation programs, extractive industries, ecotourism, and environmental movements. In mountain and rural regions, they play critical roles in managing home gardens, seeds, agroecological practices, and family farms (Lima & Cunha, 2024).

3 Barriers to Women's Participation in Biodiversity Conservation and Natural Resource Management

Despite their crucial roles as long-standing custodians of biodiversity, preserving seeds women's contributions are often overlooked and underrepresented in decision-making and governance (Ally et al., 2024; Catacora-Vargas, 2021; Jabiel, 2025; Lima & Cunha, 2024). Barriers such as lack of gender-sensitive policies, limited public incentives, and systemic inequalities hinder their full participation. Research highlights that effective biodiversity conservation and sustainable development require the meaningful inclusion of women at all levels of governance and policy-making (Ally et al., 2024; Byers & Sainju, 1994; Catacora-Vargas, 2021; Jabiel, 2025; Lima & Cunha, 2024; Shiva, 1992). Sociocultural norms and insufficient legal frameworks that do not encourage women's active participation in conservation are primarily to blame for these difficulties. Conservation results are weakened when women are excluded from environmental leadership roles. Studies suggested that gender inclusive such as women centric approaches, lead to more effective and sustainable conservation outcomes (Ally et al., 2024; Catacora-Vargas, 2021).

3.1 Sociocultural Barriers

Limited access to resources and leadership roles hindered women's potential contributions in conservation (Ally et al., 2024). These difficulties can be made worse by cultural norms and deeply rooted gender stereotypes, which serve as major obstacles that keep women from successfully utilizing their skills and abilities in agriculture (Byers & Sainju, 1994; Chhetri, 2020; Lima & Cunha, 2024; Molden et al., 2014).

3.2 Economic Barriers

Restricted access to leadership roles, financial resources, and land; difficulties in securing land ownership rights; inadequate funding; unequal distribution of labor; extended working hours; and other challenges persist. Additionally, limited participation in decision-making processes and insufficient access to essential resources further exemplify these constraints (Byers & Sainju, 1994; Chhetri, 2020; Lima & Cunha, 2024; Molden et al., 2014).

3.3 Educational and Institutional Barriers

Women are prevented from contributing fully to development by barriers such as restricted access to leadership positions, resources, and education (Ally et al., 2024). Furthermore, their prospects for career progression are limited by a lack of proper education and training (Ally et al., 2024; Byers & Sainju, 1994; Chhetri, 2020; Lima & Cunha, 2024).

3.4 Structural and Policy Barriers

Women's participation is hindered by a lack of institutional frameworks, their exclusion from decision-making processes, and failure to acknowledge women's roles. Women often face difficulties owning land, restrict their long term participation in natural resource management (Catacora-Vargas, 2021; Chhetri, 2020). Additionally, rural women and their traditional knowledge are frequently overlooked by policymakers when advocating for environmental protection and conservation policies (Byers & Sainju, 1994; Lima & Cunha, 2024; Sarker, 2017).

Meeting these challenges requires strategies for removing structural barriers and adopting gender-equitable policies that support women (Chhetri, 2020). Several studies suggested that women significantly contribute to environmental rehabilitation and natural resource conservation by participating in decision-making (Chhetri, 2020; Lima & Cunha, 2024). There are many instances of effective female-led projects promoting rural development in Sikkim in spite of these obstacles. These programs range from community-based groups supporting sustainable agriculture and ecotourism to women's cooperatives involved in organic farming and agro processing. Such initiatives demonstrate how women's empowerment can drive inclusive and sustainable development by highlighting their leadership and inventiveness in rural communities (Chhetri, 2020).

4 Need for Advancing Conservation Through Gender Equity and Economic Opportunities

Mountains cover over one-fifth of the Earth's land and support nearly 15% of the global population (Molden et al., 2014; Pradhan et al., 2021). They are rich in cultural diversity and host around half of the world's biodiversity hotspots. Mountain community's livelihood thrive on natural resources such as forests, agriculture, and biodiversity. Protecting these resources are vital for the well-being of mountain communities and humanity as a whole (Molden et al., 2014). Biodiversity loss disproportionately affects women, particularly indigenous women whose deep connection to the land supports the well-being of 476 million indigenous people globally. As ecosystems degrade due to deforestation, climate change, and pollution, these women lose access to vital resources like water, energy, food, and medicinal plants. These environmental challenges inexplicably affect women's livelihoods, health, and nutrition, threatening their well-being and household food security. The decline in biodiversity puts traditional knowledge, women's identity, and community resilience at serious risk (CDA, Collaborative Learning Projects, 2025; Mohan & Sharma, 2022; Molden et al., 2014; Resurrección et al., 2019). Given that women play a vital role in managing natural resources at both family and community levels and are among the most affected by environmental degradation. Therefore, incorporating a gender perspective is essential for developing effective and sustainable long-term solutions (Parida, 2022).

Ensuring gender equality and women empowerment is crucial for ecosystem restoration, addressing climate change, biodiversity conservation, food security, as well as Sustainable Development Goals. When women are empowered economically they are more likely to invest in their families and communities, which in turn not only uplift their agency but also fosters broader community advantages, and promoting long-term sustainability. Connecting economic opportunities with conservation efforts strengthens women's roles and enhances biodiversity protection (Ally et al., 2024; Molden et al., 2014). Women empowerment in conservation advances gender equality and strengthen the biodiversity protection law, which in turn promote social justice, equitable society, and more sustainable environmental outcomes (Ally et al., 2024). Empowering women is not only vital for gender equality but also crucial for improving the overall well-being of rural communities. It promotes greater talent, innovation, and leadership, thereby, fostering more inclusive and holistic development of rural communities. Furthermore, women's empowerment helps reduce the poverty cycle by enhancing their involvement in education, healthcare, decision-making, and income-generation activities that support their families (Chhetri, 2020).

Inclusion of women in collaborative governance processes can strengthen the effectiveness of conservation policies by fostering more comprehensive and adaptive management strategies. It promotes women empowerment and increases ecosystem resilience by promoting trust, facilitating better communication, and producing long-lasting results. Better biodiversity conservation is also supported by women's economic empowerment (Ally et al., 2024). Cross-sectional factors like age, social class, race, ethnicity, and religion must be taken into account because they exacerbate the difficulties that women encounter. Religious and cultural norms often limit women's access to financial resources and land ownership. To address these barriers, deliberate public policies must promote women's participation in decision-making (Lima & Cunha, 2024). Engaging women in decision-making strengthens conservation by combining traditional knowledge with innovative practices addressing local needs. Social-ecological theory suggested the impact of interaction between social factors, such as gender dynamics and ecological systems on environmental outcomes, as gender roles often influence how resources are managed and used, which can impact biodiversity (Ally et al., 2024; Bennett & Dearden, 2014). Increasing women participation in governance and conservation efforts is vital for ensuring sustainable development and gender equity in decision-making processes. Studies show that greater inclusion in biodiversity governance and management can improve the conservation outcomes and resource restoration. Studies reveal that women's engagement promotes holistic approaches integrating social equity and ecological sustainability (Ally et al., 2024). Despite this, significant barriers hinder women's participation and the incentives to engage in these processes (Gupta, 2015; Lima & Cunha, 2024). Once women's roles were overlooked, now there is growing recognition of their impact on the natural environment and how environmental changes affect their health and well-being (Mahour, 2016).

5 Indigenous Women in Biodiversity Conservation and Rural Livelihoods in India

In developing countries, women are largely dependent on nature for survival and play a key role in managing and conserving them. They are key contributors to food production, especially in rural areas, producing 60–80% of the food. Despite many obstacles women have continuously led environmental movements and conservation efforts. Acknowledging women contributions in environmental conservation is important for prosperous future and sustainable development (Mathur et al., 2023).

India's ethnic and tribal communities have long standing relationship with the forests and relied on forest resources for their survival, the use plants for foods, shelter such as they need forest to make their huts and leaves and seeds to cook for foods. In Northeast India, women are central to these conservation efforts, playing a vital role in maintaining ecological balance through their traditional knowledge and active stewardship (Sarker, 2017).

Women are central to rural economic activities and livelihood generation in India, where agriculture is the mainstay. They perform key roles in nearly all stages of farming, from sowing and weeding to harvesting and marketing. Yet, their significant contributions have often gone unrecognized economically and socially. The government has introduced various laws to acknowledge better and support women's participation in rural development (Dwivedy, 2012).

Tribal women conserve several plants in the forest that are used for healing wounds and other herbal treatments. Many plant species are of great economic value. By protecting and applying this knowledge, indigenous communities play an essential role in protecting the environment and preventing ecological degradation (Sarker, 2017).

5.1 Indigenous Women's Traditional Knowledge and Biodiversity Conservation in Northeast India

Northeast India, renowned for its rich diversity, is home to numerous small ethnic groups, each possessing distinctive historical, geographical, cultural, and social characteristics (Ali & Das, 2003). A large part of Northeast India is covered with flora and fauna. This region, situated within the Eastern Himalayas, also hosts various cultural, social, and linguistic traditions indigenous communities maintain. It is a valuable center of indigenous knowledge related to food, healthcare, farming, and natural resource stewardship (Sarker, 2017).

Women across Northeast India are instrumental in preparing various nutritious ethnic foods using local ingredients like soybeans, bamboo shoots, and leafy vegetables. Fermented soybean dishes are common throughout the region, with different names in each state—such as *Bekang* in Mizoram, *Aagya* in Meghalaya, *Akhuni* in Nagaland, and *Kinnema* in Sikkim. A variety of bamboo based fermented food like Hikku and Hirring are also traditionally prepared by the Apatani tribal women's of Arunachal Pradesh. These customs reflect the regions rich indigenous knowledge systems on culinary practices and cultural diversity (Parida, 2022; Sarker, 2017). The Apatani of Ziro Valley in Arunachal Pradesh are known for their innovative and sustainable hill farming system, which integrates rice-fish cultivation, animal husbandry, and social forestry. This ecologically sound and economically viable method supports food and nutritional security while conserving biodiversity. Although both men and women participate in farming, Apatani women, like many in patriarchal societies, are often excluded from decision-making despite their critical, knowledge-based contributions. Their role in upholding this organic and environmentally friendly agricultural method highlights the need for recognizing women's efforts in sustainable development (Dwivedy, 2012; Khadka & Verma, 2012).

Women from rural and tribal communities in Assam play a vital role in generating livelihoods through forest based resources. Their livelihood generation activities not only provide daily sustenance but also help preserve traditional knowledge and cultural heritage. Their dynamic skill on use of forest resources reveals a deep connection with the ecosystem (Goodwin, 2024). Tribal women from communities like Bodo, Rabha, Karbi, and Dimasa play a vital role in conserving local biodiversity. Tribal women from communities like Bodo, Rabha, Karbi, and Dimasa play a vital role in conserving local biodiversity. Bodo women, in particular, possess rich traditional knowledge of plants used for sociocultural, religious, and medicinal purposes.

A study by Bodoland University found that Bodo women actively conserve around 48 plant species for various uses. This plant-based knowledge passed down through generations, highlights women's central role in biodiversity conservation through cultural and indigenous practices (Sarker, 2017). Among tribes like the Misings and Bodos, women craft traditional textiles using natural dyes and forest-sourced fibers. These textiles are important in ceremonial dress and daily wear (Goodwin, 2024).

In Assam, various ethnic groups use a range of plant ingredients to prepare starter culture cakes for homemade alcoholic beverages, believing these ingredients offer health benefits (Yein et al., 2022). In Assam, women also use 101 varieties of herbs in culinary preparations during the vibrant festival of Rongali Bihu, which is believed to offer significant health benefits. In Assam, fermented rice known as poita bhat is an important part of the culinary tradition, primarily prepared by women. It is valued for its health benefits and also serves as a method to prevent the spoilage of leftover rice (Barooah et al., 2020). This practice highlights the role of women as environmental stewards through sustainable food practices.

In Meghalaya, Khasi women hold rich traditional knowledge of edible and non-edible plants, local food traditions, and methods of soil regeneration, all of which contribute significantly to the sustainable use of natural resources (Shangpliang, 2018). As custodians of seeds, these indigenous women play a vital role in safeguarding their community's food sovereignty. Their deep ecological understanding also enables them to gather medicinal plants and wild edibles from nearby forests. Valued as essential food producers, Khasi women are pivotal to both household sustenance and the broader community (Chitnis, 2021).

6 Initiatives to Enhance Indigenous Women's Participation in Conservation Through Economic Empowerment in Northeast India

6.1 Promoting Women's Livelihoods Through Organic Farming in Sikkim

Organic agrotourism is a key element of rural tourism and a powerful tool for rural development. It helps preserve the rural environment and local cultures, reduces migration, increases income, and shifts household economic roles. Rural tourism brings farmers financial benefits and new ideas, creating jobs and generating revenue while allowing visitors to experience local life and products. Sikkim, benefiting from favorable climate conditions, has practiced organic farming for over a decade, making it a vital livelihood source for rural communities (Chhetri, 2020). Women's empowerment is crucial for sustainable growth in Sikkim, where they are integral to rural development efforts.

Agro-tourism in Sikkim is vital in empowering women by blending ecological conservation, economic development, and social inclusion. By promoting sustainable farming practices and incorporating them into agricultural value chains and locally based tourism, this growing sector has created significant opportunities for rural women. The growing in demand for organic products and experience-based tourism has played a critical role in enhancing rural livelihoods by increasing entrepreneurship, rural income, and employment, especially among marginalized groups and women. Success stories such as Ms. Tenzing Dolma's thriving agro-tourism farm and the Sangha Women's Cooperative highlight the role of rural women in sustainable rural development underscore the need to equip women with training, resources, and support systems to promote women led rural development (Chhetri, 2024).

Furthermore, groundbreaking Organic Mission, launched by Sikkim's government in 2003 has played a significant role in promoting organic farming and agrotourism. These policies offer technical support, financial incentives, infrastructure development, and marketing support while restricting chemical inputs, thus promoting sustainable livelihoods and economic growth. Collectively, agrotourism in Sikkim empowers women economically and socially while advancing environmental sustainability (Chhetri, 2020, 2024).

6.2 Empowering Indigenous Women Through Sustainable Livelihoods in Assam

Establishment of Pirbi-Karbi Ethnic Haat in Assam

Aaranyak, an Assam-based NGO, empowers local communities, particularly women, through alternative livelihood initiatives that reduce dependence on forest resources. The Pirbi-Karbi Ethnic Haat is one of its key initiatives to foster a community-driven business model that emphasizes fair trade and reinvestment into biodiversity conservation. Launched in 2019, Pirbi exemplifies a successful partnership between community development and ecosystem management.

Pirbi-Karbi Ethnic Haat, situated near Kaziranga National Park in Assam, is a unique community-led marketplace showcasing authentic ethnic and organic products of the Karbi people. In Karbi, Pirbi, meaning "biosphere," reflects the community's deep bond with nature. The haat features traditional foods, hand-woven garments, and handicrafts, embodying Karbi's cultural heritage with value-added products. Established under Aaranyak's Community-based Natural Resources Management initiative and aligned with the IUCN biodiversity business model, Pirbi promotes sustainable, biodiversity-friendly commerce. It operates on the principles of Fair for People and Fair for Biodiversity, supporting fair trade, ethical production, and local livelihoods. Pirbi reinvests 12% of its profits into regional biodiversity and community projects and shares an additional 5% with local producers and artisans (Aaranyak, 2023; The Sentinel, 2023).

Empowering Indigenous Women Through Sustainable Livelihoods in Karbi Anglong, Assam

Aaranyak conducted a winter mushroom cultivation training program in Karbi Anglong, Assam, with a focus on empowering indigenous women; the initiative provided hands-on training in mushroom biology, cultivation methods, and the nutritional and economic benefits of mushroom farming. This sustainable livelihood option aims to enhance women's financial self-reliance while promoting eco-friendly agricultural practices. Supported by IUCN and KfW, the program reflects Aaranyak's commitment to biodiversity conservation through inclusive community development and capacity-building efforts in the Kaziranga-Karbi Anglong and Manas Landscapes (Northeast Today, 2024).

Skill Development for Conservation and Livelihood in Manas, Assam

Another initiatives of Aaranyak was tailoring training programs for marginalized women, which was initiated to empower them economically and foster a sense of ownership in forest conservation. Enabling indigenous women with practical skills empowers them in decision-making process in biodiversity conservation and sustainable resource use. This inclusive approach supports community engagement, ultimately leading to more resilient and long-term conservation efforts (Aaranyak, 2024; The Sentinel, 2024). The organization also provides training and skill development programs such as weaving, piggery farming, bee-keeping, and fishery farming as alternative livelihood options (Aaranyak, n.d.). These initiatives aim to reduce indigenous women's dependence on forest resources by equipping them with sustainable income-generating skills, thereby supporting community empowerment and environmental conservation.

6.3 Community-Conserved Areas (CCAs) for Biodiversity Conservation in SEPLS (Socio-Ecological Production Landscape or Satoyama Landscape) in Nagaland, India (Satoyama Project)

In Nagaland, 88.3% of its natural habitats are managed by individuals and clans under the guidance of village and district councils by constitutionally protected customary rights. Due to limited livelihood options, many communities rely heavily on natural resources, leading to environmental degradation. However, through traditional conservation practices, particularly establishing Community-Conserved Areas (CCAs) since the 1800s, have helped protect biodiversity. About one-third of Nagaland's villages today have CCAs, with 82% enforcing bans on tree felling and hunting.

Spanning over 1700 km^2, these areas contribute significantly to carbon storage and climate resilience, offering a sustainable, community-driven model for conservation.

A pilot project in the villages of Sukhai, Kivikhu, and Ghukhuyi in Nagaland's Zunheboto district successfully linked Community-Conserved Areas (CCAs) to promote biodiversity conservation through traditional practices and alternative livelihoods like ecotourism. Key activities included documenting indigenous Ecological Knowledge, building community capacity in biodiversity monitoring, and raising awareness about landscape conservation. The project led to the protection of 939 hectares as CCAs and bans on hunting and destructive fishing across a total landscape of 3751 hectares, resulting in increased biodiversity with 222 bird and 200 butterfly species documented.

Moreover, the program not only improved the household incomes through ecotourism among marginalized community members, youth, and women groups, but also enhanced the management of natural resources, and revitalized traditional ecological knowledge, especially among elders. Changes in community attitudes shown a greater understanding of the environmental and economic values of biodiversity conservation and increase in support for conservation measures like hunting ban. This conservation model is integrated into governance frameworks through multifaceted approach such as financial, legal, and ecological monitoring (Edake, 2021; Edake et al., 2019). Local communities were involved in the project through education, sensitization, and participatory planning. This initiative also helped expand the conservation efforts in the neighboring villages. Bans on destructive fishing and hunting were enforced across 3751 hectares, within the project areas, effectively protecting biodiversity and natural resources. Regular biodiversity surveys documented rich species diversity, including bird and butterfly species, recorded in People's Biodiversity Registers that preserve local ecological knowledge. Training local youth in biodiversity assessment, capacity building and as nature guides for ecotourism, has created alternative livelihoods, attracting tourists visiting this area to spot 'bird and butterfly species, conservation activities, and local culture. This has motivated broader community participation in conservation, linking sustainable resource use with economic benefits.

6.4 *Empowering Indigenous Women Through Gender-Inclusive Community Natural Resource Management for Sustainable Livelihoods and Biodiversity Conservation*

The North Eastern Region Community Resource Management Project for Upland Areas (NERCORMP) is a collaborative initiative jointly funded by the North Eastern Council, the Government of India, and the International Fund for Agricultural Development. The project's primary aim is to sustainably improve the livelihoods of vulnerable communities by promoting environmentally sound resource management. In 460 villages across six districts in Northeast India, NERCORMP facilitates the conservation of 10–20 hectares of community land per village as Community-Conserved Areas (CCAs) managed by gender-inclusive Natural Resource Management Groups (NaRMGs).

Recognizing women's unique relationship with forests and focusing on sustainable use, like gathering fruits, roots, and medicinal plants, NERCORMP developed a gender strategy to enhance their leadership in managing CCAs. Women in NaRMGs list non-timber forest products (NTFPs) for sustainable extraction, set traditional usage rules, determine collection seasons, and establish fair pricing and income use guidelines. They also preserve spiritual practices linked to forest conservation and define roles for women and youth, especially during men's seasonal migration.

This strategy has empowered women by enhancing their decision-making power in conservation, as seen in villages like Amsai-I and Jebalgre, where women now govern NTFP harvesting. This approach not only strengthens forest protection but also fosters gender equity and community resilience (NERCORMP, n.d.).

6.5 Community-Led and Gender-Inclusive Conservation and Sustainable Livelihoods in Nagaland

The Yaongyimchen Community Conserved Area (YCBCA), a part of the Biodiversity Peace Corridor in Nagaland, stands as a model of indigenous-led conservation that is spearheaded by local communities, including women and students. The community came together to protect its natural resources by prohibiting hunting and logging in response to deforestation rates, declining wildlife, and water scarcity (Community Conserved Areas, 2024). Central to the initiative is protecting the Amur Falcon, which has led to a resurgence in wildlife populations. In addition to conservation initiatives, the community has adopted sustainable livelihood practices that support ecological restoration and economic sustainability by establishing fruit orchards, piggeries, and ginger cultivation. Under the leadership of Nuklu Phom, the initiative has expanded to include 16 villages across four districts in a shared mission of biodiversity conservation and sustainable development. The project supports both environmental preservation and alternative livelihoods by transitioning from jhum (shifting cultivation) to sustainable, non-extractive land-use practices. It also emphasizes the revival of tribal education, enabling elders to pass traditional ecological knowledge to younger generations. The corridor fosters inter-village cooperation, allowing wildlife to move freely and communities to thrive through eco-friendly economic activities (Community Conserved Areas, 2024; Tandon, 2021; Whitley Fund for Nature, 2021).

6.6 *Empowering Women Through Livelihood-Based Conservation Initiatives in Kamrup, Assam*

In Assam, conservation biologist Dr. Purnima Devi Barman is leading a groundbreaking initiative to protect the endangered Greater Adjutant Stork (*Leptoptilos dubius*). Central to her efforts is the formation of the "Hargila Army," a group of local village women who have become key players in community-based conservation. Through awareness programs, habitat protection efforts, and community engagement, these women have taken an active role in safeguarding the stork and its nesting areas (Doley & Barman, 2025). Providing financial incentives to the women was the major driving force that not only support the livelihoods of the women but also encourage greater participation in conservation efforts. The program distributed looms, yarn, and sewing machines among women in the form of financial incentives, enabling them to weave textiles decorated with traditional Assamese motifs.

These garments are uniquely decorated with motifs of the Greater Adjutant Stork, blending cultural heritage with conservation awareness. In addition, training opportunities were arranged at a local Fashion Institute, allowing women to enroll in a Diploma in Fashion Technology. An internship program in weaving was also developed, providing rural women with hands-on experience and certification to support their future employment and income. By integrating conservation with community development, Dr. Purnima Devi Barman's initiative creates lasting change for the Greater Adjutant Stork and the people who share its habitat (Barman et al., 2020).

7 Discussion and Conclusions

Indigenous traditional ecological knowledge are rich and multifaceted. Their traditional knowledge includes sustainable agriculture practices, biodiversity conservation, forest and mountain resource management, and spiritual-ecological relationships. Indigenous peoples practice holistic ecosystem stewardship rooted in their cultural and spiritual worldview, known as cosmovision. Strengthened by community norms and governance systems, these practices often result in better conservation outcomes than in non-indigenous territories.

Traditional ecological knowledge among indigenous communities in India play a crucial role in sustainable land management and environmental conservation. This indigenous knowledge is shaped by their centuries of cultural practices and connection with nature, which has greatly influenced agricultural practices like crop rotation, organic farming, and biodiversity conservation. Notably, tribal wisdom has guided sustainable use of resources across various regions in India, with the Northeast being identified as a naturally organic zone due to its minimal chemical usage and traditional farming systems.

Women are central to the transmission and application of this knowledge. Their ecological roles are deeply embedded in daily livelihoods, from land rehabilitation and endangered plant conservation to soil health maintenance through organic recycling. Indigenous women also contribute significantly to in-situ conservation, home gardening, and local biodiversity management activities that directly influence ecosystem health and food security.

7.1 Indigenous Women's Traditional Ecological Knowledge and Their Role in Sustainable Natural Resource Management

Sustainable Agricultural Practices

Women have historically been central to agriculture, especially in seed selection, storage, preservation, and exchange. As the original cultivators, they domesticated key plant species for food, fiber, fuel, and fodder. Their tradition agricultural knowledge extend to processing plants for household use, maintaining soil health through organic practices, and preserving crop diversity for resilience and genetic vigor. By using eco-friendly methods—like animal manure for soil fertility, mixed cropping, and organic farming—women have played a vital role in promoting sustainable and resilient agriculture. Women from Sikkim, Northeast India, are extensively engaged in organic farming and contribute to agrotourism and economic development in the region (Dwivedy, 2012; Lin & Wang, 2014; Luharia & Sharma, 2019; Sahoo et al., 2022; Subramaniam & Venkat, 2009).

Sustainable Management of Biodiversity and Natural Resources

Women play a vital and multifaceted role in biodiversity conservation and natural resource management. They are responsible for situ conservation of useful wild and domesticated plants. They manage home gardens, fishing resources, and participate in the governance of forests, water bodies, and climate-related issues (Deda & Rubian, 2004; Howard, 2015; Khadka & Verma, 2012; Aguilar et al., 2010; Shiva, 1992; Sood et al., 2015). Women are active in conservation efforts, sustainable family farming, and agroecology, often collecting non-timber forest products and preserving biodiversity in protected areas. Their in-depth knowledge of their surrounding environment, forests, and species makes them effective stewards of the environment (Lima & Cunha, 2024; Ramachandran, 2008; Sahoo et al., 2022). In many indigenous and rural communities, women's practical understanding is essential for maintaining ecosystem health and resilience. In Sikkim, India, women play a vital role in the conservation of in situ genetic resources and agrobiodiversity. They take primary responsibility for homestead gardening and the upkeep of plots near their homes, where they cultivate a diverse range of plant species that enhance agricultural biodiversity. As principal custodians of seed management, they possess rich indigenous knowledge in seed selection, processing, and preservation (Khadka & Verma, 2012).

Nutrition and Health

Indigenous women's traditional knowledge of edible and non-edible plants, local food customs, and soil regeneration process influence their sustainable use of natural resources. Through ecofriendly practices they harvest various plants varieties and are custodians of seeds, and they also play an important role in maintaining food sovereignty in their communities. Indigenous rice varieties they cultivate are often superior to modern strains. Women across Northeast India are key in preparing nutritious ethnic foods using local ingredients like bamboo shoots, soybeans, and leafy greens, with fermented dishes being a regional staple. In Arunachal Pradesh, Adi women from remote communities, possess extensive knowledge of locally available nutritionally important food species and traditional products. They rely on a wide range of local plant species, managed through diverse land use systems and traditional conservation practices (Chitnis, 2021; Dwivedy, 2012; Parida, 2022; Shangpliang, 2018; Singh et al., 2020).

Sustainable Maintenance of Mountain Ecosystem

Women are instrumental in managing subsistence resources in mountain ecosystems. Women perform various functions for survival and hold diverse knowledge regarding the potential and processes of ecosystems. In mountain regions, women are involved in animal husbandry, nutrient recycling, subsistence farming, and livestock waste management, all of which are critical for maintaining productivity on nutrient-poor soils. Women often engage in informal collective practices like labor exchange and group foraging, reflecting their strong role in sustainable mountain resource management. Based on the use of natural resources, mountain women have traditionally engaged in small scale business ventures such as selling firewood, fodder, dairy products, fermented beverages, and processing non-timber forest products like herbs, resins, oils, fibers, fruits, and seeds (Byers & Sainju, 1994; Khadka & Verma, 2012; Sainju, 1993; Tyagi & Churchgate, 2006).

Despite their critical roles, women's contributions remain undervalued and often invisible in formal conservation discourse. Conventional development approaches tend to reinforce gender biases by limiting women's access to critical resources such as education, land and participation in decision-making processes (Ally et al., 2024; Catacora-Vargas, 2021; Jabiel, 2025; Lima & Cunha, 2024). Women involvement in biodiversity governance and sustainable practices are hindered by socio-cultural norms, gender stereotypes, and exclusion from environmental discussions which impede women's active participation in conservation (Byers & Sainju, 1994; Chhetri, 2020; Molden et al., 2014).

The engagement is further restrict by inadequate public incentives, lack of gender-inclusive policies, and limited legal aids that promote women empowerment. These disparities not only hinder women's participation but also limits the effectiveness of conservation efforts (Ally et al., 2024; Byers & Sainju, 1994; Chhetri, 2020; Lima & Cunha, 2024). To overcome these challenges necessitates a comprehensive approach that supports gender equality, assimilates women fully into governance and management of biodiversity, and needs targeted support to overcome structural inequalities.

7.2 Women's Participation Outcomes in Conservation

Study highlighted that women adopt sustainable conservation practices and offer distinctive perspectives on sustainable natural resource management. Because they are usually community-oriented and focus on the long-term, efficient use of natural resources, their active participation in biodiversity preservation and resource management frequently results in better environmental outcomes. Because they are usually community-oriented and focus on the long-term, efficient use of natural resources, their active participation in biodiversity preservation and resource management frequently results in better environmental outcomes. The underlying socio-economic and cultural factors significantly impact women roles, affecting both the extent and significance of their involvement. Women participation in decision making process and biodiversity governance, provide fresh viewpoints on inclusive, equitable, and prosperous environmental strategies. Evidence indicates their involvement significantly contributes to ecosystem restoration and sustainable resource use (Ally et al., 2024; Gupta, 2015; Lima & Cunha, 2024). Therefore, there is a need for designing policy frameworks that promote gender equity and integrate women's roles in environmental governance to facilitate adequate biodiversity preservation.

Despite some development there are significant challenges remain that hinders women full involvement in the governance and management of biodiversity. The major hindrance is the absence of government incentives that support gender equality and promote women's active roles in governance, biodiversity conservation, and sustainable development among the challenges. Moreover, environmental policies that overlooked women role further lessens their efficacy. Therefore, gender-sensitive policies must be integrated into natural resource management, governance, and decision-making, especially in the context of biodiversity conservation.

Meaningful and substantial inclusion of women is necessary to realize the positive impacts of their participation in biodiversity decisions.

7.3 Economic Empowerment and Gender Equality for Conservation

Empowering women and promoting gender equality are essential for achieving meaningful outcomes in climate resilience, conserving biodiversity, and Sustainable Development Goals. Since women's economic empowerment enables them to invest in the welfare of their families and communities, women's economic empowerment is essential for advancing long-term sustainability. Linking economic opportunities with conservation strengthens women's roles and Protection of biodiversity. Combining efforts of biodiversity conservation and gender equity promotes social justice, more equitable society, and sustainable outcomes. Empowering women helps break the cycle of poverty by increasing their involvement in education, healthcare, decision-making, and income generation. Moreover, their participation in collaborative governance fosters inclusive and adaptive management, strengthens trust and communication, and contributes to more effective and enduring biodiversity conservation.

In Northeast India, the conservation process gives importance to gender-inclusive and empowerment through livelihood generation programs. The livelihood generation programs revolved around skills and capacity-building programs to enhance women's participation in sustainable natural resource management and decision-making processes. Several initiatives across Northeast India illustrate how integrating sustainable livelihoods with biodiversity conservation empowers women, particularly in rural and indigenous communities.

Sikkim's organic agriculture led agrotourism have significantly influenced women's economic and social empowerment in the state. Women access to opportunities related to income generation, entrepreneurship, and leadership roles all attributed to sustainable farming and community-based tourism in the region. Government policies, like the Organic Mission (2003), have provided crucial support through training, financial incentives, and infrastructure development while promoting environmental sustainability.

In Assam, the Pirbi-Karbi Ethnic Haat fosters biodiversity-friendly commerce under fair trade and ethical production principles. It reinvests profits into community and biodiversity projects. Mushroom cultivation training initiated by Aaranyak in Karbi Anglong, skilled indigenous women with sustainable- ecofriendly agriculture practices and supports in financial independence. This initiative promotes gender equality and encourage women active participation in biodiversity conservation. Additionally, women oriented conservation initiatives by Dr. Purnima Devi Barman proved to be a successful example of gender inclusive conservation measures. The program highlights how financial incentives empowered women and encouraged their participation in species conservation efforts. By aligning conservation goals with community development and gender empowerment, the initiative demonstrates a sustainable and inclusive model for protecting biodiversity.

Other examples include the Yaongyimchen Community Conserved Area and Satoyama project in Nagaland, where communities adopt sustainable livelihoods (like orchards and piggeries), improve resource management, and revive traditional ecological knowledge. These projects also strengthen local support for conservation policies and integrate biodiversity into local governance. The NERCORMP initiative further supports vulnerable groups by linking sustainable resource management with improved livelihoods. These examples demonstrated how empowering women through sustainable livelihoods leads to ecological resilience, gender equality, and effective conservation outcomes.

Following approaches might help in enhancing participation of woman in conservation

- **Preserving our rich traditional knowledge**: The rich traditional knowledge in communities might be very useful in promoting sustainable developments
- **Integration of traditional ecological knowledge**: For conservation and natural resource management to be effective and sustainable, traditional ecological practices must be actively incorporated.
- **Inclusion of indigenous women**: Indigenous women should be meaningfully involved in all aspects of conservation efforts, as they are key environmental stewards and holders of valuable ecological knowledge.
- **Breaking the barriers**: The barriers restricting participation of woman must be removed in a systematic way.
- **Policy measures**: Gender-inclusive policies and their implementation might play an important role in strengthening involvement of woman
- **Capacity building**: Skill-enhancing training programs are necessary to strengthen indigenous women and their contributions towards conservation.
- **Financial incentives**: Empowering women economically through Livelihood generation and financial incentives might have great positive impact.
- Support from institutions: Strong institutional support can help in economic and social empowerment of woman subsequently leading to better contributions.

Although women's roles in natural resource management and conservation are significant, there is still a need for more targeted research to deepen the understanding of their contributions to biodiversity protection.

References

Aaranyak. (2023). PIRBI ethnic haat bolsters Karbi presence in Kaziranga Heritage Site. https://aaranyak.org/wp-content/uploads/2024/03/PIRBI-ethnic-haat-bolsters-Karbi-presence-in-Kaziranga-Heritage-Site-2.pdf.

Aaranyak. (2024). Aaranyak launches tailoring school for indigenous women of Manas. Retrieved May 24, 2025, from https://aaranyak.org/wp-content/uploads/2024/03/Aaranyak-launches-tailoring-school-for-indigenous-women-of-Manas.pdf

Aaranyak. (n.d.). Conservation & Livelihood—aaranyak.org. Retrieved May 24, 2025, from https://aaranyak.org/division/conservation-livelihood/.

Aguilar, L., Mata, G., & Quesada-Aguilar, A. (2010). Gender and biodiversity factsheets. IUCN. Retrieved June 23, 2025, from https://iucn.org/content/gender-and-biodiversity-factsheets.

Ali, A. N. M. I., & Das, I. (2003). Tribal situation in North East India. *Studies of Tribes and Tribals, 1*(2), 141–148.

Ally, T. A., Kumar, R., Verma, R. K., Verma, R. K., & Bhargava, R. (2024). Role and empowerment of women in wildlife conservation: The intersection of gender equality and biodiversity protection. *Pakistan Journal of Life and Social Sciences, 22*, 12918–12927.

Barman, P. D., Sharma, D. K., Cockrem, J. F., Malakar, M., Kakati, B., & Melvin, T. (2020). Saving the greater adjutant stork by changing perceptions and linking to Assamese traditions in India. *Ethnobiology Letters, 11*(2), 20–29.

Barooah, M., Bora, S. S., & Goswami, G. (2020). Ethnic fermented foods and beverages of Assam. In *Ethnic Fermented Foods and Beverages of India: Science History and Culture* (pp. 85–104).

Bennett, N. J., & Dearden, P. (2014). Why local people do not support conservation: A qualitative study from Cambodia. *Environmental Management, 54*(3), 245–258. https://doi.org/10.1007/s00267-014-0248-5

Byers, E., & Sainju, M. (1994). Mountain ecosystems and women: Opportunities for sustainable development and conservation. *Mountain Research and Development*, 213–228.

Catacora-Vargas, G. M. (2021). Indigenous women's rights in biodiversity conservation and sustainable use. In *Expert Group Meeting 'Achieving gender equality and the empowerment of all women and girls in the context of climate change, environmental and disaster risk reduction policies and programmes* (Vol. 21). EGM/ENV/EP.

CDA Collaborative Learning Projects. (2025). Biodiversity's Sisters: The essential role and challenges of Indigenous women in biodiversity Strategy processes. https://www.cdacollaborative.org/. Retrieved May 23, 2025, from https://www.cdacollaborative.org/.

Chhetri, A. (2020). Organic agro tourism of Sikkim: An initiative for sustainable rural development, women empowerment and green environmental practices. *Khoj: An International Peer Reviewed Journal of Geography, 7*(1), 67–73.

Chhetri, A. (2024). Sustainable rural renaissance: Fostering women's empowerment through organic agro-tourism. *Current Agriculture Research Journal, 12*(2).

Chitnis, R. (2021). The keepers of seeds and land, the Khasi Women of Meghalaya. Sprf. https://sprf.in/photo-archive/the-keepers-of-seeds-and-land-the-khasi-women-of-meghalaya/.

Colchester, M. (1994). Sustaining the forests: The community-based approach in the south and Southeast Asia. *Development and Change, 25*(1), 69–100. https://doi.org/10.1111/j.1467-7660.1994.tb00510.xb

Community Conserved Areas. (2024). Yaongyimchen CCA (Yaongyimchen Community Biodiversity conservation Area). https://communityconservedareas.org/data/show/985.

Deda, P., & Rubian, R. (2004). Women and biodiversity: The long journey from users to policy-makers. *Natural Resources Forum, 28*, 201–204.

Dhakal, T. D., & Leduc, B. (2010). Women's role in biodiversity management in Himalayas. In *Sustainable mountain development*. International Centre for Integrated Mountain Development (ICIMOD).

Doley, D. M., & Barman, P. (2025). The role of biodiversity communication in mobilizing public participation in conservation: A case study on the greater adjutant stork conservation campaign in Assam, Northeast India. *Journal for Nature Conservation, 84*, 126847.

Dwivedy, N. (2012). Suggestions for female farmer's areas of gainful employment-a study of rural area of Sikkim in north-eastern India. *Research Journal of Humanities and Social Sciences, 3*(3), 304–315.

Edake, S., Sethi, P., & Lele, Y. (2019). Mainstreaming Community-Conserved Areas (CCAs) for biodiversity conservation in SEPLS—A case study from Nagaland, India. In *Understanding the multiple values associated with sustainable use in socio-ecological production landscapes and seascapes (Satoyama initiative thematic review)* (Vol. 5, pp. 169–179).

Edake, S. (2021). Implementing Satoyama Initiative through Community-Conserved Areas (CCAs)—A success story from Nagaland, India. FFTC Agricultural Policy Platform (FFTC-AP). https://ap.fftc.org.tw/article/1592.

Gabriel, A. G., & Mangahas, T. L. S. (2017). Indigenous people's contribution to the mitigation of climate variation, their perception, and organizing strategy for sustainable community based forest resources management in Caraballo Mountain, Philippines. *Open Journal of Ecology, 7*(02), 85. https://doi.org/10.4236/oje.2017.72007

Gabriel, A. G., De Vera, M., & B. Antonio, M. A. (2020). Roles of indigenous women in forest conservation: A comparative analysis of two indigenous communities in the Philippines. *Cogent Social Sciences, 6*(1), 1720564.

Gämperli Krauer, U., Wymann von Dach, S. & Bieri, S. (2017). Mountain women's future. In *Issue brief on Sustainable Mountain Development*. Centre for Development and Environment.

Goodwin, L. L. (2024). Women and forestry in Assam and their contribution to sustainability and conservation. In *Gender, society and environment: Contemporary perspectives*

Gupta, H. (2015). Women and climate change: Linking ground perspectives to the global scenario. *Indian Journal of Gender Studies, 22*, 408–420. https://doi.org/10.1177/0971521515594278

Howard-Borjas, P. (2001). *Women in the plant world: The significance of women and gender bias for biodiversity conservation.* Wageningen Universiteit.

Howard-Borjas, P., & Cuijpers, W. (2002). Gender relations in local plant genetic resource management and conservation. Encyclopedia for Life Support Systems: Biotechnology.

Howard, P. L. (2015). Gender relations in biodiversity conservation and management. In *The Routledge Handbook of Gender and Development* (pp. 117–128). Routledge.

IPBES (Intergovernmental Science-Policy Platform on Biodiversity and Ecosystem Services). (2019). Summary for policymakers of the global assessment report on biodiversity and ecosystem services. IPBES.

Jabiel, S. (2025). Guardians of nature: Indigenous women and biodiversity. https://www.activesustainability.com/sustainable-development/women-indigenous-biodiversity/

Khadka, M., & Verma, R. (2012). Gender and biodiversity management in the Greater Himalayas: Towards equitable mountain development (pp. vi+-66).

Lima, H. S. D. M., & Cunha, H. F. A. (2024). The role of women and the obstacles to biodiversity conservation in developed and developing countries. *Environment, Development and Sustainability*, 1–25.

Lin, Y., & Wang, H. H. (2014). Natural resources management: An exploration of the Tayal model. *Taiwan Journal of Indigenous Studies, 4*, 139e172.

Luharia, J., & Sharma, H. (2019, November). Role of Indigenous women in securing sustainable livelihoods in Western Himalayan Region, India. In *Research Association for Interdisciplinary Studies (RAIS) Conference Proceedings* (No. 6–7, pp. 77–82).

Mahour, K. (2016). Role of women in environment conservation. *Journal of Advanced Laboratory Research in Biology, 7*(1), 17–26.

Mathur, R., Katyal, R., Bhalla, V., Tanwar, L., Mago, P., & Gunwal, I. (2023). Women at the forefront of environmental conservation. *Current World Environment, 18*(2), 706.

Mohan, R., & Sharma, P. (2022). The role of women in the conservation of environment with special reference to Assam. *Journal of Positive School Psychology, 6*(3), 5849–5855.

Molden, D., Verma, R., & Sharma, E. (2014). Gender equality as a key strategy for achieving equitable and sustainable development in mountains: The case of the Hindu Kush-Himalayas. *Mountain Research and Development, 34*(3), 297–300.

NERCORMP. (n.d.). NERCORMP gender strategy for community conserved areas. NERCORMS. Retrieved May 24, 2025, from https://nercorms.org/nercormp-gender-strategy-for-community-conserved-areas/.

Northeast Today. (2024). Assam: Aaranyak facilitates mushroom cultivation training In Karbi Anglong. Northeast Today. https://northeasttoday.in/2024/07/23/assam-aaranyak-facilitates-mushroom-cultivation-training-in-karbi-anglong/.

Parida, A. (2022). Dimensions of women's empowerment. Women empowerment and their role in conservation of biodiversity in India. SRF Publication.

Pilgrim, S., & Pretty, J. (2010). Nature and culture: An introduction. In S. Pilgrim & J. Pretty (Eds.), *Nature and culture. Rebuilding lost connections* (pp. 1–20). Earthscan.

Prasad, C., & Singh, R. P. (1992). Farm women: A precious resource. In R. K. Punia (Ed.) *Women in agriculture*. Education, training and development (Vol. 2). Northern Book Centre.

Pradhan, M.S., Rai-Paudyal, B., Rai, A., Bai, Y., Hengsuwan, P., Bun, P., & Yongzom, D. (2021). Exploring the role of gender equality in addressing climate change, biodiversity conservation, ecosystem restoration and food security. In B. Thapa, C. Fu, & L. Zhang (Eds.), *Gender equality and sustainable development in the Mountain Areas of Asia*. LI-BIRD.

Ramachandran, B. (2008). Gender and bio-diversity management in a food-gathering tribal society. In *Primitive tribal groups in India: Tradition*. Development & Transformation" Seraialas Publications.

Resurrecci6n, B. P., Goodrich, C. G., Song, Y., Bastola, A., Prakash, A., Joshi, D., et al. (2019). In the shadows of the Himalayan mountains: persistent gender and social exclusion in development. In *The Hindu Kush Himalaya assessment: Mountains, climate change, sustainability and people* (pp. 491–516).

Sahoo, G., Wani, A. M., Swamy, S. L., Mishra, A., & Mane, S. P. (2022). Indigenous people activities on ecosystems and sustainable development—A paradigm shift. In *Indigenous people and nature* (pp. 3–27). Elsevier.

Sainju, M. (1993). *Mountain ecosystems and women: Opportunities for sustainable development and conservation*. UN International Consultation on Advancing Women in Ecosystem Management.

Sarker, T. (2017). Biodiversity: The role of women in North East India. *Quest Journals Journal of Research in Humanities and Social Science, 5*(7), 71–74.

Shangpliang, R. M. (2018). Livelihood and gender: Marginalisation of Khasi rural women in land and forest rights.

Shiva, V. (1992). Women's indigenous knowledge and biodiversity conservation. *India International Centre Quarterly, 19*(1/2), 205–214.

Singh, R. K., Kumar, A., Singh, A., & Singhal, P. (2020). Evidence that cultural food practices of Adi women in Arunachal Pradesh, India, improve social-ecological resilience: Insights for Sustainable Development Goals. *Ecological Processes, 9*(1), 1–19.

Sood, M., Gupta, V., & Jan, A. (2015). Role of women in conserving plant genetic resources and related traditional knowledge for food security. In *Plant genetic resources and traditional knowledge for food security* (pp. 237–253).

Subramaniam, T., & Venkat. (2009). Contribution of tribal wisdom in making of Indian culture. In *Sundeep prakashan* (pp. 60–66).

Tandon, A. (2021). Indian "Green Oscar" winner Nuklu Phom envisions a Biodiversity Peace Corridor. Mongabay-India. https://india.mongabay.com/2021/05/indian-green-oscar-winner-nuklu-phom-envisions-a-biodiversity-peace-corridor/

The Sentinel. (2023). Pirbi ethnic haat bolsters Karbi presence in Kaziranga Heritage Site of Assam. https://www.sentinelassam.com/; https://www.sentinelassam.com/north-east-india-news/assam-news/pirbi-ethnic-haat-bolsters-karbi-presence-in-kaziranga-heritage-site-of-assam-641216

The Sentinel. (2024). Aaranyak launches tailoring school for indigenous women of Manas Tiger Reserve. https://www.sentinelassam.com/; https://www.sentinelassam.com/north-east-india-news/assam-news/aaranyak-launches-tailoring-school-for-indigenous-women-of-manas-tiger-reserve

Tyagi, R., & Churchgate, M. (2006). *Role of mountain women in environment governance in India*. The Indian Society of Ecological Economics.

Whitley Fund for Nature. (2021). Nuklu Phom|India|2021 Whitley Award winner. Whitley Award. https://whitleyaward.org/winners/establishing-a-biodiversity-peace-corridor-in-nagaland/

Yein, J., Pegu, B. K., Sanong, J., & Pegu, N. R. K. B. (2022). Traditional ethnomedicinal practice in the homemade alcoholic beverage of mising community of Majuli district, Upper Assam. *Biosciences Biotechnology Research Asia, 19*(2), 497.

Zhou, L., & Huo, Z. (2020). Indigenous knowledge and biodiversity conservation: A case study of women's roles in sustainable practices. *Journal of Environmental Studies and Sciences, 10*(2), 149–161. https://doi.org/10.1007/s13412-020-00629-0

Voices from the Himalayas: Indigenous Women in a Glocalized World

Dharmendra Singh

Abstract In an age dominated by global forces such as modernization and globalization, the relevance of local cultures and identities lies in their ability to resist and incorporate the externally enforced changes. The Himalayan regions, place to sundry indigenous cultures and communities, have witnessed glaring transformations in social, economical, and cultural landscape due to transnational powers. These powers have inflicted a dual burden on indigenous women, requiring them to equalize traditional roles such as care-giving, farming, and handicraft production with modern responsibilities, including wage labour and participation in the tourism industry. This paper attempts to uncover the myriad impacts of globalization, modernization and cultural shifts on indigenous women of the Bhotia, Buksa, Jaunsari, Raji, and Tharu tribes of a Himalayan region, centering on their status, identity, food habits and empowerment. By examining these aspects, this study aims to bring to the surface a complex interplay of the global forces and local realities, shedding lights on the lived experience of these women in an ever evolving world.

Keywords Tribal women · Tourism · Modernization · Cultural shift and hybridity

The persistent ballooning of modernity, hurled by the signage of neoliberal economic paradigms, has accelerated fervent metamorphosis in the social, cultural, and environmental fabric of indigenous people across the Himalayas. Within this broad region, the tribal populations of Uttarakhand form a particularly unfortified and vulnerable demographic. As tradition and modernity collide under the pressures of modernization, commodification, tourism, and state-led development, what emerges is a liminal "Third Space" (Bhabha, 1994)—a zone of in-betweenness marked by hybridity and contested identities. In this context, the existence and indigenous knowledge systems of tribal communities, particularly those of women, face systemic erasure and reconstitution. This research investigates the dialectical oscillation between cultural preservation and forced assimilation by critically examining and evaluating the sundry

D. Singh (✉)
Department of English, PGDAV College, University of Delhi, New Delhi, India
e-mail: dharmendra.singh@pgdav.du.ac.in

N. Moran et al. (eds.), *The Participatory Role of Indigenous Women*,
https://doi.org/10.1007/978-3-032-24059-0_8

impacts of globalization on the Bhotia, Buksa, Jaunsari, Raji, and Tharu tribes of Uttarakhand state, with a critical focus on the gendered dimensions of cultural attrition, agricultural displacement, and changing food practices in the wake of tourism as a dominant industry in the region.

Mass tourism, a byproduct of globalization, engendered the commercialization of indigenous tribal cultural artifacts as exemplified by the fetishization of Bhotia costumes. The Tharu and Raji tribes wrestled with the empirical threat of industrial encroachment, which not only disrupted their symbiotic relationship with the forest biotics but also altered their unique socio-cultural identities through commingling discourses of growth. The Jaunsari, known as keepers of a complex agro-pastoralist landscape, braving the disintegration of their ethnobotanical knowledge systems, as market driven monocrop agriculture practices has emerged as a new norm.

At the core of this analysis is the epistemic violence thrusted upon the tribal women, whose physical and emblematic labour have succoured the existence of these tribes for long yet remained strikingly absent from the dominant existing developmental knowledge. The Buksa and Tharu women maneuvered a janus faced reality wherein their handmade and agricultural prowess are both exoticized for tourist consumption and rendered economically non-viable in a cash-centric modern economy. The nutritional sovereignty of these tribes is eroded by the joint forces of climate change and market dependency, as evidenced in the alarming food insecurity paralyzing remote tribal areas.

1 Literature Review

A review of literature explored the varied textualities of cultural preservation and commercialization, developmentalist discourse and cultural erasure, and the impact of modernization and globalization on indigenous women in the Himalayas, with a specific focus on the five tribes of Uttarakhand examining how these processes, often under the guise of "progress" and "empowerment," have paradoxically entrenched ostracization and marginalized women's roles as cultural custodians and economic agents.

The Bhotias, traditional pastoralists of the Himalayas, exemplify the challenges of cultural perpetuation amidst state and market interventions. Hoon (1996), in *Living on the Move*, meticulously documented how sustainable migratory practices were altered by external forces. The urbanization of tribal areas, as detailed by Cultural Survival (2023), revealed how infrastructural projects and sedentarization policies impeded transhumance, thereby increasing the burden on tribal women. They face diminishing access to pasture and struggle to negotiate commercial livelihoods (Pandey et al., 2017). Historically, the Bhotiyas' economy thrived on Indo-Tibetan trade, exchanging Indian goods for Tibetan wool, salt, and borax (Chatterjee, 1976).

The 1962 Sino-Indian War and subsequent border closures severely disrupted these networks, leading to a decline in pastoralism and economic instability (Pandey et al., 2017). This forced the Bhotiyas to adapt, shifting towards agriculture, wage labour, and handicrafts (Negi, 2007). Despite these challenges, their traditional woolen handloom industry, producing items like thulma (woolen blanket), shawls, and carpets, continues to reflect their rich cultural heritage (Kumar et al., 2022). This industry faced significant hurdles, including limited access to high-quality wool, intense competition from machine-made products, and a lack of modern technology.

Bhotia women are central to both household and community economic activities, actively participating in weaving, livestock rearing, and agriculture, thereby significantly contributing to family income (Pande & Tripathi, 2016). Despite these crucial contributions, they encounter numerous obstacles, including limited access to education, healthcare, and broader economic opportunities. Literacy rates among Bhotiya women remain low, often due to families prioritizing boys' education (Pande & Tripathi, 2016).

The migration of men to urban centers for employment has disproportionately increased the domestic and economic responsibilities borne by women (Pandey et al., 2017). While globalization and modernization have brought some opportunities, such as government initiatives like "Ek Ikrit Hathkargha Vikas Yojana" and the establishment of urban haats (marketplaces) for handloom products (Kumar et al., 2022), they have also posed threats. The influx of cheap, machine-made textiles and the decline of traditional markets have undermined the viability of handloom weaving (Kumar et al., 2022). A growing disinterest in traditional occupations among the younger generation, influenced by modern education, further threatens intergenerational knowledge transfer (Kumar & Satti, 2023). The Bhotiyas' cultural identity is also under threat from assimilation pressures and the erosion of traditional practices, including medicinal knowledge and sustainable resource management (Samal et al., 2010), highlighting the urgent need for revitalization efforts.

The Buksa and Jaunsari tribes similarly grapple with the spectral presence of cultural erasure amidst dominant developmentalist narratives. The Jaunsari, known for their polyandrous kinship structures and animistic rituals, experience ontological dissonance as tourism commodifies their festivals (Uttarakhandi, 2023; The Hindu, 2025). The photographic gaze, as critiqued in exhibitions on Jaunsari life, often reduces their cosmology to exoticized vignettes, thereby alienating women from the very traditions they are expected to perpetuate (The Hindu, 2025). This commodification undermines the authentic cultural experience and often marginalizes the agency of indigenous women.

Modern agricultural techniques have further reconfigured tribal epistemologies, particularly for the Tharu and Raji tribes. The Tharu's symbiotic relationship with the Terai forests has been fractured by industrialization (Village Square, 2023), leading to land alienation and the promotion of monoculture farming. This shift has displaced women's crucial roles as custodians of seed diversity. The Raji, already marginalized by state neglect, now face "food apartheid", where market dependencies have eroded traditional foraging practices, leaving women disproportionately vulnerable to nutritional precarity. The transition from subsistence agriculture to cash-crop economies, often presented as 'empowerment,' has, in reality, entrenched patriarchal labor hierarchies, devaluing women's agrarian expertise in favor of mechanized, male-dominated production.

Tourism, while lauded as a panacea for economic stagnation, functions as a double-edged sword. While it generates revenue, it simultaneously imposed a homogenizing aesthetic on tribal cultures, prioritizing performative authenticity over lived experience. The Tharu's loss of linguistic and ritual autonomy underscores how development-induced displacement fractures intergenerational knowledge transmission. In this context, women bear the significant affective labour of preserving culture amidst assimilationist pressures.

In synthesis, the tribal women of Uttarakhand occupy a liminal space—simultaneously hypervisible as cultural signifiers and invisibilized as agential subjects. Modernity, under the guise of progress, has engendered a paradoxical reality where empowerment is contingent upon the very structures that perpetuate their subjugation. The glocalization paradigm, rather than forming genuine hybridity, has often replicated colonial hierarchies, rendering indigenous women both the targets and casualties of neoliberal modernity.

2 Research Methodology

This study followed a methodological framework rooted in decolonial feminist epistemology, relying exclusively on secondary sources to critically examine the dialectics of glocalization and the lived experiences of indigenous women in Uttarakhand's tribal communities. Discourse analysis techniques are employed to deconstruct hegemonic representations that frame tribal identities as either exoticized or romanticized traditions or obstacles to development. The commodification of Bhotia textiles and the gastronomic dispossession of Tharu food systems are interpreted through frameworks including Bourdieu's theory of cultural capital (Bourdieu, 1986) and the political ecology of food sovereignty (Patel, 2009). The current analysis also deliberated upon visual and digital cultural artefacts, such as Instagram-based tribal archives like @bhuli.art, as counter-narratives to dominant portrayals of indigeneity. Additionally, theoretically grounded upon Appadurai's concept of 'scapes' (Appadurai, 1996) and Mohanty's critique of neoliberal feminism (Mohanty, 2003), the study mapped the gendered contours of cultural transformation, positing that tribal women occupy a paradoxical position as both subjects of dispossession and agents of resilience within the shifting terrains of tradition and modernity.

Sacred groves are forest patches preserved through socio-religious practices, playing a vital role in conserving endemic and medicinal species in the Garhwal Himalaya. These groves, protected by taboos against tree felling and hunting, act as reservoirs of biodiversity. Studies reveal that trees like *Ficus religiosa*, *Ocimum sanctum*, *Azadirachta indica*, and *Cedrus deodara* are commonly found in these groves and are revered for their spiritual and ecological value (Anthwal et al., 2006). For Bhotia women, who are traditional custodians of health and healing, sacred groves are a primary source of medicinal plants. These groves also sustain soil fertility, regulate microclimates, and provide non-timber forest products essential for livelihood and medicine (Anthwal et al., 2010). Their conservation is a product of both ecological necessity and cultural reverence.

Bhotia women have historically maintained systems such as *khals* (small percolation pits), *tals* (ponds), and *chals* (channels) in upland regions, demonstrating a nuanced understanding of hydrology. These community-engineered systems ensure spring recharge and control soil erosion (Anthwal et al., 2010). Such water structures are intricately linked with the agro-pastoral livelihoods of Bhotia communities. Women, being primary water gatherers and land cultivators, assume responsibility for seasonal desilting, ritual maintenance, and allocation of water resources. Their practices demonstrate both eco-engineering skills and social governance in managing scarce mountain water systems.

Bhotia women's role in agroforestry is profound, particularly in cultivating multipurpose trees on farms and managing alpine medicinal species like *Saussurea obvallata* and *Origanum vulgare*. They are deeply knowledgeable about the uses of trees for food, fuelwood, fodder, and medicine (Anthwal et al., 2006). Their ethnobotanical practices often include both cultivation and wild harvesting, conducted in harmony with seasonal and lunar cycles. This dual engagement with cultivated and wild species supports conservation through sustainable use and contributes to in situ biodiversity preservation. Their role is not merely operational but strategic—they are decision-makers in what to plant, harvest, and preserve.

The Bhotia, historically semi-nomadic pastoralists, now confronted the commodification of their cultural patrimony, as sartorial traditions reductively aestheticized for touristic gaze and consumption, severing vestimentary practices from their cosmological significations. The aestheticization of Bhotia attire for commercial consumption (Behance, 2023) demonstrated the fetishization of indigeneity, where cultural symbols being detached from their ontological moorings and refashioned for global market consumption. This performative commodification, while economically enabling, engendered a discursive violence wherein women, as primary bearers of textile traditions, are both empowered and entrapped within neoliberal narratives of authenticity. The Behance exposition on Bhotia costumes epitomized this fetishization, wherein intricately woven textiles, once imbued with socio-ritualistic meaning, are transmuted into decontextualized artifacts for the cosmopolitan gaze, eroding the hermeneutic link between material culture and collective memory.

Similarly, the transhumant rhythms of Bhotia pastoralism are being destabilized by sedentarization policies and ecological precarity, forcing women—who historically managed migratory herds—into precarious wage labour, thereby disrupting matrilineal knowledge systems tied to seasonal mobility (Hoon, 1996). This alimentary dispossession revealed how market-driven agricultural shifts have precipitated nutritional insecurity among Bhotia women, as monocropping displaces biodiverse subsistence practices, rendering traditional food sovereignty obsolete.

Among the Buksa, digital platforms like Instagram perform a dual-edged semiotic function: while @bhuli.art ostensibly celebrated feminine artisanal labour through curated visual narratives, it inadvertently participates in the neoliberal co-optation of indigeneity, wherein tribal women's creative labour is algorithmically valorized yet economically marginalized. The Instagram page @bhuli.art, despite its veneer of empowerment, epitomized the extractive logic of digital capitalism, wherein indigenous aesthetics are appropriated as 'authentic' commodities while the material conditions of Buksa women remain structurally unaltered (Bhuli Art, 2023; @buksa_tribe, 2023).

The Raji, one of Uttarakhand's most marginalized tribes, the absence of digital or institutional representation signaled not preservation but erasure—their oral histories and agroecological knowledge remained unarchived, rendering them hypervisible as exoticized subjects in developmentalist discourse yet invisible as epistemic agents (Pande & Tripathi, 2016). The Tharu's predicament epitomized the necropolitical dimensions of industrial encroachment, as their communally managed forests—once the bedrock of matrilocal kinship systems—are expropriated for infrastructural projects, displacing women from their roles as custodians of biodiversity.

These transformations are not merely cultural but deeply epistemic. Indigenous women emerged as both casualties and resistors of glocalization's epistemic violence—their bodies inscribed as sites of cultural authenticity even as their agential voices systematically silenced. Women were central to the handloom industry, yet their labour rendered invisible within official development narratives. The dialectic between touristic commodification and industrial erasure exposed a fundamental tension: indigeneity is either hyper-commodified as 'heritage' or obliterated as 'backwardness,' with women navigating this double bind through fragmented acts of resilience—be it the Bhotia weavers adapting motifs for commercial markets or Tharu women clandestinely preserving seed varieties.

The food habits of these communities further unearthed this tension. The Bhotia's dietary practices, including fermentation techniques and high-altitude preservation strategies, rapidly declining as market penetration introduced new dietary regimes (Tamang, 2022). Such shifts not only alter nutritional landscapes but disrupt the symbolic and ritualistic dimensions of food, often governed by women. The decline of traditional food sovereignty, as Patel (2009) argued, undermined ecological and cultural autonomy. Similarly, Buksa women's displacement from ancestral land due to coerced land transfers disrupted their access to traditional crops and medicinal flora, increasing dependency on state-run schemes and market provisions.

Urbanization and youth migration exacerbated these dynamics, particularly among Buksa and Raji women, who experience a cultural dichotomy between traditional roles and modern aspirations (Verma & Yadav, 2010). The large-scale migration of men from hill villages to urban labor markets has led to a feminization of agriculture and local governance in Uttarakhand. Women have become the de facto heads of households and land managers, overseeing farming, water sourcing, and forest resource collection (Thapliyal, 2018; Medium, 2024). This economic reconfiguration has expanded their practical authority, while also intensifying their labor burden.

According to narratives documented on platforms like *Voices of Rural India*, women adapt traditional farming and weaving skills while engaging with digital storytelling to shape development dialogues (Kalyanwala, 2025). This dual role—as ecological stewards and content creators—reflects a broader transformation where indigenous women assert both epistemic and political agency. Raji women displayed paradoxical forms of agency; while structurally marginalized, they wield significant domestic power, often determining financial expenditures and familial decisions (Pande & Tripathi, 2016). Such agencies are increasingly circumscribed by neoliberal incursions that prioritized market logic over the customary laws.

Globalization—through market integration, tourism, and the media—has reshaped conservation in Uttarakhand by creating economic incentives that align with ecological and cultural aegis. A prominent example of this transformation is the rise of community-based homestay tourism, which allows Himalayan villagers to engage with global tourism markets while preserving their natural and cultural heritage.

Sarmoli, a small village in the Pithoragarh district, exemplifies how homestay tourism can reverse rural depopulation and foster environmental conservation. The village's homestay initiative, led largely by women, integrates ecotourism with Van Panchayat forest governance. Hosts offer rooms in traditional homes, organize nature walks, and educate tourists on the local ecology of the Greater Himalayas. The model channels 80% of earnings directly to hosts and reinvests 5% into village-level conservation (The Hindu, 2013).

As Malika Virdi, a prominent conservationist and founder of Himalayan Ark, explains, the program aims not just to earn money but to cultivate "shared responsibility for the landscape" (The Hindu, 2013). This initiative has led to afforestation, forest fire prevention, and a decline in outmigration, while simultaneously increasing household incomes and women's leadership in conservation decision-making.

In the Tons River Basin (Uttarkashi district), homestay tourism has provided supplementary income for agricultural households while preserving local ecological knowledge and architectural traditions. Sati and Banergee (2025) report that homestay owners serve local cuisine, maintain deodar-wood homes, and offer cultural performances rooted in Jaunsari and Garhwali traditions. Despite limited infrastructure, over 70 homestay units across villages like Sankari, Dharkot, and Osla hosted 114 tourists in 2023, generating substantial income (Sati & Banergee, 2025).

Globalization has not only altered material conditions but has also problematized the symbolic and semiotic terrain of indigenous identity. Appadurai's (1996) 'scapes' framework aptly mapped how the flows of media, capital, and technoscience reconfigure tribal imaginaries, often privileging consumable aesthetics over lived realities. Bourdieu's (1986) notion of cultural capital was useful in understanding how tribal knowledge systems were appropriated, commodified, and circulated within hegemonic circuits of value, particularly in tourism and artisan industries. Chandra Talpade Mohanty's (2003) critique of neoliberal feminism further illuminated how indigenous women were repositioned as 'entrepreneurs' or 'beneficiaries' within state-sanctioned development schemes, thereby displacing their epistemic authority with technocratic interventions.

To resist these incursions, tribal women deployed micro-resistances that sustained cultural memory and ecological balance. These include clandestine seed saving, informal ritual instruction, oral transmission of healing practices, and selective adaptation of global media to articulate indigenous aesthetics. Yet, as Spivak (1988) warns, such resistances often remain inaudible within the dominant episteme unless mediated through sanctioned academic or developmental discourses. Hence, the urgent task remains the decolonization of knowledge production—to recenter indigenous women not as ethnographic subjects but as sovereign architects of their futures.

Access to education has been both a catalyst and consequence of changing gender roles. While female literacy in Uttarakhand has improved (69.34% as per Census 2011), deep inequalities remain in rural and indigenous areas. Education empowers women not only through literacy but by expanding their civic capability—the confidence and competence to navigate bureaucratic and political systems (Ghosh et al., 2015). Women like those in the Maati Collective in Munsiari have effectively combined oral traditions, ecological knowledge, and formal education to produce alternative media narratives that challenge dominant development discourses (Kalyanwala, 2025). Their storytelling affirms women's status as memory keepers and knowledge holders of sustainable living—repositioning them as environmental experts rather than passive beneficiaries.

At the heart of this upheaval are the indigenous women of the Himalayas, whose lives embody both the ruptures and the continuities of a rapidly shifting world. From the commodification of Bhotia textiles to the erosion of Buksa biradari networks, from the displacement of Tharu agroecological practices to the erasure of Raji food traditions, and the ritual fragmentation among the Jaunsaris, each case reveals the uneven terrain upon which women navigate the paradoxes of tradition and modernity.

Far from being passive recipients of these shifts, tribal women emerged as complex agents—resisting, adapting, and innovating within the constraints imposed by state policy, market logic, and patriarchal structures. Yet their agency often unfolded within a context of structural precarity and epistemic invisibility. While tourism, education, and digitization offered new avenues for self-expression and economic participation, they simultaneously risked entrenching reductive representations and extractive relationships that devalued indigenous knowledge systems. The increasing disconnect between state-led developmental narratives and the lived realities of tribal women marked a fundamental epistemological crisis—one in which cultural heritage is simultaneously fetishized and endangered, preserved in form but hollowed in meaning.

What resulted is a profound need to recalibrate both academic inquiry and policy praxis: to move beyond tokenistic inclusion and toward genuine epistemic justice. This requires centering the narratives, knowledges, and aspirations of indigenous women—not as supplementary voices but as foundational to any sustainable and equitable vision of development. Their perspectives are indispensable in reimagining futures that resist homogenization, affirm cultural sovereignty, and reinstate the relational ontologies that have long sustained Himalayan life-worlds. Without such a shift, the march of Glocalization will continue to script a future where indigenous heritage survives not as lived practice, but as curated relic—musealized, marketable, and ultimately, mute.

References

Anthwal, A., Gupta, N., Sharma, A., Anthwal, S., & Kim, K.-H. (2010). Conserving biodiversity through traditional beliefs in sacred groves in Uttarakhand Himalaya, India. *Resources, Conservation and Recycling, 54*(11), 962–971. https://doi.org/10.1016/j.resconrec.2010.02.003

Anthwal, A., Sharma, R. C., & Sharma, A. (2006). Sacred groves: Traditional way of conserving plant diversity in Garhwal Himalaya, Uttaranchal. *The Journal of American Science, 2*(2), 35–38.

Appadurai, A. (1996). *Modernity at large: Cultural dimensions of globalization*. University of Minnesota Press.

Bhabha, H. K. (1994). The location of culture. Routledge.

Bourdieu, P. (1986). The forms of capital. In J. Richardson (Ed.), *Handbook of theory and research for the sociology of education* (pp. 241–258). Greenwood.

Chatterjee, B. B. (1976). The Bhotias of Uttarakhand. *India International Centre Quarterly, 3*(1), 3–16.

Tamang, J. P. (2022). Dietary culture and antiquity of the Himalayan fermented foods and alcoholic fermented beverages.*Journal of Ethnic Foods, 9*(30).

Ghosh, R., Chakravarti, P., & Mansi, K. (2015). Women's empowerment and education: Panchayats and women's self-help groups in India. *Policy Futures in Education, 13*(3), 294–314. https://doi.org/10.1177/1478210315571214

Hoon, V. (1996). *Living on the move: Bhotias of the Himalayas*. Sage Publications.

Kalyanwala, K. (2025). Marginalised women's voices in the Indian environmental justice movement: Stories from a Himalayan community. *Frontiers in Communication, 10*, 1536392. https://doi.org/10.3389/fcomm.2025.1536392

Kumar, D., Sharma, N., & Rawat, A. (2022). Reviving Bhotiya weaving traditions in Uttarakhand. *Textile Heritage Journal, 15*(2), 284–290.

Kumar, S. (2022). Challenges and opportunities: Tribal community and handlooms in Uttarakhand, India. *Journal of Mountain Research, 17*(2), 288–293.

Kumar, S., & Sati, V. (2023). Highland-lowland interaction in mountain regions: A study of changing patterns among Bhotiya tribes of Garhwal Himalaya. *Human Geographies, 17*, 87–105.

Mohanty, C. T. (2003). *Feminism without borders: Decolonizing theory, practicing solidarity*. Duke University Press.

Negi, C. S. (2007). Declining transhumance and subtle changes in livelihood patterns and biodiversity in the Kumaon Himalaya. *Mountain Research and Development, 27*(2), 114–118.

Pande, P., & Tripathi, M. K. (2016). Vanraji women: Status, challenges and solutions. *The Indian Journal of Political Science, 77*(2), 247–252. https://doi.org/10.2307/26575709

Pandey, A., et al. (2017). Withering of traditional institutions? An institutional analysis of the decline of migratory pastoralism in the rangelands of the Kailash Sacred Landscape, western Himalayas. *Environmental Sociology, 3*(1), 87–100.

Patel, R. (2009). Food sovereignty. *The Journal of Peasant Studies, 36*(3), 663–706. https://doi.org/10.1080/03066150903143079

Samal, P. K., Dhyani, P. P., & Dollo, M. (2010). Indigenous medicinal practices of Bhotia tribal community in Indian Central Himalaya. *Indian Journal of Traditional Knowledge, 9*(1), 140–144.

Sati, V. P., & Banergee, S. (2025). Sustainable homestay tourism for enhancing rural livelihood in the Uttarakhand Himalaya: A study of the Tons River Basin. *BIO Web of Conferences, 151*, 03001. https://doi.org/10.1051/bioconf/202515103001

Spivak, G. C. (1988). Can the subaltern speak? In C. Nelson & L. Grossberg (Eds.), *Marxism and the interpretation of culture* (pp. 271–313). University of Illinois Press.

Verma, S. C., & Yadav, G. S. (2010). Food habits among the educated youth of Tharu and Buksa tribes. *International Journal of Sociology and Anthropology, 2*(4), 66–70.

Behance. (2023). Bhotia textiles exposition. https://www.behance.net/

Bhuli Art. (2023). @bhuli.art [Instagram profile]. https://www.instagram.com/bhuli.art/

Cultural Survival. (2023). The urbanization of tribal areas: Impact on indigenous communities. https://www.culturalsurvival.org/

Medium. (2024, May 3). Part II: Women's role—Why traditional farming practices are diminishing in hilly regions of Uttarakhand. https://medium.com/@khetibadi/part-ii-womens-role-why-traditional-farming-practices-are-diminishing-in-hilly-regions-of-b778f734df19

Thapliyal, N. M. (2018). Innovative practices by grassroot women of Uttarakhand. *Alleviation: An International Journal of Nutrition, Gender & Social Development, 5*(5). https://alvt.apcjournals.com/FrmInnovativePracticesbyGrassrootWomenofUttarakhandDetails.aspx

The Hindu. (2013, February 9). Homestay with a difference. *The Hindu*. Retrieved from https://www.thehindu.com/features/magazine/homestay-with-a-difference/article4396757.ece

The Hindu. (2025, January 28). Mumbai: A photo exhibition sheds light on the Jaunsari tribe of Uttarakhand. https://www.thehindu.com/news/cities/mumbai/photo-exhibition-jaunsari-tribe-of-uttarakhand/article69149614.ece

Uttarakhandi. (2023, July 15). Polyandry and animistic rituals: The cultural fabric of the Jaunsari tribe. https://www.uttarakhandi.com

Village Square. (2023, September 20). Industrialization ruptures symbiotic relationship: The Tharu and the Terai forests. https://www.villagesquare.in

Policy Frameworks and Institutional Supports for Indigenous Women in India

Nazia Talat

Abstract This chapter explores the inter-relationships between policy frameworks and institutional supports available for ensuring the rights of indigenous women in India. Constitutional provisions and legislative acts such as the Forest Rights Act (FRA) and the Panchayats (Extension to Scheduled Areas) Act (PESA), as well as the national policies designed to address the unique challenges faced by tribal women are also analysed in this chapter. For the implementation of these legislations the roles of institutions such as the Ministry of Tribal Affairs, the National Commission for Scheduled Tribes, and grassroots organizations is also explored. The long-drawn challenges of land alienation, gender-based violence, and socio-economic marginalization are highlighted with an emphasis on the need for culturally sensitive and context-specific interventions. The chapter evaluates the efficacy of existing institutional mechanisms while advocating for better policy implementation, enhanced institutional coordination, and the active participation of indigenous women in shaping their own development, ultimately contributing to a more equitable and just future.

Keywords Indigenous women · Forest Rights Act (FRA) · Panchayats (Extension to Scheduled Areas) Act (PESA) · National Commission for Scheduled Tribes · India

1 Introduction

The terms 'Adivasi', 'tribal communities', or 'Indigenous People' are used to describe people considered to be the original inhabitants of India. While there are 5653 distinct communities in India, 635 are recognized as tribes or Adivasis (Nepram, 2017). Indigenous women in India, often referred to as Adivasi women, are defined by their historical continuity with pre-colonial societies and their distinct cultural identities. They are part of the broader Indigenous communities, which are constitutionally

N. Talat (✉)
Shyama Prasad Mukherji College for Women, University of Delhi, New Delhi, India
e-mail: nazia@spm.du.ac.in

N. Moran et al. (eds.), *The Participatory Role of Indigenous Women*,
https://doi.org/10.1007/978-3-032-24059-0_9

recognized as ‘Scheduled Tribes’ in India. This recognition aims to support their development following historical marginalization. The constitutional recognition of Indigenous communities as ‘Scheduled Tribes’ is intended to aid their development and protect their rights, although gaps remain between policy and practice (Thomas et al., 2021). Indigenous women play a crucial role in their communities, yet they face unique challenges due to the intersectionality of gender and Indigenous identity, which results in double marginalization. Indigenous women are integral to the cultural and economic fabric of Adivasi communities, often serving as cultural custodians and economic contributors (Toppo et al., 2024). They are part of communities that strive to preserve their ethnic identity and cultural patterns, as outlined by international agreements such as the United Nations Declaration on Indigenous Peoples’ Rights (Nepram, 2017). However, they still face certain socio-economic challenges, like low literacy rates, poor health and lesser access to resources (Manna, 2024). Gender equality and empowerment is further marred by the limited participation of indigenous women in the processes of decision making, owing to patriarchy (Manna, 2024). They also face inequalities in health and nutrition-based parameters, showing poor nutritional outcomes, with high prevalence of chronic energy deficiency and nutritional inadequacies, particularly in iron, calcium, and vitamin A (Agrawal & Agrawal, 2010; Kapoor et al., 2022).

The indigenous population in India, comprises approximately 8.6% of the national demographic, and is predominantly located in states like Jharkhand, Madhya Pradesh, Chhattisgarh, Odisha, and Maharashtra, Kerala and North East India (Agrawal & Agrawal, 2010; Thomas et al., 2021). The language, traditions and ecological knowledge of the indigenous women, is apparent despite their marginalisation within their communities. Choudhary et al. (2023) emphasise on the need to preserve linguistic heritage as a means of cultural identity for indigenous women. Indigenous languages often carry unique ecological knowledge, which is vital for biodiversity conservation, and are considered as the custodians of cultural practices (Das & Hossain, 2023). The intersection of gender and indigenous identity leads to a unique form of subalternity, where indigenous women face compounded challenges of cultural extinction and socio-economic marginalization (Das & Hossain, 2023).

The sustainable practices and belief system of indigenous people in India have maintained an intimate relationship with the land, forests, and natural resources, thereby conserving biodiversity (Challa et al., 2023; Kumar Palita et al., 2023). However, socio-economic and political changes have led to the marginalization of these communities, separating them from their traditional lands and livelihoods, resulting in ecosystem degradation and the loss of biodiversity (Challa et al., 2023). The modifications in property rights have further disadvantaged the indigenous communities as their livelihood dependent heavily on these resources (Kjosavik & Shanmugaratnam, 2007). The significance of indigenous communities in the maintenance of a resilient and healthy planet cannot be overemphasized. Still, they find themselves neglected in the environmental and developmental policies (Challa et al., 2023). Therefore, it is imperative to recognise their role in biodiversity conservation and sustainable development and include them in legal frameworks and conservation efforts in India (Challa et al., 2023).

Historically, Adivasi populations have faced perpetual cycles of violence, disposability, and exploitation linked to colonization (Dodgson & Struthers, 2005; Drever, 2022). The impact of colonisation was limited not only to the physical control of indigenous people but also to the erasure of the identities of these people, paving the path for gendered violence that continues even today (Smith & Ross, 2004). More recently, Adivasi women have been the victims of systemic violence, including sexual violence, which is exacerbated by societal stereotypes that render them "disposable" (Drever, 2022). The marginalisation and victimisation of these women is further exacerbated by the frequent failure of criminal justice system to protect these women (Drever, 2022).

In the above context, this chapter attempts to analyse the policy frameworks (constitutional, legislative) designed to protect and empower indigenous women, explore the institutional mechanisms (government bodies, schemes, NGOs) that provide support and critically assess the effectiveness and challenges in implementation. In doing so it also seeks to offer a way forward for strengthening these supports.

This chapter is divided into seven sections. Section 1 introduces the chapter by focusing on the meaning of Indigenous Women in India, their diversity, intersectionality of their vulnerability, and the objectives of this chapter. Section 2 elaborates on the socio-economic, land rights issues, violence and discrimination and other challenges faced by indigenous women in India. Section 3 lays the Groundwork for Indigenous Women's Rights in India. Section 4 discusses the institutional support available to indigenous women in India. Section 5 discusses the inter-relationships between policies and institutions to alleviate the challenges faced by indigenous women in India. Section 6 attempts to evaluate the efficacy of the existing policy and institutional mechanisms in India for indigenous women. Section 7 concludes the chapter.

2 Understanding the Challenges Faced by Indigenous Women

2.1 Socio-Economic Disparities

There are several socio-economic, cultural and geographic factors responsible for the lower literacy and educational attainment of indigenous women in India, particularly the Scheduled Tribes, as compared to their male counterparts and the general population. Poverty prohibits the families from focusing on education, lower enrolment and higher dropout rates among tribal women (Dey & Maiti, 2024; Rathore, 2011). Tribal women tend to be part of the labour force in order to contribute to the family income, limiting their ability to send their kids to school (Naik, 2016). Access to education of tribal women is also impacted adversely by strong patriarchal norms (Naik, 2016; Nayak, 2025b) and conservative cultural practices, further limiting their educational opportunities (Wani & Majid, 2024). Despite such hurdles some states have shown better literacy rates among tribal women because of community engagement and progressive policies (Dey & Maiti, 2024; Rathore, 2011).

The health and nutrition indicators, like maternal mortality, infant mortality, malnutrition etc., show significant disparities with worse outcomes among indigenous women in contrast to non-indigenous populations (Akkiraju, 2022; Madankar et al., 2024). The reasons for these disparities could be socio-economic disadvantages, limited access to healthcare, and adherence to harmful traditional practices (Madankar et al., 2024). Indigenous women also exhibit poor nutritional outcomes showing deficits in essential nutrients such as iron, calcium, and vitamin A, because of low dietary diversity and inadequate intake of fruits, vegetables, and protein sources (Kapoor et al., 2022). The poor health outcomes are ultimately determined by poverty and marginalisation as barriers to accessing healthcare (Agrawal & Agrawal, 2010; Das et al., 2023) leading to poorer health outcomes (Agrawal & Agrawal, 2010). Policies addressing food security often overlook the specific needs of women, placing the burden of household food security solely on them (Rao et al., 2017). A lack of gender-transformative approaches in policy frameworks further perpetuates the cycle of poverty and food insecurity among indigenous women (Rao et al., 2017).

As of 2019, the participation of women in labour force stands very low at 20.3% of women engaged in work force, including both formal and informal sectors (3ie, 2021). The participation of indigenous women is predominantly seen in the informal sector where their contributions are often unrecognized and undervalued (Zaidi, 2022), because of low literacy rates, limited access to productive resources, and systemic violence (Manna, 2024); cultural norms, historical gender roles perpetuating economic disparities, restricting women's access to asset ownership and financial decision-making (Veena, 2024). Even at the time of Covid 19, it was observed that the decline in female labour force (14%) was much higher as compared to the decline in the male labour force (1%) (Nikore et al., 2021). Manna (2024) notes that the existing gaps between policy and practice prohibits women from getting access to employment and financial opportunities. There is an urgent need for affirmative action, gender-sensitive labour laws and support for childcare, to enhance women's economic empowerment (Veena, 2024).

2.2 *Land and Resource Rights*

Adivasis have been frequently displaced from their land due to development projects and economic policies, undermining their access to resources and livelihoods. This has contributed to long-term impoverishment and livelihood insecurity, which became particularly severe during crises such as the COVID-19 pandemic (Kujur, 2023; Behera & Padhi, 2022). Despite making up only 8% of the population, STs account for almost 40% of the displaced (Mitra & Rao, 2009). Alienation of Adivasis from their land is a gradual process, which is often framed as a voluntary migration, complicating the understanding of displacement (Mitra & Rao, 2009). The displacement impacts women disproportionately, leading to disempowerment and loss of livelihood, as resettlement policies often overlook their specific needs (Bala, 2006; Bisht, 2009). Women have traditionally lacked ownership and control over property, particularly land, which is a key determinant of economic security and social status (Agarwal, 1994). This lack of property rights increases their vulnerability, including in contexts such as displacement, where compensation mechanisms are often tied to formal ownership (Panda & Agarwal, 2005). Only 11% of landowners in India are women, highlighting a significant gender gap in land ownership, despite legal provisions, because of prevalence of patriarchal norms (Licart, 2023). Without land women are not recognised as farmers, restricting their access to credit and government benefits (Dubochet, 2013), increasing their vulnerability and perpetuating cycles of poverty and violence (Jain et al., 2023; Sircar, 2016). The resettlement and rehabilitation policies have been criticised for being blind to gender, failing to address the particular difficulties displaced women face (Bala, 2006), for example, the Land Acquisition Act of 1894 perpetuates inequalities by not recognizing common property resources, further marginalizing women (Jhala, 2013).

The impact of forest conservation policies on the livelihoods and traditional practices of indigenous women in India is profound and multifaceted. Forest conservation policies frequently restrict access to forest resources, which are vital for the economic sustenance of indigenous women. There are several examples where forest policies have restricted the life and livelihoods of indigenous communities. Khasi women in Meghalaya face challenges in managing inherited forest resources due to restrictive policies (Shangpliang, 2012). Adivasi women in the Nilgiri Biosphere region have experienced hardships due to the denial of traditional rights, leading to economic marginalization and a loss of their ecological knowledge (Dudley, 1999). Traditional practices, such as the reverence for sacred groves among the Karbi women, are threatened by forest degradation and modern policies that overlook indigenous knowledge systems (Hansepi & Laisram, 2022). The cultural transition observed in West Bengal indicates a shift from self-sustained livelihoods to increased participation in external economic activities, undermining traditional practices (Bit & Banerjee, 2013). Indigenous women tend to be disproportionately affected by deforestation as compared to indigenous men, because of their primarily responsibilities related to gathering of resources like fuel and water. This has led to increased labor burdens and health issues due to longer travel distances (Rawat, 1995). The erosion of women's status in tribal cultures, exacerbated by external influences, further complicates their ability to engage in forest management and conservation efforts (Dudley, 1999). The Forest Rights Act (FRA) 2006 aimed to empower tribal women by granting joint titles and promoting their involvement in decision-making has improved women's political participation, however, the economic gains remain limited due to entrenched patriarchal structures (Raj & Anitha, 2024).

2.3 *Violence and Discrimination*

The violence and discrimination faced by Indigenous women in India is a multi-faceted issue deeply rooted in historical, social, and political contexts. Indigenous women experience violence not only from external societal structures but also from within their communities, often exacerbated by colonial legacies and systemic discrimination. Indigenous women experience heightened discrimination because of their ethnicity and gender, leading to unique experiences of violence and inequality (Nayak, 2025a). The Partition of India in 1947 serves as a significant historical backdrop, where mass abductions and sexual violence against women were rampant, reflecting a broader pattern of gender-based violence during communal conflicts. The narratives of these women have been largely silenced, with their experiences often overshadowed by nationalistic discourses that prioritize collective honour over individual trauma (Parsloe & Campbell, 2021). The experiences of Indigenous women are not acknowledged or supported in both academic and public discourses contributing to their ongoing victimization and marginalization (Smith & Ross, 2004).

2.4 *Customary Laws and Patriarchy*

Customary laws are very common among tribal communities. Their structure is patriarchal in nature often resulting in discriminatory practices that restrict women's rights and resources (Yanthan, 2018). These customary laws can also conflict with modern legal frameworks, leading to systemic discrimination against women, particularly in land rights and inheritance (Singh & Bhatnagar, 2024). Indigenous women are vulnerable in the face of these customary laws as they face compounded disadvantages due to both gender and ethnic discrimination (Sarin & Gopalakrishnan, 2022).

3 Policy Frameworks: Laying the Groundwork for Indigenous Women's Rights

3.1 *Constitutional Provisions as the Foundation*

The constitutional framework for indigenous women in India is shaped by a combination of legal recognition, cultural autonomy, and ongoing challenges. The Indian Constitution acknowledges Scheduled Tribes, aiming to empower these communities through affirmative action and self-governance. This framework is crucial for indigenous women, who often face unique socio-cultural barriers. Indigenous women are recognized as part of Scheduled Tribes, which aims to address historical injustices and promote their rights (Thomas et al., 2021).

(A) ***Fundamental Rights***: Articles 14 (Equality before law) ensures that tribal women have the same legal rights as any other citizen. It prevents discriminatory laws or practices that deny them equal protection. However, customary laws in tribal communities sometimes override constitutional provisions, limiting their access to land and inheritance rights.[1] Article 15 (Prohibition of discrimination), prohibits discrimination based on religion, race, caste, sex, or place of birth. It allows the government to make special provisions for women and Scheduled Tribes. The Constitution provides for affirmative action under Article 15(3), allowing special provisions for women, including those from indigenous communities (Nayar et al., 2023). Despite this, tribal women often face exclusion from mainstream legal protections due to customary laws.[2] Article 16 (Equality of opportunity in public employment) guarantees equal access to government jobs for tribal women. It provides for the reservations for Scheduled Tribes, ensuring representation. However, social and economic barriers still limit their participation in formal employment.[3]

(B) ***Directive Principles of State Policy***: Article 46 (Promotion of educational and economic interests of STs and protection from exploitation) of the Indian Constitution directs the state to promote the educational and economic interests of Scheduled Tribes (STs) and protect them from social injustice and exploitation. For tribal women, this provision has significant implications in terms of educational empowerment (through scholarship programs, reservation in educational institutions, skill development initiatives), economic upliftment (through livelihood support, microfinance and entrepreneurship, employment reservations), and protection from exploitation (through legal safeguards, land rights and forest protection, social welfare schemes etc.) (Besra & Sarangi, 2024; Shah et al., 2019).

(C) ***Special Provisions for Tribal Areas***: India has special provisions for tribal areas under the Fifth and Sixth Schedules of the Constitution, ensuring autonomy, protection, and development for Scheduled Tribes.[4] Tribal regions on the Indian mainland are included by the Fifth Schedule (Article 244 (1)). It provides special governance mechanisms through Tribal Advisory Councils allowing for states to modify laws to suit tribal customs and traditions. It also ensures protection of land to prevent alienation from tribal communities. Sixth Schedule (Article 244(2)) applies to tribal areas in Meghalaya, Assam, Mizoram, and Tripura. It creates Autonomous District Councils (ADCs), which have the authority to enact laws and administer justice. It also grants self-governance in matters like land, forests, and local administration, protecting tribal identity and culture through special legal provisions. Special protections for Scheduled and Tribal

[1] https://ebooks.inflibnet.ac.in/hrdp05/chapter/indian-constitutional-framework-articles-14-15-and-16-of-the-indian-constitution-and-womens-rights/ (Accessed on 30-05-2025).

[2] https://www.barandbench.com/columns/succession-rights-of-tribal-women-time-to-make-them-gender-equal (Accessed on 30-05-2025).

[3] https://www.tscld.com/tribal-women-equal-succession (Accessed on 30-05-2025).

[4] https://www.nextias.com/blog/tribals-in-india/ (Accessed on 30-05-2025).

Areas are provided by the Indian Constitution's Fifth and Sixth Schedules, influencing local governance and customary law. While the Panchayat Act extension to Schedule V areas has facilitated tribal women's participation in natural resource management in some regions, customary practices in Schedule VI areas continue to limit women's control over resources (Krishna, 2001). There are political and constitutional obstacles to the establishment of local self-governance in Sixth Schedule territories, and there are no special provisions for Panchayati Raj (Borah & Bhuyan, 2023). Despite constitutional protections, Scheduled Tribes disproportionately face displacement and landlessness due to development projects (Wahi & Bhatia, 2018). Urban development in Fifth and Sixth Schedule areas, such as Greater Ranchi and New Shillong Township, raises concerns about land procurement and its impact on tribal rights and governance (Xaxa, 2021). These issues highlight the complex interplay between constitutional provisions, customary practices, and development pressures in tribal areas.

(D) ***Political Representation***: Articles 243D and 243T of the Indian Constitution mandates reservation for women, including those from Scheduled Tribes (STs), in Panchayats and Municipalities to ensure their political representation.[5] Seats reserved for Scheduled Tribes (STs) in Panchayats, under Article 243D, in proportion to their population in the area. It requires that women from ST communities be given at least one-third of these reserved seats. The position of Chairpersons in Panchayats also have reserved quotas for STs and women, ensuring leadership opportunities. Under Article 243T seats reserved for STs in urban local bodies. It is mandatory for at least one-third of these reserved seats to be allocated to women from ST communities.

Articles 330 and 332 of the Indian Constitution ensure political inclusion for Scheduled Tribes (STs) by setting aside seats in the State Legislative Assemblies and the Lok Sabha. In each state or territory, seats are reserved for Scheduled Tribes (STs) based on their population under Article 330 (Reservation in Lok Sabha). It ensures tribal representation in national policymaking. Although it originally meant only for 10 years, but it has been extended multiple times (currently up to 2030). Article 332 (Reservation in State Assemblies) mandates reservations of seats for STs in state legislatures. It helps participation of tribal communities in governance at the state level, ensuring political empowerment and influence on policy making. These provisions aim to integrate tribal voices into mainstream politics, but challenges remain in effective representation and leadership development.,[6,7]

[5] https://www.constitutionofindia.net/articles/article-243d-reservation-of-seats/ (Accessed on 30-05-2025).

[6] https://iasorigin.com/part-xvi-of-the-indian-constitution-articles-330-to-342/ (Accessed on 30-05-2025).

[7] https://indiankanoon.org/doc/818841/ (Accessed on 30-05-2025).

(E) ***Constitutional Oversight***: The Indian Constitution's Article 338A (National Commission for Scheduled Tribes (NCST)) creates the NCST as a special agency tasked with protecting the rights of Scheduled Tribes (STs). In order to concentrate solely on tribal welfare, it was established by the Constitution (89th Amendment) Act, 2003, which separated the NCST from the National Commission for Scheduled Castes and Tribes.[8] It guarantees the application of STs' legal and constitutional protections, addresses grievances related to the deprivation of tribal rights, provides recommendations for socio-economic development and tribal welfare, submits annual reports on the status of ST rights and policy effectiveness, and functions with civil court authority, allowing it to summon individuals and demand evidence in investigations.[9] The NCST plays the role of a critical watchdog, ensuring that tribal communities receive fair treatment in governance, land rights, education, and employment. It also works to prevent exploitation and displacement, advocating for stronger legal protections.[10]

3.2 Key Legislations Addressing Unique Challenges

(A) ***The Scheduled Castes and Scheduled Tribes (Prevention of Atrocities) Act, 1989*** along with its amendments (2015, 2018), is an essential legal structure created to shield indigenous women from systematic oppression, assault, and prejudice.,[11,12] The Act criminalizes sexual violence against tribal women, ensuring strict penalties for offenders. It also prevents illegal eviction of tribal women from their land and homes. It prohibits denial of access to resources, employment, and services based on caste or tribal identity. Fast-track courts under this act ensure speedy trials for cases and Special Public Prosecutors are appointed to handle cases involving atrocities against tribal women. It also ensures compensation, rehabilitation and witness protection for the victims.

To shield underprivileged groups from prejudice and violence, the Scheduled Castes and Scheduled Tribes (Prevention of Atrocities) Act, 1989, was passed (Acharya & Acharya, 2020). Despite its intentions, the Act's effectiveness has been questioned, with concerns about misuse and the Supreme Court labeling it as an instrument of "blackmail" (Acharya & Acharya, 2020). The 2015 Amendment aimed to strengthen the Act, but a 2018 Supreme Court ruling introduced procedural safeguards that significantly weakened its implementation (Saxena,

[8] https://www.constitutionofindia.net/articles/article-338a-national-commission-for-scheduled-tribes/ (Accessed on 30-05-2025).

[9] https://indiankanoon.org/doc/197519278/ (Accessed on 30-05-2025).

[10] https://ncst.nic.in/?lang=en (Accessed on 30-05-2025).

[11] https://www.indiacode.nic.in/bitstream/123456789/1920/1/a1989-33.pdf (Accessed on 30-05-2025).

[12] https://blog.ipleaders.in/scheduled-castes-and-scheduled-tribes-prevention-of-atrocities-act-1989/ (Accessed on 30-05-2025).

2018). The Mahajan judgment of 2018 further diminished the Act's effectiveness by making it more difficult to file FIRs and arrest accused individuals, potentially enabling upper-caste perpetrators to violate the law with impunity (Nawsagaray, 2018). These developments have raised concerns about the Act's ability to protect marginalized communities and ensure social justice.

(B) ***The Panchayats (Extension to Scheduled Areas) Act (PESA), 1996***: The Panchayats (Extension to Scheduled Areas) Act (PESA), 1996 is a landmark legislation that empowers Gram Sabhas (village assemblies) in tribal areas by granting them autonomy in governance, resource management, and decision-making.[13] PESA empowers Gram Sabhas by ensuring tribal communities retain control over their affairs, preventing exploitation, upholding traditional governance systems, and integrating tribal customs into local administration. It ensures the participation of indigenous women in decision making through mandatory representation, giving them leadership positions, and involving them in community led initiatives. PESA also ensures Natural Resource Management by giving the authority over land and resources to the Gram Sabhas involvement of tribal women in forest-based enterprises like minor forest produce collection, and incorporating traditional ecological knowledge into sustainable agriculture and biodiversity conservation. PESA acknowledges indigenous governance systems, allowing communities to follow their traditional dispute resolution methods. It also ensures tribal land rights, preventing forced evictions and loss of cultural heritage, encourages traditional medicine, crafts, and agricultural techniques, preserving tribal identity.

Gram Sabhas in tribal areas were intended to be empowered by the Panchayats (Extension to Scheduled Areas) Act (PESA) of 1996, but its implementation has been difficult. While PESA was designed to protect cultural identities and promote decentralized natural resource management (Sisodia & Dalapati, 2018), its effectiveness has been limited. Indigenous women's participation in decision-making remains poor, with traditional political structures often excluding them. The Act's compliance varies across states, with some yet to implement state-specific guidelines (Kindo & Bhowmick, 2024). Despite PESA's potential to safeguard tribal customs and promote self-governance, Adivasi participation in Gram Sabhas is disappointing due to non-Adivasi domination and bureaucratic influence (Sisodia & Dalapati, 2018). More than two decades after implementation, the panchayat raj system in scheduled areas still struggles to fully realize PESA's objectives of preserving tribal identities and empowering local governance (Sisodia & Dalapati, 2018).

(C) ***The Forest Rights Act (FRA), 2006*** is a landmark piece of legislation in India that recognizes the rights of forest-dwelling Scheduled Tribes and other traditional forest dwellers. The FRA acknowledges the historical injustices faced by forest communities whose rights were often overlooked (Gochhayat, 2011). It aims to rectify the colonial legacy that stripped these communities of their

[13] https://pubadmin.institute/rural-local-governance/pesa-act-1996-panchayati-raj-scheduled-areas (Accessed on 30-05-2025).

customary rights, thereby restoring their connection to the land (Das, 2013). It grants them legal recognition over individual forest land for habitation and cultivation, community forest rights for access to resources like minor forest produce, grazing, and traditional practices and the right to protect and conserve forests, ensuring sustainable management by local communities. A crucial aspect of the FRA is its mandate for joint ownership of land titles between spouses. It empowers tribal women by ensuring legal claims to land, reducing dependency on male relatives, enhancing economic security, allowing women to participate in decision-making regarding land use and resource management and challenging patriarchal norms. The Act mandates joint titles for land ownership, which empowers women by ensuring their legal recognition in land rights alongside men (Zaidi, 2019). This provision is significant in promoting gender equality, as it raises women's social standing and economic independence by enabling them to take part in decisions about land usage and management (Raj & Anitha, 2024). By securing land rights, the FRA enables tribal women to access resources necessary for their livelihoods, such as forest produce, which is vital for their sustenance and economic stability (Khanna, 2024). Studies indicate that women beneficiaries of the FRA experience increased autonomy in agricultural production and improved self-confidence, contributing to their overall empowerment (Raj & Anitha, 2024)

(D) ***Inheritance Rights and Customary Law***: The Hindu Succession Act, 1956, and its 2005 Amendment were pivotal in advancing gender equality in inheritance rights. However, tribal women remain largely excluded due to the Act's Section 2(2), which declares that Scheduled Tribes are exempt from it unless the government notifies them.,[14,15,16] Because of this exclusion, tribal women's rights to inherit are controlled by customary laws, which often favour male heirs. Tribal communities have autonomy over customary laws, impacting marriage and inheritance rights. However, these laws can sometimes disadvantage women, leading to legal challenges (Uma, 2023). Courts have occasionally intervened, but the application of Hindu law to tribal cases complicates the inheritance rights of women, often undermining their tribal identity (Uma, 2023). Despite constitutional protections, indigenous women encounter numerous socioeconomic obstacles, such as restricted access to healthcare and education (Thomas et al., 2021). The need for disaggregated data on health outcomes for indigenous women highlights ongoing disparities (Thomas et al., 2021). The exclusion of indigenous women from inheritance rights has come under increased scrutiny from courts. In Kamla Neti v. Special Land Acquisition Officer (2022), the Supreme Court stressed the need of gender equality in tribal succession rules. While tribal autonomy is important, legal experts argue

[14] https://www.drishtiias.com/daily-updates/daily-news-analysis/exclusion-of-st-women-from-hindu-succession-act (Last accessed on 30-05-2025).

[15] https://www.tscld.com/tribal-women-equal-succession (Last accessed on 30-05-2025).

[16] https://www.barandbench.com/columns/succession-rights-of-tribal-women-time-to-make-them-gender-equal (Last accessed on 30-05-2025).

that customary laws should evolve to align with constitutional principles of equality.

Other Relevant Legislations: One important piece of legislation, the Protection of Women from Domestic Violence Act, 2005 (PWDVA), aims to protect women against economic, emotional, and physical abuse, among other types of domestic violence. Compared to earlier legislation like Section 498A of the IPC and Section 125 of the CrPC, this Act offers a more complete framework for protection and remedies such protection orders, residency orders, and monetary compensation (Kalyani, 2013; Lalremruati, 2024). However, its effectiveness is often questioned due to societal attitudes and enforcement challenges (Hornbeck et al., 2007; Karanjawala & Chugh, 2009). Unlike previous laws, the PWDVA extends protection to various categories of women, including sisters, widows, and mothers, not just wives (Kalyani, 2013). The Indian Penal Code (IPC) Sections 354 and 376 protect women from sexual assault, the Prohibition of Child Marriage Act, 2006 addresses early marriages that are common in some tribal tribes, and the Dowry Prohibition Act, 1961 protects women from violence related to dowries.

4 Institutional Supports: Implementing Policies and Delivering Aid

4.1 Government Ministries and Nodal Bodies

(A) **Ministry of Tribal Affairs (MoTA)** plays a pivotal role as the central nodal ministry for the overall policy, planning, and coordination of programs for the development of Scheduled Tribes (STs) in India. Established in 1999, its primary objective is to provide a focused approach to the integrated socio-economic development of tribal communities, who are among the most under-privileged sections of Indian society.[17] It functions in the areas of policy formulation, planning and coordination, financial assistance, monitoring and evaluation, social security and welfare, promoting voluntary efforts, and implementation of Acts like the Protection of Civil Rights Act, 1955, and the Scheduled Castes and the Scheduled Tribes (Prevention of Atrocities) Act, 1989.[18] MoTA, often in collaboration with its public sector undertaking, the National Scheduled Tribes Finance and Development Corporation (NSTFDC), and the Tribal Cooperative Marketing Development Federation of India (TRIFED), implements several schemes to economically empower tribal women.[19] For

[17] https://www.pib.gov.in/PressReleasePage.aspx?PRID=2090883 (Last accessed on 30-05-2025).

[18] https://trifed.tribal.gov.in/ministry-of-tribal-affairs (Last accessed on 30-05-2025).

[19] https://www.prnewswire.com/in/news-releases/empowering-tribal-communities-A-journey-of-sustainable-development-302189717.html (Last accessed on 30-05-2025).

example, the Adivasi Mahila Sashaktikaran Yojana (AMSY)[20] is an exclusive concessional loan scheme specifically designed for the economic development and financial independence of eligible Scheduled Tribe women. The Pradhan Mantri Janjatiya Vikas Mission (PMJVM) aims to facilitate and promote livelihood opportunities and entrepreneurship among the tribal population by establishing an efficient and transparent mechanism for the production, procurement, processing, and marketing of tribal produce/products. Van Dhan Yojana (VDY): A critical component of PMJVM, it focuses on setting up Van Dhan Vikas Kendras (VDVKs) and Van Dhan Producer Enterprises (VDPEs). These centers provide common-use facilities for value addition, skill development, and market linkages, transforming tribal gatherers into entrepreneurs.[21] TRIFED (Tribal Cooperative Marketing Development Federation of India Ltd.) another nodal agency, is a crucial organization under MoTA, dedicated to the socio-economic development of tribal communities by promoting tribal product marketing.[22] It markets tribal products through its "TRIBES INDIA" retail outlets (both physically and online (via www.tribesindia.com) and through exhibitions like Aadi Mahotsav. MoTA also implements various schemes focused on improving educational outcomes for tribal children, with a specific focus on bridging the literacy gap for tribal girls and women.[23] Eklavya Model Residential Schools (EMRS) provide quality middle and high-level education to Scheduled Tribe students in remote tribal areas, reducing dropout rates and improving access to educational opportunities.[24] There are several other scholarships and fellowship schemes for children belonging to ST category.

(B) **National Commission for Scheduled Tribes (NCST)**: The National Commission for Scheduled Tribes (NCST) is a constitutional body established under Article 338A of the Indian Constitution, specifically created to safeguard and promote the rights and interests of Scheduled Tribes (STs). It acts as a watchdog, investigator, monitor, and advisor to the government on all issues pertaining to the growth and welfare of tribes.[25] The NCST's mandate inherently includes a strong focus on issues affecting tribal women, who often face layered vulnerabilities due to their gender and tribal identity. A chairperson, vice-chairperson, and three other members make up the NCST's membership, with the crucial provision that at least one of these members must be a woman. The presence of a woman member ensures that the unique perspectives, challenges, and aspirations of tribal women are adequately represented within the Commission. It

[20] https://pib.gov.in/PressReleseDetail.aspx?PRID=1813196 (Last accessed on 30-05-2025).

[21] https://www.pib.gov.in/PressReleseDetail.aspx?PRID=2090883 (Last accessed on 30-05-2025).

[22] https://en.vikaspedia.in/viewcontent/social-welfare/scheduled-tribes-welfare/national-institutes-for-st-welfare (Last accessed on 30-05-2025).

[23] https://www.pib.gov.in/PressReleasePage.aspx?PRID=2090883 (Last accessed on 30-05-2025).

[24] https://www.pib.gov.in/PressNoteDetails.aspx?NoteId=153421&ModuleId=3 (Last accessed on 30-05-2025).

[25] https://vajiramandravi.com/national-commission-for-scheduled-tribes-ncst/ (Last accessed on 30-05-2025).

provides a direct voice for tribal women's concerns in policy discussions and decision-making.[26]

(C) **Ministry of Women and Child Development (MoWCD)**: Through a variety of programs and initiatives, the Ministry of Women and Child Development (MoWCD) significantly contributes to the advancement of women's safety, security, and empowerment in India. While these schemes are designed for all women, their reach and effectiveness for indigenous women often require targeted strategies and a deep understanding of their unique socio-cultural contexts.[27] Several broad women-centric schemes like Indigenous women are supported by One-Stop Centers, Women Helplines, Beti Bachao Beti Padhao, and other organizations. Through services like medical aid, legal aid and guidance, psycho-social counseling, police assistance, and temporary shelter, One Stop Centers (OSCs) offer integrated support and assistance to women impacted by violence in both public and private settings.[28] Women Helplines (e.g., 181 National Women's Helpline, 112 National Emergency Helpline) provide a 24/7 toll-free number for women facing violence or distress to seek immediate help, information, and referral to relevant services. One of the main initiatives to combat the falling Child Sex Ratio (CSR) is Beti Bachao Beti Padhao (BBBP) It also promotes the education and empowerment of the girl child. It's a multi-sectoral program involving advocacy, community mobilization, and convergence of services.[29]

(D) **Other Ministries**: While the Ministry of Tribal Affairs (MoTA) is the nodal ministry for tribal welfare, other ministries also play a crucial role in improving the well-being and livelihoods of tribal communities, including tribal women. The Ministry of Health and Family Welfare (MoHFW) and the Ministry of Rural Development (MoRD) are particularly significant in this regard.[30] Maternal and Child Health (MCH) Programs (e.g., Janani Shishu Suraksha Karyakram—JSSK, Pradhan Mantri Surakshit Matritva Abhiyan—PMSMA, Janani Suraksha Yojana—JSY), are some of the programs of M0HFW to reduce infant and maternal mortality.[31] MoHFW has relaxed population norms for setting up Sub-Centres, Primary Health Centres (PHCs), and Community Health Centres (CHCs) in tribal and hilly areas (e.g., PHC for 20,000 population in tribal areas vs. 30,000 in plain areas). They also have flexible norms for ASHA workers (one ASHA per habitation in tribal areas) and provide more

---

[26] https://www.galaxyclasses.co.in/details?res_type=ca&res_id=7520 (Accessed on 30-05-2025).

[27] https://www.ncw.gov.in/publications/women-centric-schemes-by-different-ministries-of-government-of-india-goi/ (Accessed on 30-05-2025).

[28] https://www.pib.gov.in/PressReleasePage.aspx?PRID=2110882 (Accessed on 30-05-2025).

[29] https://wcdhry.gov.in/schemes-for-women/beti-bachao-beti-padhao/ (Accessed on 30-05-2025).

[30] https://www.business-standard.com/economy/news/health-and-welfare-of-tribal-communities-leaving-no-one-behind-123121000242_1.html (30-05-2025).

[31] https://pmc.ncbi.nlm.nih.gov/articles/PMC11020602/ (Last accessed on 30-05-2025).

Mobile Medical Units (MMUs) to improve outreach.[32] Under the National Health Mission (NHM) outreach camps are frequently organized in tribal areas to increase awareness of MCH services and facilitate community mobilization. The scheme for promoting menstrual hygiene among adolescent girls (including tribal girls) aims to increase awareness, access to sanitary products, and safe disposal practices.[33] The Ministry of Rural Development: Impact on Tribal Women's Livelihoods (MoRD) implements large-scale rural development programs that significantly impact the livelihoods and economic well-being of tribal women by providing employment, housing, and promoting self-help groups.[34] The Mahatma Gandhi National Rural Employment Guarantee Act (MGNREGA) provides a crucial safety net and a source of regular, albeit seasonal, income for tribal women, many of whom are landless laborers or rely on precarious informal work (Thapar-Bjorkert et al., 2019). Primitive Tribal Groups (PVTGs) and legally released bonded laborers are automatically included as beneficiaries under PMAY-G (Pradhan Mantri Awaas Yojana-Gramin), ensuring coverage for some of the most vulnerable tribal women.[35]

4.2 Local Governance and Community-Based Structures

(A) **Panchayati Raj Institutions**: In contrast to other regions of India, Panchayati Raj Institutions (PRIs) in Scheduled Areas are governed by a different set of laws. The provisions of the Panchayats (Extension to the Scheduled Areas) Act, 1996 (PESA) largely regulate this system. With some adjustments and exclusions to take into consideration the distinct sociocultural, administrative, and economic traits of tribal populations, PESA was passed in order to apply the provisions of Part IX of the Constitution pertaining to Panchayats to the Scheduled Areas.[36] The 73rd Amendment mandates a minimum one-third reservation for women in all PRIs, including those belonging to STs. PESA further strengthens this by ensuring that if a seat is reserved for STs, at least one-third of *those* seats are also reserved for tribal women.[37] This reservation has led to an unprecedented number of tribal women becoming elected representatives

[32] https://nhm.gov.in/index1.php?lang=1&level=2&sublinkid=1221&lid=188 (Last accessed on 30-05-2025).

[33] https://nhm.gov.in/index1.php?lang=1&level=3&sublinkid=1021&lid=391 (Last accessed on 30-05-2025).

[34] https://www.pib.gov.in/PressReleasePage.aspx?PRID=1647439 (Last accessed on 30-05-2025).

[35] https://www.aavas.in/blog/pmay-gramin#:~:text=Households%20without%20shelter.,adult%20member%20aged%2016%E2%80%9359 (Last accessed on 30-05-2025).

[36] https://www.undp.org/india/publications/panchayat-raj-extension-scheduled-areas-act-1996-policy-brief (Accessed on 30-05-2025).

[37] https://secforuts.mha.gov.in/73rd-amendment-of-panchayati-raj-in-india/ (Accessed on 30-05-2025).

(Sarpanches, Panchayat members). This gives them an official forum to take part in the decision-making process, which has historically been dominated by men.[38]

(B) **Self-help Groups (SHGs)**: In India, Self-Help Groups (SHGs) have become a potent grassroots movement that has a big impact on women's lives, especially in rural and tribal areas. These are tiny, unofficial groups of women, usually ten to twenty, who gather together on a monthly basis to save a little sum of money, which is then used to create a common fund. This fund provides collateral-free loans to members for various purposes, often at reasonable interest rates decided by the group.[39] Government programs such as the Ministry of Rural Development's Deendayal Antyodaya Yojana-National Rural Livelihoods Mission (DAY-NRLM) have aggressively encouraged the establishment and growth of SHGs, including those in tribal areas. These programs provide initial capital, revolving funds, capacity building, and linkage to banks for larger loans.[40]

4.3 Role of Civil Society and Grassroots Organizations:

Community-Based Organizations (CBOs) and Non-Governmental Organizations (NGOs) are extremely important and frequently necessary in advancing welfare and development, especially among underserved groups like tribal populations. They act as crucial intermediaries, bridges, and catalysts, filling critical gaps in areas where government reach might be limited or where a more localized, culturally sensitive approach is required. They give a voice to marginalized tribal groups, who often lack the means or platforms to articulate their grievances and needs to policymakers.[41] They conduct research, publish reports, and converse with government agencies to have an impact on the creation of policies and ensure that tribal perspectives are incorporated into national and state-level development plans. This includes advocating for stronger implementation of acts like PESA and the Forest Rights Act.[42] Access to justice remains a significant challenge for many tribal individuals, particularly women, due to illiteracy, poverty, remote locations, and lack of awareness of

---

[38] https://plutusias.com/empowering-tribal-communities-the-impact-of-the-pesa-act/#:~:text=Additionally%2C%20the%20mention%20of%2014,representatives%2C%20along%20with%20states%20increasing (Accessed on 30-05-2025).

[39] https://knowledgehub.unsse.org/knowledge-hub/womens-self-help-groups/ (Accessed on 30-05-2025).

[40] https://www.myscheme.gov.in/schemes/day-nrlm (Last accessed on 30-05-2025).

[41] https://www.ijraset.com/research-paper/role-of-ngos-in-the-upliftment-of-tribals-in-jharkhand (Last accessed on 30-05-2025).

[42] https://uppcsmagazine.com/the-role-of-non-governmental-organizations-in-policy-formulation/ (Last accessed on 30-05-2025).

legal rights. NGOs and CBOs step in to bridge this gap.[43] Some examples of civil society and Grassroot organisations are listed below:

A. **Self-Help Groups (SHGs) in Tribal Areas**

i. ***R. Kavitha(Adilabad, Telangana)***: This is a direct success story highlighted by NSTFDC. R. Kavitha, from the Lambada ST community, was abandoned by her husband. As a member of a Self-Help Group, she availed a loan of ₹50,000 (and later more, including refinance from NSTFDC) to start a Kirana (grocery) store named DhanaLaxmi Kirana in Utnoor, Adilabad. Today, she successfully runs the shop, earning a substantial monthly income and providing for her children's education. This exemplifies how SHG membership provided her with a financial base, confidence, and entrepreneurial understanding, leading to significant economic independence.[44]

ii. ***Pangi Valley, Himachal Pradesh (CEVA and NABARD collaboration)***: In the remote Pangi Valley, a non-profit organization called Collective Efforts for Voluntary Actions (CEVA) collaborated with NABARD to encourage tribal women to form 75 SHGs. These groups initially focused on traditional handloom and handicraft products made from tribal wool (socks, scarves, caps). With NABARD's training and CEVA's focus on marketing, they created the "Pangi Hills" brand. They expanded to organic products like hazelnuts and medicinal herbs. Women members now earn substantial incomes (e.g., ₹25,000–₹30,000 in an exhibition or ₹2 lakh–₹2.5 lakh per annum per group). This showcases how SHGs, with NGO and institutional support, can transform traditional skills into sustainable livelihoods and create market linkages.[45]

iii. ***Van Dhan Vikas Kendras (VDVKs)and TRIFED***: While specific individual names might be harder to pinpoint in public reports, the overall impact of VDVKs under PMJVM is a major case study. For example, in Khunti, Jharkhand, TRIFED organized a "Mahila SHG Sammelan" where President Droupadi Murmu interacted with tribal women from SHGs/VDVKs. These women showcased and demonstrated value-added products like lac jewelry, lac bangles, Kharswan Haldi (turmeric), Sabai Grass products, and tribal paintings. This demonstrates how the VDVK model, built around SHGs, is enabling tribal women to add value to minor forest produce, diversify their income, and access larger markets through TRIFED's marketing channels.

iv. ***"Ambe Mahila Mandal" (Gujarat)***: This SHG, formed with technical advice from Krishi Vigyan Kendra, started by purchasing raw materials to make soap, detergent powder, Vicks, and Vaseline. This shows how

---

[43] https://www.probonoinst.org/2024/11/01/breaking-the-cycle-addressing-domestic-violence-in-indigenous-communities/ (Last accessed on 30-05-2025).

[44] https://nstfdc.tribal.gov.in/ (Last accessed on 30-05-2025).

[45] https://securehimalaya.org/pangi-hills-brand/ (Last accessed on 30-05-25).

simple, locally relevant micro-enterprises can be initiated through SHGs, contributing to women's economic stability.[46]

B. **NGOs/CBOs in Advocacy, Legal Aid, and Capacity Building**

i. **PRADAN (Professional Assistance for Development Action)**[47]: PRADAN is a prominent NGO working extensively with tribal communities, particularly women. PRADAN has facilitated the formation of "Nari Adalat" or women's courts in some tribal areas. These are informal justice mechanisms run by tribal women themselves to resolve local disputes, particularly those related to domestic violence, marital issues, and minor land disputes, providing accessible and culturally sensitive legal aid and mediation services. This example highlights their role in fostering collective agency and grassroots justice delivery. PRADAN has also helped tribal women diversify livelihoods beyond subsistence agriculture, focusing on poultry, goat rearing, and vegetable cultivation, linking them to markets.

ii. **SEWA (Self-Employed Women's Association), Gujarat**[48]: While not exclusively tribal, SEWA has a strong presence in tribal areas of Gujarat. SEWA organizes women in the informal sector, including tribal women involved in minor forest produce collection, agriculture, and handicrafts. It provides them with access to microfinance, skill training, and market linkages. SEWA advocates for the rights of self-employed women, including tribal women, guaranteeing social security benefits, improved working conditions, and equitable pay. They have successfully negotiated with local authorities and markets on behalf of their members.

iii. **ActionAid India**[49]: This NGO works with various marginalized communities, including tribals. To assist indigenous groups in asserting their rights under the Forest Rights Act (FRA) and fending off land alienation, they offer legal assistance and awareness-raising initiatives. They support communities in documenting their claims and engaging with legal processes.

iv. **Gram Vikas (Odisha)**[50]: Gram Vikas works with tribal communities in Odisha on various development initiatives, including water, sanitation, and livelihood. They empower local communities, including women, to take ownership of development projects, providing them with technical skills and management training. This capacity building ensures that projects are sustainable and meet local needs. They conduct extensive awareness

[46] https://www.researchgate.net/publication/262764811_Enterprising_Farm_Women_in_Rajasthan_and_Gujarat (Last accessed on 30-05-25).

[47] https://www.pradan.net/our-impact/ (Last accessed on 30-05-25).

[48] https://www.sewa.org/ (Last accessed on 25-05-2025).

[49] https://world-habitat.org/world-habitat-awards/winners-and-finalists/restoration-of-dignity-and-human-rights-of-indigenous-tribal-community-in-karnataka/ (Last accessed on 25-05-2025).

[50] https://www.gramvikas.org/project/odisha-tribal-empowerment-livelihoods-programme/#:~:text=The%20programme%20aims%20to%20improve,livelihood%20support%20and%20food%20security (Last accessed on 25-05-2025).

campaigns on health, hygiene, and the importance of education, leading to behavioral changes and improved well-being among tribal women and their families.

5 Inter-relationships: How Policies and Institutions Converge (or Diverge)

The inter-relationships between policies and institutions regarding indigenous women in India reveal a complex interplay of support and conflict, particularly in the context of land rights, health, and cultural identity. These dynamics often converge and diverge, reflecting the multifaceted challenges faced by indigenous women. Tribal women in India historically enjoyed communal land ownership, but patriarchal structures and customary laws often undermine their rights (Singh & Bhatnagar, 2024). The Indian judiciary plays a crucial role in balancing tribal autonomy with the need for gender justice, highlighting the tension between traditional practices and modern legal frameworks (Singh & Bhatnagar, 2024). Divergence in policy effectiveness for indigenous women can across India, for example indigenous women in Kerala, despite the state's overall success in health and education, face significant gaps in outcomes (Thomas et al., 2021). The lack of disaggregated data on health outcomes for tribal women further complicates the assessment of their needs and the effectiveness of existing policies (Thomas et al., 2021). Indigenous women face compounded discrimination based on gender, ethnicity, and socioeconomic status, necessitating a holistic approach to policy-making that addresses these intersecting identities (Celorio, 2022).

5.1 Policy-Institution Linkages

The translation of constitutional mandates into legislative acts for indigenous women in India involves a multi-layered process that encompasses legal frameworks, political representation, and targeted government schemes. This process is crucial for addressing the historical marginalization of indigenous women and ensuring their rights and empowerment. The Indian Constitution, particularly through Articles 15(3) and the 73rd and 74th Amendments, mandates affirmative action for women, including indigenous populations. These provisions allow for special measures to increase women's involvement in decision-making and governance (Nayar et al., 2023; Raman, 2002). To safeguard women's rights and advance gender equality, laws like the Protection of Women from Domestic Violence Act (2005) and the Sexual Harassment of Women at Workplace Act (2013) have been passed. These laws are designed to address specific issues faced by women, including those from indigenous communities (Saxena, 2024). Government institutions are tasked with implementing these laws through various schemes aimed at empowering indigenous women. For instance, initiatives in Kerala have shown success in improving health and education outcomes for tribal women, in line with the United Nations Sustainable Development Goals (Thomas et al., 2021). Although there has been a lot of progress, there are still obstacles in the way of successfully converting these directives into real advantages for indigenous women, highlighting the need for ongoing reforms and societal change to ensure comprehensive empowerment.

5.2 Synergies and Success Stories

Following are some examples where strong policy provisions, coupled with effective institutional support have positively impacted indigenous women:

A. **Forest Rights Act (FRA) and Women's Land Titles**

   a. **Podochuanpadar village, Odisha**[51]: A case study highlights the successful implementation of FRA in this village. The Gram Sabha played a crucial role in granting land titles to 25 households for Individual Forest Rights (IFR) and 200 acres to the community under Community Forest Rights (CFR). This secure land tenure empowers women, enhancing their decision-making power and financial security.
   b. **Southern Rajasthan:** Research indicates that FRA implementation, while facing challenges, has improved the social standing of tribal women (Zaidi, 2019). Women who received land titles reported increased confidence and a stronger ability to assert their rights.
   c. **Narmada district, Gujarat**: Following FRA enactment, an organization helped tribal communities claim tenure security (Haque, 2020). A significant number of IFR claims were approved, often with joint titles in both the husband and wife's names. This joint ownership is a crucial step towards gender equality in land rights.
   d. **Jharkhand**: NGOs have worked to increase awareness of FRA and women's rights,[52] which has led to increased assertion of rights by women in these communities.

B. **PESA and Women's Empowerment in Gram Sabhas**: PESA empowers Gram Sabhas, enabling women to participate in decision-making processes related to natural resource management and local development. As a result, there are now more women in leadership positions and a greater focus on addressing gender-specific issues.

   a. **Himachal Pradesh**: A case study showcases a woman from a marginalized community who, despite facing societal challenges, became a successful

[51] https://nfs.inroad.in/best-practice-detail?id=101065 (Last accessed on 30-05-2025).
[52] Landesa (2016).

Panchayat President.[53] Her leadership led to positive changes in the community.

b. **Odisha**: Elected tribal women in Odisha have played a critical role in monitoring MGNREGA works, ensuring fair wages, and helping tribal families access housing and pension schemes.[54]

5.3 Disconnects and Implementation Gaps

The policy framework in India for the empowerment of indigenous women is robust, with a number of programs designed to enhance their socioeconomic circumstances. However, institutional failures such as lack of coordination, resource scarcity, bureaucratic hurdles, and corruption significantly hinder effective implementation. These challenges are evident in various programs and policies designed for tribal communities and women, highlighting the gap between policy intent and on-ground reality. The execution of the Protection of Women from Domestic Violence Act (PWDVA) is hampered by poor coordination among agencies, inadequate funds, and ineffective monitoring mechanisms, which limits its effectiveness in safeguarding women from domestic violence (Parvati, 2015). Tribal development programs such as the Tribal Sub-Plan and Vanbandhu Kalyan Yojana suffer from administrative inefficiencies and lack of awareness, leading to uneven impact and limited success in improving socio-economic conditions (Bhatnagar, 2024a, 2024b). The Development of Women and Children in Rural Areas (DWCRA) face operational challenges due to insufficient technical training and political commitment among grassroots bureaucrats, resulting in high rates of program disuse (Sharma, 2011). Corruption is a significant barrier, with instances of misallocation of funds in programs like DWCRA, where local officials divert resources meant for beneficiaries, undermining the program's objectives (Sharma, 2011). The economic empowerment of women is further constrained by systemic issues in policy design that fail to address social and patriarchal constructs, affecting the overall effectiveness of government programs (Mitra, 2024). While the policy framework for indigenous women in India is comprehensive, these institutional failures highlight the need for improved coordination, resource allocation, and accountability mechanisms. Addressing these issues is crucial for translating policy into tangible benefits for indigenous women, ensuring their empowerment and socio-economic development.

[53] https://securehimalaya.org/pangi-hills-brand/ (Last accessed on 30-05-25).

[54] https://www.gramvikas.org/project/odisha-tribal-empowerment-livelihoods-programme/#:~:text=The%20programme%20aims%20to%20improve,livelihood%20support%20and%20food%20security (Last accessed on 25-05-2025).

6 Evaluating Efficacy and Advocating for a Just Future

In India, empowering indigenous women is essential to attaining inclusive development and gender equality. The socioeconomic difficulties that tribal women experience have been addressed by a number of policy frameworks, although their efficacy varies. Table 1 highlights the key policy frameworks, their objectives, and their impact on indigenous women in India, citing relevant contexts.

The Forest Rights Act (FRA) 2006 is a significant policy aimed at empowering tribal communities, including women, by granting them land rights. The Act recognizes the importance of land as a resource for subsistence, social status, and identity in tribal communities. Studies have shown that women's land rights under the FRA have boosted productivity and empowerment, advancing gender equality (Raj & Anitha, 2024). However, challenges such as limited economic gains and persistent patriarchal norms remain, restricting women's decision-making power and mobility (Raj & Anitha, 2024).

Women were granted reservations in Panchayati Raj Institutions (PRIs), which are local self-governance bodies, by the 73rd Constitutional Amendment Act of 1992. This policy has been instrumental in increasing the political participation of tribal women. The 73rd Amendment's application in Telangana has revolutionized the state by dispelling gender stereotypes and promoting an inclusive society (Indira, 2023). However, in states like Arunachal Pradesh, Tribal women continue to be underrepresented in panchayat politics, with limited representation in higher decision-making bodies (Yadav, 2017).

The collection and management of non-timber forest products (NTFPs), which are vital to their livelihoods, are mostly the responsibility of tribal women. However, they often face exploitation by middlemen due to illiteracy and lack of access

Table 1 Key policy frameworks and their impact on indigenous women in India

Policy framework	Objective	Impact
Forest Rights Act (FRA) 2006	Grant land rights to tribal communities, including women	Boosted productivity and empowerment, advancing gender equality (Raj & Anitha, 2024)
73rd Constitutional Amendment	Increase women's representation in local self-governance bodies	Transformed gender stereotypes, fostering inclusivity in Telangana (Indira, 2023)
Self-Help Groups (SHGs)	Provide platform for economic independence and leadership development	Enhanced financial independence and socio-economic development (Verma & Sharma, 2024)
Vanbandhu Kalyan Yojana	Promote tribal development through economic and social empowerment	Improved education and economic opportunities, though implementation varies (Bhatnagar, 2024a, 2024b)

Source Compiled by the author from the respective studies

to markets. Policies aimed at enhancing their skills and providing better market access are necessary to improve their economic empowerment (Jagati, 2024). Tribal women now have a way to become financially independent and improve their leadership abilities through self-help groups (SHGs), which helps them advance socioeconomically (Verma & Sharma, 2024).

Tribal women still face several obstacles in spite of different governmental initiatives, such as limited access to healthcare and education and gender-based violence. The National Family Health Survey (2019) revealed higher rates of violence against tribal women compared to other groups. To protect their safety and wellbeing, policies that combat gender-based violence and advance gender equality are crucial (Nayak & Alam, 2023).

In Assam, government-initiated development programs have improved the socio-economic conditions of tribal women, but regional variations persist. The Kamrup district's Dimoria Development Block emphasizes how these initiatives affect women's empowerment and the necessity of ongoing assessment and development (Devee, 2023). On the other hand, Dhalbhumgarh block's tribal women's involvement in local administration shows how democratic institutions contribute to their empowerment (Paul & Samal, 2023).

The implementation of policies aimed at empowering tribal women often faces challenges such as administrative inefficiencies, lack of awareness, and inadequate implementation. Strengthening policy execution, enhancing administrative efficiency, and increasing awareness and participation are crucial for improving the socio-economic conditions of tribal women (Bhatnagar, 2024a, 2024b). In order to guarantee justice and equality for indigenous women, it is also crucial to address the interaction of gender, customary rules, and land rights (Singh & Bhatnagar, 2024).

7 Conclusion

Indigenous women or Adivasi women have their unique cultural identities and historical connections to pre-colonial communities. Other than being the economic providers, they also act as the cultural stewards of their communities. However, the junction of their gender and indigenous identity puts them into double marginalisation. The participation of these women in decision making is inhibited by low literacy rates, ill health, restricted access to resources, and patriarchal systems. Indigenous women exhibit poor nutritional outcomes because of inadequate intake of vital minerals like iron, calcium, and vitamin A, as well as a high prevalence of chronic energy insufficiency. Indigenous women have faced continued discrimination, systemic violence, including sexual violence, since the colonial times, because of societal assumptions that make them "disposable." Their abuse and marginalization are exacerbated by the criminal justice system's frequent failure to provide them with protection.

Despite the above disadvantages, indigenous women exhibit strong and enduring ties to the land, forests, and other natural resources, playing a vital role in the preservation of biodiversity. However, social and political changes have disenfranchised them, and they have been isolated from their ancestral lands and methods of survival, leading to environmental degradation and biodiversity loss. Since property rights have evolved, they are much more disadvantaged.

In India, empowering indigenous women necessitates a multifaceted strategy that takes social justice, economic empowerment, political engagement, and land rights into account. Several policy frameworks have achieved great strides, yet there are still issues. Long-term improvements for India's indigenous populations require consistent policy reform, inclusive development plans, and strong monitoring and assessment systems.

References

3ie. (2021). Swashakt evidence programme. Enabling women's economic empowerment through collective enterprises in rural India. https://www.3ieimpact.org/sites/default/files/2021-11/Swashakt-PO-brief.pdf

Acharya, P., & Acharya, P. (2020). An analysis of the Scheduled Castes and Scheduled Tribes (Prevention of Atrocities) Act, 1989. *SSRN Electronic Journal*. https://doi.org/10.2139/ssrn.3732709

Agarwal, B. (1994). Gender and command over property: A critical gap in economic analysis and policy in South Asia. *World Development, 22*(10), 1455–1478. https://doi.org/10.1016/0305-750X(94)90031-0

Agrawal, P. K., & Agrawal, S. (2010). To what extent are the indigenous women of Jharkhand, India living in disadvantageous conditions: Findings from India's National Family Health Survey[1]. *Asian Ethnicity, 11*(1), 61–80. https://doi.org/10.1080/14631360903506778

Akkiraju, V. S. (2022). A comparative study of maternal and child health indicators of tribal and non-tribal areas of selected states in India. *International Journal of Health Sciences and Research, 12*(12), 17–24. https://doi.org/10.52403/ijhsr.20221203

Bala, M. (2006). Indian National Policy on resettlement and rehabilitation and the marginalization of women. *SSRN Electronic Journal*. https://doi.org/10.2139/ssrn.942913

Behera, D. K., & Padhi, R. K. (2022). Mining, displacement and tribal livelihoods in India. https://doi.org/10.1177/0972558X221096265

Besra, S., & Sarangi, P. (2024). Tribal women in India: A framework of empowerment. *International Journal of Science and Research (IJSR), 13*(2), 1766–1773. https://doi.org/10.21275/SR24228124813

Bhatnagar, A. (2024a). Political frameworks for tribal empowerment in India: Successes, challenges, and recommendations. *ShodhKosh: Journal of Visual and Performing Arts, 5*(5). https://doi.org/10.29121/shodhkosh.v5.i5.2024.1612

Bhatnagar, S. (2024b). Patriarchy and the Indian Society. *SSRN Electronic Journal*. https://doi.org/10.2139/ssrn.4806415

Bisht, T. C. (2009). Development-induced displacement and women: The case of the Tehri Dam, India. *The Asia Pacific Journal of Anthropology, 10*(4), 301–317. https://doi.org/10.1080/14442210903271312

Bit, J., & Banerjee, S. (2013). Livelihood dependence, traditional knowledge conservation and sustainable forest management—A study on Birbhum Forest Division, West Bengal. *Arthaniti:*

Journal of Economic Theory and Practice, 12(1–2), 1–27. https://doi.org/10.1177/0976747920130101
Borah, P. P., & Bhuyan, A. J. (2023). Local self-governance in sixth scheduled areas of Assam: Constitutional and political challenges. *Indian Journal of Public Administration, 69*(3), 692–697. https://doi.org/10.1177/00195561231166361
Brodsky, G. (2016). Indian Act Sex discrimination: Enough inquiry already, just fix it. *Canadian Journal of Women and the Law, 28*(2), 314–320. https://doi.org/10.3138/cjwl.28.2.314
Celorio, R. (2022). *Women and international human rights in modern times: A contemporary casebook*. Edward Elgar Publishing. https://doi.org/10.4337/9781800889392
Challa, K., Sharma, S. S., Singh Dari, S. S., & Khubalkar, D. (2023). Recognizing role of indigenous communities in biodiversity conservation in India: An analysis. *Russian Law Journal, 11*(1S). https://doi.org/10.52783/rlj.v11i1S.366
Choudhary, N., Premkumar, L., Singh, C., Mondal, S., Priya, S., Sudarshan, B., Perumal Samy, P., & Mohan, S. (2023, July 3). Bharatavani Project—Reviving linguistic diversity and cultural heritage in India: A case study. In *Proceedings of the International Workshop on Digital Language Archives: LangArc-2023*. International Workshop on Digital Language Archives: LangArc-2023. https://doi.org/10.12794/langarc2114300
Das, B., & Hossain, S. M. D. (2023). Ecofeminist concerns and subaltern perspectives on 'Third World' indigenous women: A study of selected works of Mahasweta Devi. *Journal of International Women's Studies, 25*(2), 2. https://vc.bridgew.edu/jiws/vol25/iss2/2
Das, S. K. (2013). India's rights revolution: Has it worked for the poor? *Oxford University Press*. https://doi.org/10.1093/acprof:oso/9780198081661.001.0001
Das, U., Kar, N., Chaplot, B., & Kumar, P. (2023). Child and maternal health vulnerability among the indigenous population in India: Based on cross-sectional data NFHS 2015-16. *Cogent Public Health, 10*(1), 2253573. https://doi.org/10.1080/27707571.2023.2253573
Devee, D. (2023). Government initiated development programmes and its impact on the empowerment of tribal women: A case study of Dimoria Development Block, Kamrup (M), Assam. *International Journal For Multidisciplinary Research, 5*(6), 9172. https://doi.org/10.36948/ijfmr.2023.v05i06.9172
Dey, T., & Maiti, J. (2024). A study of the status of tribal women of Purulia district of West Bengal reference to education. *International Journal of Multidisciplinary Trends, 6*(2), 15–18. https://doi.org/10.22271/multi.2024.v6.i2a.374
Dodgson, J. E., & Struthers, R. (2005). Indigenous women's voices: Marginalization and health. *Journal of Transcultural Nursing, 16*(4), 339–346. https://doi.org/10.1177/1043659605278942
Drever, M. (2022). Gendered disposability of indigenous women across time and space. *INvoke, 7*. https://doi.org/10.29173/invoke49009
Dubochet, L. (2013). When women farm India's Land: How to increase ownership? https://policy-practice.oxfam.org.uk/publications/when-women-farm-indias-land-how-to-increase-ownership-304918
Dudley, A. M. (1999). Indigenous forest use practices and sustainability: A case study of the Adivasis of the Nilgiri Biosphere region, South India. https://library2.smu.ca/handle/01/22334
Gochhayat, S. A. (2011). Project on FRA ACT 2006: The scheduled tribes and other traditional forest dwellers (Recognition of Forest Rights) Act 2006. *SSRN Electronic Journal*. https://doi.org/10.2139/ssrn.1957149
Hansepi, L., & Laisram, R. (2022). Karbi women and environmental conservation. *Grassroots Journal of Natural Resources, 5*(3), 1–17. https://doi.org/10.33002/nr2581.6853.050301
Haque, T. (2020). Securing forest rights and livelihoods of tribals. Challenges and Way Forward, SRSC Policy Paper Series-2/2020, S. R. Sankaran Chair (Rural Labour) National Institute of Rural Development and Panchayati Raj (Ministry of Rural Development, Government of India) Rajendranagar, Hyderabad. www.nirdpr.org.in

Hornbeck, A., Johnson, B., LaGrotta, M., & Sellman, K. (2007). The protection of women form domestic violence Act: Solution or mere paper tiger? *The Loyola University Chicago International Law Review, 4*(2), 273. https://lawecommons.luc.edu/cgi/viewcontent.cgi?article=1081&context=lucilr

Indira, B. (2023). Empowerment of tribal women through the 73rd amendment: A study of Telangana. *International Journal of Creative Research Thoughts, 11*(7), 448.

Jagati, P. (2024). Issues and challenges with empowerment of tribal women in India: A forest-based study. *Asia Minor Studies, 12*(2), 213–220. https://doi.org/10.17067/asm.1445005

Jain, C., Saxena, D., Sen, S., & Sanan, D. (2023). Women's land ownership in India: Evidence from digital land records. *Land Use Policy, 133*, Article 106835. https://doi.org/10.1016/j.landusepol.2023.106835

Jhala, S. (2013). Displaced by development: Confronting marginalisation and gender injustice. *Madhya Pradesh Journal of Social Sciences, 18*(2), 122. https://www.questia.com/library/journal/1G1-412685679/displaced-by-development-confronting-marginalisation

Kalyani, V. P. (2013). Protection of women from Domestic Violence Act, 2005: A critical appraisal. *Social Science Research Network.* https://papers.ssrn.com/sol3/papers.cfm?abstract_id=2342113

Kapoor, R., Sabharwal, M., & Ghosh-Jerath, S. (2022). Diet quality, nutritional adequacy and anthropometric status among indigenous women of reproductive age group (15–49 years) in India: A narrative review. *Dietetics, 2*(1), 1–22. https://doi.org/10.3390/dietetics2010001

Karanjawala, T., & Chugh, S. (2009). The legal battle against domestic violence in India: Evolution and analysis. *International Journal of Law, Policy and the Family, 23*(3), 289–308. https://doi.org/10.1093/lawfam/ebp006

Khanna, S. (2024). Forest rights and the forest rights act. In P. Cullet, L. Bhullar, & S. Koonan (Eds.), *The Oxford Handbook of environmental and natural resources law in India* (1st ed., pp. 487–506). Oxford University Press. https://doi.org/10.1093/oxfordhb/9780198884682.013.28

Kindo, D. P., & Bhowmick, P. K. (2024). Assessing the compliance of the Local Self-Governance Act for PESA in Jharkhand: A policy review after two-and-a-half decades of its inception. *Journal of Rural Development*, 40–59. https://doi.org/10.25175/jrd/2023/v43/i1/173020

Kjosavik, D. J., & Shanmugaratnam, N. (2007). Property rights dynamics and indigenous communities in highland Kerala, South India: An institutional–historical perspective. *Modern Asian Studies, 41*(6), 1183–1260. https://doi.org/10.1017/S0026749X06002617

Krishna, S. (2001). Gender, tribe and community control of natural resources in north-east India. *Indian Journal of Gender Studies, 8*(2), 307–321. https://doi.org/10.1177/097152150100800210

Kujur, J. (2023). Development-induced displacement and tribal livelihoods in India. *Journal of Development Studies.* https://doi.org/10.1177/09731741221129352

Kumar Palita, S., Panda, D., & Kumar Nayak, J. (2023). Indigenous communities and biodiversity conservation: An Indian perspective. *Science and Culture, 89*(September–October). https://doi.org/10.36094/sc.v89.2023.Indigenous_Communities_and_Biodiversity_Conservation.Palita.347

Lalremruati, P. (2024). Conceptual analysis of the protection of women from Domestic Violence Act, 2005. *International Journal For Multidisciplinary Research, 6*(1), 13571. https://doi.org/10.36948/ijfmr.2024.v06i01.13571

Landesa. (2016). *India: Gender in a forest tenure context.* Landesa Rural Development Institute.

Licart, T. (2023). Women's land ownership and patrilocality in India. In C. Z. Guilmoto (Ed.), *Atlas of gender and health inequalities in India* (Vol. 16, pp. 75–87). Springer. https://doi.org/10.1007/978-3-031-47847-5_7

Madankar, M., Kakade, N., Basa, L., & Sabri, B. (2024). Exploring maternal and child health among tribal communities in India: A life course perspective. *Global Journal of Health Science, 16*(2), 31. https://doi.org/10.5539/gjhs.v16n2p31

Manna, A. (2024). Empowerment and gender equality among tribal women. *Integrated Journal for Research in Arts and Humanities, 4*(1), 11–17. https://doi.org/10.55544/ijrah.4.1.2

Mitra, A., & Rao, N. (2009). Displacing gender from displacement: A view from the Santal Parganas, Jharkhand. In *Displaced by development: Confronting marginalisation and gender injustice* (pp. 34–58). SAGE.

Mitra, S. (2024). Policy framework for women's economic empowerment: What is Amiss? In I. Hirway (Ed.), *India social development report 2023* (1st ed., pp. 79–92). Oxford University Press. https://doi.org/10.1093/oso/9780198885979.003.0005

Naik, R. B. (2016). Social, economic and educational status of tribal women in India: Some issues. *International Journal of Scientific Research, 5*(7). https://www.worldwidejournals.com/international-journal-of-scientific-research-(IJSR)/recent_issues_pdf/2016/July/July_2016_1468998269__176.pdf

Nawsagaray, N. (2018). Misuse of the Prevention of Atrocities Act: Scrutinising the Mahajan judgment, 2018. *Economic & Political Weekly, 53*(22).

Nayak, K. V., & Alam, S. (2023). Change in gender relations: Re-visiting gender-based violence in tribal communities of India. In D. A. Karras, S. K. Oruganti, & S. Ray, *Interdisciplinary perspectives on sustainable development* (1st ed., pp. 257–261). CRC Press. https://doi.org/10.1201/9781003457619-52

Nayak, K. V. (2025a). Indigenous women's evaluation of anti-Indigenous violence: An intersectional analysis. *AlterNative: An International Journal of Indigenous Peoples, 21*(1), 42–53. https://doi.org/10.1177/11771801241309360

Nayak, K. B. (2025b). Gender inequality in educational status of tribals in India—An extensive review of studies. *International Journal For Multidisciplinary Research, 7*(1), 35014. https://doi.org/10.36948/ijfmr.2025.v07i01.35014

Nayar, D. S., Shah, D. V., Sharma, U., & Pandya, D. M. (2023). From tradition to transformation: Legal frameworks for women's rights in India. *Journal of Survey in Fisheries Sciences, 10*(1), 3747–3752. https://doi.org/10.53555/sfs.v10i1.1719

Nepram, B. (2017). Indigenous women of Northeast India at the forefront of a strong non-violent peace movement. In *Indigenous peoples' rights and unreported struggles: Conflict and peace.* The entire volume is available in Academic Commons at https://doi.org/10.7916/D82R5095

Nikore, M., Gupta, M., Prabhu, P., & Narang, V. (2021). India's missing working women: How COVID-19 pushed women out of formal labour markets. *International Journal of Social Science and Economic Research, 6*(4), 1335–1355. https://doi.org/10.46609/IJSSER.2021.v06i04.015

Panda, P., & Agarwal, B. (2005). Marital violence, human development and women's property status in India. *World Development, 33*(5), 823–850. https://doi.org/10.1016/j.worlddev.2005.01.009

Parsloe, S. M., & Campbell, R. C. (2021). Folks don't understand what it's like to be a native woman: Framing trauma via #MMIW. *Howard Journal of Communications, 32*(3), 197–212. https://doi.org/10.1080/10646175.2021.1871867

Parvati, P. (2015). Implementing the PWDVA: Safeguarding women from domestic violence. https://oxfamilibrary.openrepository.com/handle/10546/582546

Paul, R. K., & Samal, A. (2023). Role of democratic structures in empowerment of tribal women of Dhalbhumgarh block: A case study. *International Journal For Multidisciplinary Research, 5*(4), 5693. https://doi.org/10.36948/ijfmr.2023.v05i04.5693

Raj, N., & Anitha, V. (2024). Empowerment of tribal women in India: Impact of the Forest Rights Act 2006. *Forum for Development Studies*, 1–20. https://doi.org/10.1080/08039410.2024.2432302

Raman, V. (2002). The implementation of quotas for women: The Indian experience. https://www.quotaproject.org/CS/CS_India.pdf

Rao, N., Pradhan, M., & Roy, D. (2017). Gender justice and food security in India: A review. *Social Science Research Network.* https://papers.ssrn.com/sol3/papers.cfm?abstract_id=2908520

Rathore, K. S. (2011). Inclinations in literacy rates and schooling among the scheduled tribe women in India, Madhya Pradesh. *Journal of Social Sciences, 16*(1), 65–80.

Rawat, A. S. (1995). Deforestation and forest policy in the lesser Himalayan Kumaun: Impacts on peasant women and tribal populations. *Mountain Research and Development, 15*(4), 311. https://doi.org/10.2307/3673807

Sarin, M., & Gopalakrishnan, S. (2022). Gender issues, including gender-based violence, among Scheduled Tribes 1. In *Tribal Development Report* (pp. 11–55). Routledge.

Saxena, P. (2024). Gender justice and legal reforms: A review of women's rights legislation in India. *Indian Journal of Law*, *2*(5), 1–6. https://doi.org/10.36676/ijl.v2.i5.50

Saxena, K. B. (2018). Legislature proposes, judiciary disposes: Supreme court's ruling on the Atrocities Act. *Social Change, 48*(2), 275–282. https://doi.org/10.1177/0049085718775369

Shah, A. Q., Dar, B. A., & Bhasir, S. (2019). Tribal women in India and social inclusion: A study of constitutional safeguards. *Journal of Emerging Technologies and Innovative Research (JETIR), 6*(1).

Shangpliang, R. M. (2012). Forest legislations and livelihood strategies: Khasi women in rural Meghalaya. *Sociological Bulletin, 61*(3), 479–493. https://doi.org/10.1177/0038022920120306

Sharma, S. (2011). "Empowerment Was Never Conceptualized as Entitlement": Problems in Operationalizing a "Feminist" Program. In S. Sharma, *"Neoliberalization" as Betrayal* (pp. 147–180). Palgrave Macmillan US. https://doi.org/10.1057/9780230119208_6

Singh, M. A. (2025). A critical study on the protection of women from Domestic Violence Act, 2005. *International Journal For Multidisciplinary Research*, *7*(1). https://doi.org/10.36948/ijfmr.2025.v07i01.34830

Singh, J., & Bhatnagar, K. (2024). Perspectives on the intersection of gender, customary laws and land rights in India. *International Journal on Minority and Group Rights, 32*(2), 195–214. https://doi.org/10.1163/15718115-bja10164

Sircar, A. (2016). Women's right to agricultural land: Removing legal barriers for achieving gender equality. https://oxfamilibrary.openrepository.com/handle/10546/618600

Sisodia, Y. S., & Dalapati, T. K. (2018). Decentralised governance and implementation of PESA in tribal areas: Evidences from the western tribal belt of India. In V. Srinivasa Rao (Ed.), *Adivasi rights and exclusion in India* (pp. 332–347). Routledge.

Smith, A., & Ross, L. (2004). Introduction: Native women and state violence. *Social Justice, 31*(4.98), 1–7. http://www.socialjusticejournal.org/archive/98_31_4/98_01Intro.pdf

Thapar-Björkert, S., Maiorano, D., & Blomkvist, H. (2019). Empowerment mechanisms—Employment guarantee, women and Dalits in India. *Contemporary South Asia, 27*(4), 486–501. https://doi.org/10.1080/09584935.2019.1686750

Thomas, S. T., Thomas, E. T., McLean, M., & Titus, T. T. (2021). Paving the way to achieving the United Nations Sustainable Development Goals for women from Indigenous communities: Lessons from Attappady, India. *Discover Sustainability, 2*(1), 4. https://doi.org/10.1007/s43621-021-00009-y

Toppo, A. O. M., Kachhap, S. S., & Minj, N. S. (2024). Locating Indian Adivasi women's struggles in relation to global Indigenous feminisms. *International Feminist Journal of Politics, 26*(5), 1001–1024. https://doi.org/10.1080/14616742.2024.2428789

Uma, S. (2023). Between the Devil and the Deep Sea: Tribal women's inheritance rights in India. *Indian Journal of Gender Studies, 30*(3), 309–329. https://doi.org/10.1177/09715215231183623

Veena, R. S. (2024). Unveiling the gap: Exploring gender inequality in economic empowerment across India. *Journal of Humanities and Education Development*, *6*(2), 65–70. https://doi.org/10.22161/jhed.6.2.9

Verma, P., & Sharma, V. K. (2024). From constrain to capability: Factors driving tribal women empowerment through SHGs. *International Journal For Multidisciplinary Research, 6*(4), 24606. https://doi.org/10.36948/ijfmr.2024.v06i04.24606

Wahi, N., & Bhatia, A. (2018). The legal regime and political economy of land rights of scheduled tribes in scheduled areas of India. Available at SSRN 3759219.

Wani, M. A., & Majid, T. (2024). Educational status of tribals of Jammu and Kashmir: A case study of Women (Gujjars) in District Budgam. *International Journal For Multidisciplinary Research, 6*(3), 23484. https://doi.org/10.36948/ijfmr.2024.v06i03.23484

Xaxa, A. (2021). Transformation of tribal hinterlands into urban spaces in India's Fifth and Sixth schedule areas: Emerging issues and contestations. *Social Change, 51*(3), 362–378. https://doi.org/10.1177/00490857211032952

Yadav, A. K. (2017). Tribal women experiencing Panchayati Raj Institution in India with special reference to Arunachal Pradesh. *IOSR Journal of Humanities and Social Science, 22*(01), 46–50. https://doi.org/10.9790/0837-2201024650

Yanthan, E. C. (2018). Gendered practice of Naga customary laws: A critical analysis. *Journal of Education, Society and Behavioural Science, 36*(8), 36–44. https://doi.org/10.9734/jesbs/2023/v36i81246

Zaidi, M. (2022). Work and women's economic empowerment in tribal Rajasthan, India. In E. Samantroy & S. Nandi, *Gender, unpaid work and care in India* (1st ed., pp. 147–164). Routledge. https://doi.org/10.4324/9781003276739-12

Zaidi, M. (2019). Tribal women's empowerment through the Forest Rights Act, 2006 in southern Rajasthan. *Asian Journal of Women's Studies, 25*(1), 47–75. https://doi.org/10.1080/12259276.2019.1565637

Women and the Wild: Understanding the Gender Dynamics of Traditional Ecological Knowledge Among Indigenous Communities of India

Chandranshu Tiwari, Kanhaiya Shah, and Anupam Sunny

Abstract Traditional ecological knowledge is a repository of collective wisdom accumulated through generations of experience. The treasure trove of cultural knowledge exists in almost symbiotic relationship with the surrounding biodiversity, passed down through collective memories and socio-cultural norms. Due to their role in sustaining livelihoods, women in traditional communities often have a more direct relationship with the environment. This close relationship puts women in a unique position where they simultaneously have the highest stake in preventing environmental degradation but minimal authority to exercise against it. Gender inequalities and social exclusions remain a persistent theme of conservation research, particularly in the Global South. Power dynamics and gender perceptions in traditional societies are a product of socio-cultural contexts, where customary laws dictate women's status and authority in decision-making. While applauded for their significance in conservation, indigenous institutions often fail to include the voices of women and other socially marginalized groups. This review examines the literature to ask if a more gender inclusive governance system would improve biodiversity conservation efforts. What factors influence the supposed "relationship" between women's socioeconomic background and attitudes towards the natural environment? What factors enhance or inhibit women's participation in participatory conservation?

Keywords Collaborative governance · Community-based conservation · Gender norms · Indigenous beliefs and taboos · Intersectionality · Participatory management · Traditional knowledge system

---

C. Tiwari
Department of Environmental Studies, Shyama Prasad Mukherji College for Women, University of Delhi, Delhi, India

K. Shah · A. Sunny (✉)
Department of Environmental Studies, Satyawati College, University of Delhi, New Delhi, India
e-mail: anupamsunny@satyawati.du.ac.in

N. Moran et al. (eds.), *The Participatory Role of Indigenous Women*,
https://doi.org/10.1007/978-3-032-24059-0_10

1 Introduction

Research in recent decades has raised much-needed attention to the repositories of environmental knowledge in the Global South. Traditional Ecological Knowledge (TEK) of indigenous communities has emerged as an increasingly crucial tool for sustainability managers to manage resource bases and reduce adverse impacts of environmental crises. In India alone, there are approximately 84 million indigenous people, representing 461 distinct ethnic groups (Singh et al., 2022). The deep-rooted relationship between these communities and their surrounding bioresources, shaped by generations of traditional knowledge, highlights the enduring presence of bio-folklore within these cultures.

Resource-dependent communities often possess intimate knowledge of their local biodiversity; however, this knowledge tends to vary significantly across gender and age groups. Numerous studies have highlighted the differential impacts of environmental change on men and women, particularly in mountain regions, where women are disproportionately affected. This disparity arises from their socially assigned roles, responsibilities, and limited access to resources and decision-making power (Singh et al., 2013, 2020; Tripathi et al., 2020). Women have long been custodians of biodiversity through their everyday practices building homes, cultivating crops, breeding livestock, gathering food and medicine, and nurturing their communities (Agarwal, 1992; Chhetri et al., 2024; Das, 2024; Munshi, 2001; Si et al., 2024; Singh & Teron, 2017; Venkatasubramanian & Ramnarain, 2018). Through these interactions, they have not only conserved biodiversity but have also passed down intricate cultural and environmental knowledge of sustainable resource management across generations, thereby contributing to local economy and livelihood security of indigenous communities (Fig. 1).

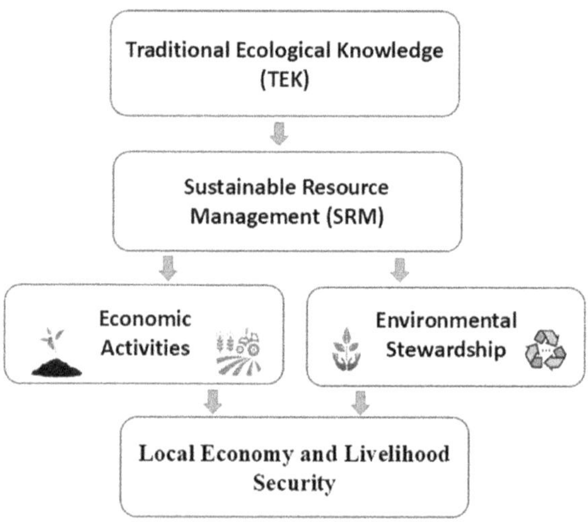

Fig. 1 Illustration depicting the role of TEK (Traditional Ecological Knowledge) in achieving the local economy and livelihood security (illustrative figure by author)

The Convention on Biological Diversity (CBD), a key international agreement promoting biodiversity conservation and sustainable use, explicitly acknowledges "*the vital role that women play in the conservation and sustainable use of biodiversity*" and underscores "*the need for the full involvement of women at all levels of policy-making and implementation for biodiversity conservation*" (UNEP, 1993: Preamble) (also see Tripathi et al., 2020).

The increasing frequency and severity of natural disasters further compound the challenges faced by women. Natural disasters lead to the loss of assets, income, and social support systems, threatening women's livelihoods and overall well-being. Gender plays a critical role in determining access to and control over resources, participation in decision-making, and opportunities for empowerment (Nosheen et al., 2024). Moreover, TEK among indigenous communities in India is inherently gendered. Women are not just passive transmitters of knowledge (or TEK) but acts as active agents shaping ecological resilience and consequently, community livelihood. Thus, recognizing their role is essential to any conversation on sustainability programs, policy-making, and indigenous knowledge governance.

With this background, the current review examines the gender dynamics of traditional ecological knowledge among indigenous communities in India. We begin by explaining TEK and its importance in preserving biodiversity, and we explain the differential impacts on genders in increasingly degraded environmental conditions. We explore the various ways in which indigenous institutions and communities exclude socially marginalized groups, including women. We conclude the review by arguing the importance of more inclusive governance models and how they might improve the climate and social resilience of indigenous societies.

2 The Nature of Traditional Ecological Knowledge

Going from the perspective of research and support to "poor" and "developing" countries of the Global South, communities in Global South are now at the forefront of finding sustainable solutions for the environmental problems of twenty-first century owing to their insights and conventional ways of resource utilization and management built over the years (Priyadarshini & Abhilash, 2019; Sunder, 2007). This "cultural environmentalism" does not come without its baggage. Despite the underlying issues, the rise of awareness regarding traditional ecological knowledge (TEK) has brought the much needed attention towards the environmental resilience of traditional communities in the developing countries with their long history of sustainable existence and its potential for managing global environmental issues (Sunder, 2007; Kelkar, 2007; Pilgrim et al., 2007; Rajamohan, 2007; Rao & Ramana 2007; Chhetri et al., 2024).

Traditional Ecological Knowledge (TEK) represents a culmination of experience gathered by a community over generations of trials and errors. TEK incorporates not just "a library of information" to rely on for everyday problems but also the religious and ethical worldviews as experienced by that particular group in that region (Mukhopadhyay, 2008; Rajamohan, 2007; Singh et al., 2013, 2020, 2022; Tiwari & Sunny, 2024). TEK serves as a connecting link between generations, enhancing community resilience and promoting sustainable growth in regions increasingly ravaged by climate change and habitat degradation (Tiwari & Sunny, 2024). A combination of local knowledge, social institutions, and prolonged cultural internalization, TEK exhibits an intimate connection to the biodiversity of any region (Rajamohan, 2007; Rao & Ramana 2007; Singh et al., 2013, 2020; Tynsong et al., 2020; Singh et al., 2022; Chhetri et al., 2024). As a collective of knowledge of practices regarding food, hunting, medicine, agriculture, and other similar activities essential for sustaining the local population, TEK provides essential knowledge for careful use and management of the regional biodiversity for sustained usage, often over generations at a time (Pilgrim et al., 2007) (also see Table 1).

As the world becomes increasingly globalized and younger generations replace older ones, the invaluable knowledge once possessed by our ancestors is slowly fading, a significant loss for humanity. However, traditional practices for utilizing and conserving natural resources have not disappeared entirely. Many ethnic communities across the globe continue to uphold these traditions, serving as living repositories of indigenous knowledge. Researchers emphasize that the resilience of social-ecological systems is essential for ensuring the sustainability of natural resources, ecosystem services, and, ultimately, a stable environment for human well-being. Today, global environmental change is occurring at an unprecedented rate, driven largely by current patterns of production and consumption and the resulting climate change (Aswani et al., 2018; Priyadarshini & Abhilash, 2019; Das et al., 2022). Furthermore, sociopolitical and economic processes operating at various scales, such

Table 1 A list of various types/components of TEK (Traditional Ecological Knowledge)

Type	Focus area	Relevant examples
Factual/empirical	Observation of nature	Weather patterns, migration timing of local wildlife, knowledge about soil types and water resources, etc.
Practice-based	Techniques and tools	Fire management, planting strategies, Water conservation, irrigation practices, etc.
Beliefs and Worldviews	Spiritual and moral systems	Sacred species, stewardship of the land, rituals tied to ecological sustainability, etc.
Ethnobiological	Species knowledge and classification	Indigenous taxonomy system of flora and fauna, medicinal plants, animal behaviour, etc.
Resource management systems	Community rules and governance	Hunting restrictions, land-use zoning, conflict resolution mechanisms, etc.
Cultural knowledge transmission	Knowledge sharing methods	Storytelling, rituals, mentorship, intergenerational teaching, etc.

as intensified logging, agribusiness expansion, and large-scale land acquisitions, are diminishing the capacity of rural communities to access and sustainably manage both formally owned and informally used resources (Ruiz-Mallén & Corbera, 2013; Tiwari & Sunny, 2024).

In many parts of the world, climate change is also altering ecosystems and species distributions by affecting rainfall and temperature patterns, thereby challenging the foundations of community-based conservation efforts (Singh et al., 2022; Tiwari & Sunny, 2024). In developing and least-developed countries, Traditional Knowledge remains crucial for food security and public health, often providing the only accessible and affordable healthcare to marginalized populations. Unfortunately, such knowledge, practices, and innovations are rapidly disintegrating for two key reasons. First, there is a declining dependence on endemic medicinal plants and animals, as modern alternatives have supplanted traditional uses.

Second, local communities are increasingly losing control over their natural resource base to the State and corporations (Jasmine et al., 2016; Pilgrim et al., 2007; Singh et al., 2010). Acknowledging the importance of traditional knowledge, of indigenous communities, the Convention on Biological Diversity (CBD) adopted the Strategic Plan for Biodiversity 2011–2020. This includes the 20 Aichi Biodiversity Targets, one of which is dedicated specifically to the preservation and promotion of Traditional Knowledge (Jasmine et al., 2016).

3 Gendered Relationships with the Environment

Tiger in the village

In 2005, the Sariska Tiger Reserve in Rajasthan became a symbol of failure in India's tiger conservation efforts. Despite substantial financial investment by the Indian government to protect the endangered Bengal tiger (*Panthera tigris tigris*), the entire tiger population in Sariska was wiped out by that year. Responding swiftly, the government initiated the world's first tiger reintroduction program in 2008 and 2009, bringing tigers back to the reserve. This crisis prompted significant shifts in both national and global tiger conservation strategies. Domestically, it led to quicker assessments, better resource distribution, and stronger conservation policies. Additionally, the success of reintroducing tigers in Sariska established reintroduction as a viable conservation strategy in India. Globally, it inspired similar plans in tiger habitats across Cambodia, Central Asia, South China, and Southeast Asia.

Given that over 70 percent of India's tiger habitats are in forests shared with human populations (Doubleday, 2020), understanding the social implications of reintroduction is just as crucial as the ecological outcomes. Concerns about these social effects stem from a history of marginalization and the continuing negative impacts on communities near protected areas. While conservation initiatives can offer local benefits, such as tourism-related jobs, income, sustainable resource use, community participation, and preservation of cultural heritage, these efforts also carry significant, often overlooked, costs.

Doubleday (2020) highlighted the intersection of gender dynamics, labor, patriarchy, and large predator conservation in Sarisk National Park. Specifically, it suggests that women may experience increased inequality due to the complex interplay of patriarchal norms, gender-specific labor roles, and the presence of reintroduced tigers. The research identifies that (1) gender-based differences in human wildlife violence emerge as an unrecognized cost of women's labor in conservation zones like Sariska, and (2) this issue is less about direct human–wildlife conflict and more a reflection of deeply embedded patriarchal structures shaping human–environment interactions.

In rural India, especially among tribal communities, women play a significant role in the forestry sector (Rajamohan, 2007; Shah & Shah, 1995; Shiva, 1992; Si et al., 2024; Venkatasubramanian & Ramnarain, 2018). Women often work as wage laborers in various forestry-related activities. Their responsibilities as gatherers include collecting firewood for household needs and market sales, fodder for animals, and numerous NTFPs (Non-Timber Forest Products) such as edible items, medicinal plants, seeds, leaves, and construction materials (Shah & Shah, 1995; Sarin, 1995; Zweifel, 1997; Lahiri-Dutt, 2001; Rao & Ramana 2007; Ellena & Nongkynrih, 2017; Carrin, 2022; Dhall, 2024). A significant portion of their livelihood comes from small-scale industries that rely on NTFPs. These include making ropes from different grasses, crafting baskets, rearing tasar silk cocoons, cultivating lac, creating bamboo products, and extracting oils. Field research consistently shows that women heavily rely on NTFPs for both sustenance and income. However, it also highlights that they earn extremely low wages for their efforts (Rao & Ramana 2007).

While the status of tribal women varies across different regions, they generally enjoy relatively higher social standing compared to women in mainstream Indian society (Shah & Shah, 1995). Nevertheless, even among tribal women, two key forms of inequality persist: access to property and participation in political life. For example, among the tribes of Jharkhand in eastern India, land ownership follows male inheritance lines, and customary laws typically deny women land rights. Politically, the traditional village council (*panchayat*) remains male-dominated, comprising mostly male household heads, with women participating only under rare circumstances. Transmission of knowledge between genders remains largely limited, and women are expected to abide by more social norms and taboos compared to their male counterparts. Despite their significant contribution in preserving and strengthening the cultural fabric of their society, women in traditional communities are expected to be dependent on their male family members (Carrin, 2022; Tyagi & Das, 2020).

Unlike scientific knowledge, TEK is more heterogeneous in nature, often unique to location but sometimes also along the lines of intersectional classes (Tiwari & Sunny, 2024). Some important case studies depicting the heterogeneous nature of TEK, namely, spiritual, intergenerational, ritualistic, communal, and symbolic-ritualistic and key roles of women in TEK are given in Table 2. Moreover, due to their distinct roles in both productive work and reproductive responsibilities, men and women engage with different environments, acquire varied skill sets, and develop unique knowledge systems shaped by gender. The local use of gendered symbols, myths, customs, and beliefs further entrenches this gendered knowledge by defining what is considered acceptable for individuals to learn or know based on their gender identity (Chhetri et al., 2024). Numerous studies have highlighted how gender-specific labor patterns, unequal rights and access to resources, cultural norms around masculinity and femininity, and belief systems all play a crucial role in shaping how knowledge about natural resources is constructed (Agarwal, 2009; Doubleday, 2020; Kelkar,

2007; Nosheen et al., 2024; Sugden & Silva, 2014; Tripathi et al., 2020; Tynsong et al., 2020; Zweifel, 1997).

Across much of the world, women are responsible for both economic production and domestic duties, yet they often lack an equal claim to the household income generated from their labor (Hansepi & Laisram, 2022; Shah & Shah, 1995; Sugden & Silva, 2014). They shoulder a disproportionate share of responsibilities (childcare, food preparation, fetching water, and collecting firewood) they perform alongside agricultural work. Despite their contributions to farming, they typically do not control the income that their labor helps generate (Kelkar, 2007; Oakley & Momsen, 2005; Sugden & Silva, 2014; Tyagi & Das, 2020).

Women's close interaction with nature, as they rely on it daily for household survival, makes them particularly vulnerable when natural resources are overexploited (Tiwari & Rani, 2025). This is evident in many regions across India. In drought-prone areas, for instance, excessive depletion of water sources and deforestation have forced women to walk long distances and spend hours collecting water (Mukhopadhyay, 2008; Tiwari & Rani, 2025). Similarly, rampant deforestation has severely impacted tribal women, making it increasingly difficult for them to gather essential firewood for daily use (Tiwari & Rani, 2025; Tripathi et al., 2020). Historically, women among forest-dwelling communities, like those of the Baiga and Bhil community of Central India, Kondh community from Odisha, and Toda people of Tamilnadu had been consistently marginalized by not just the state but their community. Although recognized today for their vast inventory of traditional medicine and soil conservation strategies, Baiga women had exerted a tenuous control over their rightful social capital (Basu, 1990; Sebestianus Lakra, 2019; Tyagi & Das, 2020; Behera, 2021; Dhall, 2024; Mathew et al., 2024).

Since the 1980s, forest governance has undergone a global shift towards decentralization, with greater rights and responsibilities being transferred to communities that rely on forests for their livelihoods (Agarwal, 2009; Elias et al., 2020; Tripathi et al., 2020). This shift has led to the rise of collaborative governance models that emphasize shared authority, accountability, and benefits between state authorities and resource users, the latter being increasingly involved in decision-making processes (Elias et al., 2020; Sarker, 2017; Sawakar, 2017). These models aim to make natural resource management not only more efficient but also fairer and more inclusive.

However, as the potential and pitfalls of these governance systems have become more apparent, so has the issue of women's limited representation in the forest user groups responsible for co-managing forest resources. The ability of individuals or groups to impact forest management decisions is largely determined by their social identity, including gender, caste or ethnicity, age, and economic status (Agarwal, 2009; Resurrección & Elmhirst, 2008; Resurrección, 2017; Tyagi & Das, 2020). In particular, social norms and structures that shape gender roles and relationships often restrict women's meaningful involvement in local governance. As a result, even when women are officially included in user groups, their participation tends to be minimal, symbolic, and largely without real influence (Kelkar, 2007; Mathew et al., 2024; Sugden & Silva, 2014; Tiwari & Rani, 2025; Tyagi & Das, 2020).

Table 2 Some important case studies depicting the heterogeneous nature of TEK (Traditional Ecological Knowledge), key role of women, and the challenges faced by indigenous communities

Indigenous community	Key role of women in TEK	Nature of TEK	Challenges faced	Key aspects identified
Dongria Kondh, Odisha	Foraging, seed conservation, and medicinal plant knowledge	Spiritual–embodied, generational, forest-based	State-led mining, forest access control	Women as guardians of biodiversity, excluded from formal forest rights but central to daily ecological practice
Mizo, Mizoram	Seed selection, soil management, and shifting cultivation	Intergenerational–agro-ecological, seasonal	Loss of land titles, invisibility in land rights, modernization	Knowledge embedded in women's cooperative work, marginalization in state discourse
Irular, Tamil Nadu	Snake tracking, herbal medicine, and ecosystem monitoring	Ritualistic–experiential, animal behavior-based,	Environmental degradation, caste-based marginalization	TEK as survival and resistance; women's cosmological ties to land often dismissed in mainstream policy
Bhil, Madhya Pradesh	Collective harvesting, forest products, and food security	Communal–sustainable harvesting techniques	Forest department restrictions, exclusion from forest governance	Communal TEK underpinned by kinship networks; gendered exclusion from forest policies
Apatani, Arunachal Pradesh	Water management, seed storage, and ritual keeping	Symbolic-ritualistic–agro-engineering	Loss of cultural practices, tourism pressure	TEK as ritual ecology; women's roles declining due to external economic pressures

4 Conservation, Inequality, and Social Exclusion

India is marked by immense diversity, with numerous intersecting social axes shaping communities and individual identities. Among these, caste remains a deeply rooted and enduring marker of identity, both within India and, to some extent, among the Indian diaspora (Jayaram, 2012; Sharma, 2012). Despite decades of constitutional guarantees and governmental initiatives aimed at reducing caste- and gender-based exclusion, the resilience of caste hierarchies and discrimination continues to be significant (Nightingale, 2006; Sharma, 2012).

Environmental discourse in India often overlooks the role of marginalization or fails to acknowledge the distinct ecological perspectives of these communities. Recognizing these perspectives would enrich our understanding of environmental politics by foregrounding voices that challenge mainstream narratives. Tribals and women from rural communities, whose livelihoods often rely heavily on access to natural resources, have a significant stake in environmental issues. Their environmental concerns are closely linked to broader struggles for social and political recognition (Lahiri-Dutt, 2001; Westermann et al., 2005).

It is now widely recognized that social capital plays a crucial role in fostering and sustaining collective action, which is essential for achieving meaningful and long-term improvements in natural resource management (NRM) (Agarwal, 2000; Westermann et al., 2005). Women are central to managing natural resources. Their relationship with the environment is often examined through the lens of poverty, as many rural women face heightened dependence on natural resources due to economic hardship. This dynamic, commonly referred to as the "feminization of poverty," is where women, especially those heading households, are disproportionately represented among the poor and are therefore more reliant on shared or common-pool resources (Nightingale, 2006; Westermann et al., 2005; Sugden & Silva, 2014).

At the same time, the gendered division of labor within households places the responsibility of securing vital resources like water and firewood largely on women. This dual dependency on common property resources not only increases their reliance on nature but also makes them especially susceptible to the adverse impacts of environmental degradation on rural livelihoods (Krishna, 2005; Sugden & Silva, 2014; Tripathi et al., 2020; Zweifel, 1997). Although gender dynamics influence decisions and outcomes in NRM, gender considerations have often been overlooked in discussions about social capital (Nightingale, 2006; Resurrección & Elmhirst, 2008; Resurrección, 2017; Rocheleau et al., 2013). Yet, research shows that men and women tend to possess different forms and degrees of social capital, shaped by variations in their social networks, collaborative tendencies, and their approaches to conflict and conflict resolution.

The nature and legal recognition of resource tenure often mirror existing gendered power dynamics. Rights to natural resources, particularly environmental and land rights, can exist either in a ***de jure*** form (formalized through legislation or legal precedent) or a **de facto** form (established through social practice and custom). Typically, men are more likely to hold *de jure* rights, while women tend to rely on de facto access. This division significantly influences the security and authority individuals have over resource tenure, often to the disadvantage of women (Rocheleau et al., 2013).

In regions such as Africa and parts of Asia, the coexistence of statutory and customary legal systems can further entrench gender inequalities in resource access. Frequently, customary laws, especially those governing family and marriage, are applied to assess women's claims to land and resources, whereas men's claims are processed under modern statutory legal frameworks. This dual legal structure tends to skew customary norms, reinforcing male dominance over resources. Although rules governing inheritance and marriage differ across cultures and are subject to change, they continue to reflect gendered distributions of rights (Basu, 1990; Krishna, 2005; Mathew et al., 2024; Resurrección & Elmhirst, 2008; Resurrección, 2017; Rocheleau et al., 2013; Zweifel, 1997).

Full ownership, often linked to wealth and male privilege, typically grants exclusive control. In contrast, women and lower-income men are more frequently relegated to the use of land or resources owned by others. This dynamic is echoed in customary arrangements where control and allocation powers lie with senior men, who decide how resources are distributed to women and younger men (Agarwal, 1992, 1998, 2000; Shiva, 1992). In many societies, women's access is mediated through male-dominated institutions, such as lineages, clans, cooperatives, or local governance bodies, meaning that their rights are often contingent and secondary to those of men (Agarwal, 1998, 2000; Nightingale, 2006).

The dynamics between those who use, own, manage, or rely on natural resources can involve conflict, collaboration, mutual support, or parallel existence, highlighting the importance of examining these relationships through the lens of power and gender (Agarwal, 2000; Rocheleau et al., 2013). Women's economic dependence is reinforced by cultural norms that legitimize their subordination to men. These norms shape not only the nature and recognition of women's labor but also their interactions within the village community throughout different stages of life. In high-caste households, a common strategy to reconcile the contradiction between the ideal of female seclusion and the practical need for women's participation in agriculture is to redefine the boundaries of the domestic space (Agarwal, 2000; Lahiri-Dutt, 2001; Rocheleau et al., 2013).

By framing family-owned fields as part of the household domain, women can engage in agricultural work without challenging norms that discourage their involvement in public or monetary labor (Sugden & Silva, 2014; Tripathi et al., 2020). This redefinition effectively means that women's agricultural contributions, though essential, are excluded from formal recognition and social prestige. Their work remains associated with subsistence activities and the non-monetized sphere, reinforcing the perception that they are not true cultivators or land managers. Meanwhile, men maintain legal ownership of land and have easier access to agricultural inputs and services, further entrenching gender disparities in agrarian settings (Agarwal, 1998, 2000; Krishna, 2005). As a result, women are often seen as laborers rather than decision-makers, a view that reflects their perceived role as endlessly adaptable and supportive, a sentiment captured in the saying, "*women are our arms and legs.*" (Lahiri-Dutt, 2001; Rocheleau et al., 2013).

5 Indigenous Institutions and the Question of Inclusion

Historically, forest-dependent communities have managed and utilized forest resources through practices such as slash-and-burn cultivation, controlled grazing, the use of fire, and maintaining sacred groves. These methods were developed to ensure the sustainable use and availability of diverse natural resources. However, colonial and post-colonial governance increasingly curtailed such local practices, often by designating protected areas under the belief that human presence would destabilize forest ecosystems (Nightingale, 2006; Krishnakumar et al., 2012; O'Neill et al. 2017). Today, global environmental changes are occurring at an unprecedented rate, driven primarily by unsustainable production and consumption patterns that contribute to climate change. Broader sociopolitical and economic dynamics such as commercial logging, the expansion of agribusiness, and land appropriation further restrict rural communities' access to and management of both legally and traditionally used natural resources (Ruiz-Mallén & Corbera, 2013).

Indigenous communities that rely on natural resources often possess in-depth knowledge about local ecosystems and factors influencing resource quality and availability (Singh et al., 2013; Tynsong et al., 2020). Some important case studies depicting various indigenous communities of India and women's contribution in environmental sustainability through traditional ecological knowledge (TEK) are presented in Fig. 2. This shared knowledge can evolve into a rich, community-held body of ecological understanding. There is increasing advocacy for integrating traditional ecological knowledge (TEK) with scientific insights, rather than merely comparing the two. Combining TEK with scientific research and traditional experimentation in collaboration with local communities can result in culturally attuned, effective, and accepted resource management strategies (Rist et al., 2010; O'Neill et al. 2017). While TEK can address information gaps and suggest innovative management approaches, its limitations, like those of science must be acknowledged.

Community-based conservation (CBC) initiatives aim to preserve biodiversity while generating local benefits, often economic (Singh et al., 2010, 2013; Lele et al., 2010). CBC efforts range widely from self-managed community conservation areas to co-managed protected areas and reflect varying degrees of local participation and governance structures (Priyadarshini & Abhilash, 2019). In self-governed initiatives, such as sacred forests and community-managed fisheries, indigenous groups and local communities hold authority and use their social norms and practices to adapt to environmental change (Kelkar, 2007; Rist et al., 2010). In contrast, co-management models involve collaborations between government bodies, NGOs, and communities, often developed in response to the shortcomings of centralized conservation approaches (Jayaram, 2012; Lele et al., 2010; Ruiz-Mallén & Corbera, 2013).

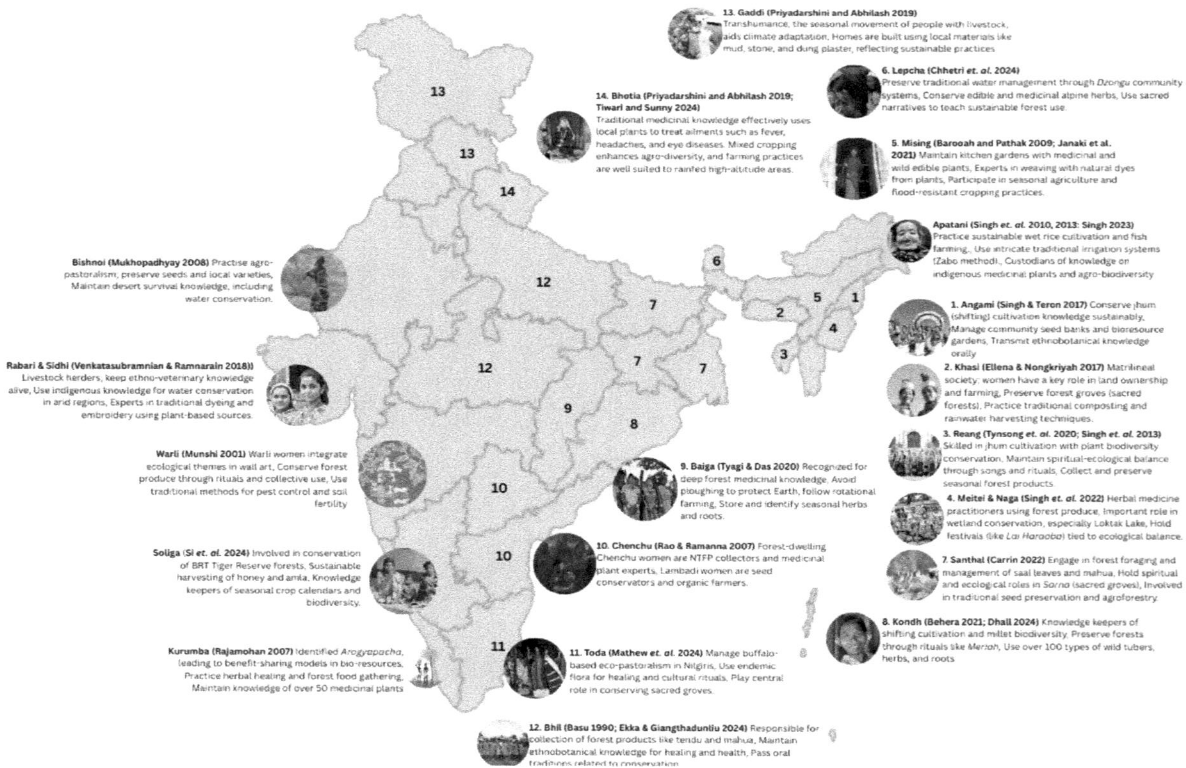

Fig. 2 Women from indigenous communities of India and their contribution in sustainability through TEK (Traditional Ecological Knowledge) (illustrative figure created by author 2025)

These collaborative models typically establish complex frameworks for shared governance and benefit from combining TEK with scientific knowledge to create specific conservation strategies (Lele et al., 2010; Ruiz-Mallén & Corbera, 2013). Moreover, integrating TEK into these arrangements often enhances resilience and adaptability. However, rigid institutional structures may impede adaptive responses to environmental shifts by being slow to learn from past experiences or resistant to innovation (Rist et al., 2010; Ruiz-Mallén & Corbera, 2013). Although formal regulations are crucial for aligning with global and national conservation goals, their implementation can lead to stakeholder conflicts and unintended harm to both communities and ecosystems. Such tensions are particularly evident in indigenous areas, where co-management frameworks have frequently contributed to the failure of integrated conservation and development projects (Lele et al., 2010; Ruiz-Mallén & Corbera, 2013).

Many communities are internally divided, with deep-rooted political conflicts playing out across various levels (Jayaram, 2012; Sharma, 2012). In some cases, even local NGOs tasked with facilitating implementation can inadvertently hinder community empowerment. One of the most significant challenges facing community-based conservation (CBC) initiatives is the fragile and often incomplete set of rights and decision-making powers granted to local communities by the state. Despite the rhetoric of participatory governance, meaningful shifts in authority over access and management of protected areas are rare, with government agencies frequently retaining control over critical decisions.

6 Towards Gender-Inclusive Conservation Governance

Women and Grassroots Conservation in Northeast India

In many mountainous regions, women play a pivotal role in ensuring food, medicinal, and nutritional security, thanks to their lifelong informal learning and place-based knowledge (Figure 2). Their close relationship with the natural environment, coupled with their active involvement in household responsibilities, equips them with a deeper understanding of local plant species, including those used for food, healing, and cultural practices, than their male counterparts. As a result, women often take the lead in preserving both biodiversity and the cultural traditions linked to it, a relationship known as biocultural diversity. This gender-based knowledge gap widens further across generations, as women of different age groups contribute in unique ways to conservation efforts. Elderly women, in particular whether from mountain areas like Arunachal Pradesh or other traditional communities possess extensive knowledge and skills critical for sustaining biodiversity and associated cultural heritage (Barooah & Pathak, 2009; Singh et al., 2013; Ellena & Nongkynrih, 2017; Sarker, 2017; Singh & Teron, 2017; Singh et al., 2020; Janaki et al., 2021; Singh, 2023; Ekka & Giangthandunliu, 2024).

The *bari* system of farming, developed over generations in Northeast India, holds significant value in terms of biodiversity conservation, household sustenance, and resource management (Barooah & Pathak, 2009; Janaki et al., 2021). A *bari* refers to a multifunctional farming unit where diverse crops, including trees, are cultivated alongside the rearing of livestock, poultry, and sometimes fish, all aimed at fulfilling the daily needs of rural families. This system is an integral part of the landscape in *Thengal-Kachari* households. Typically located adjacent to the main residence, the *bari* is accessed through a *batchor* or entrance gate that connects to the house via a narrow pathway. A traditional homestead generally includes not only living quarters for extended family members but also vegetable and fruit gardens, various trees, bamboo clusters, spaces for threshing, sheds for animals and poultry, and water bodies like ponds.

Women possess intricate knowledge regarding the growth patterns and various uses of different plants. They skillfully organize planting in limited spaces by considering each species' tolerance to conditions like waterlogging, shade, intense sunlight, and drought. These plants serve multiple functions, such as sources of food, medicine, fuel, timber, and even for sericulture and income generation. While men typically handle labor-intensive activities like digging ponds, constructing fences, preparing beds, and harvesting large trees, women are primarily responsible for the everyday maintenance of these systems. Their roles include weeding, training climbing plants, managing pests and diseases, and harvesting crops like vegetables, spices, and medicinal herbs. Additionally, women manage post-harvest processes, completing about 90% of these tasks, and serve as the main caretakers of livestock and poultry.

Elderly women, even those no longer actively involved in physical labor, continue to play an essential role by sharing their knowledge with younger generations, particularly in areas like indigenous plant use and natural pest control (Singh et al., 2013, 2020). Women's daily responsibilities create sustained interaction with natural environments. Through regular dependence on land and water for essential needs such as food, fuel-wood, medicine, and household goods, women accumulate valuable ecological knowledge. This deep understanding of biodiversity, resource management, and sustainable practices is carefully preserved and passed on within communities.

Managing common property natural resources is a complex and evolving process that involves balancing the diverse interests of various social groups within a community. Effective protection of shared resources is only possible when community livelihoods are secure and local populations feel a sense of ownership and benefit in conserving those resources. Ensuring secure usufruct rights, particularly for both men and women, is essential to gain widespread support for conservation efforts (Agarwal, 1992; Shah & Shah, 1995; Zweifel, 1997). The shift toward state control and privatization of communal lands has gradually eroded traditional community-based systems of managing and accessing natural resources. For rural women and children, especially from tribal, landless, or smallholder backgrounds, access to common property like village commons has historically served as a vital source of sustenance, offering some degree of economic independence from male-dominated land ownership systems (Agarwal, 1992).

Government-led afforestation efforts often fall short, largely due to a focus on commercial monoculture plantations that replace biodiverse forests, the appropriation of land used by locals, and the top-down nature of these interventions. As a result, such schemes have not only failed to support the rural poor but have also sometimes displaced their existing rights and access, prompting local resistance. Furthermore, women are frequently excluded from meaningful participation in these initiatives, often relegated to limited roles like tending nurseries without involvement in decision-making or species selection. On the other hand, community forestry projects frequently face challenges due to internal economic disparities and resulting distrust, especially among the poor, who fear they may not have fair access to the benefits of the trees once grown (Agarwal, 1992, 1998, 2000; Shah & Shah, 1995).

Recent research increasingly highlights that rural communities are not unified entities with a single set of concerns or priorities (Leach & Green, 1997; Rocheleau et al., 2013; Resurrección & Elmhirst, 2008; Resurrección, 2017). Individuals within these communities, across gender and other social divisions, often hold both overlapping and conflicting interests when it comes to natural resource use and management. The central issue, therefore, is not to idealize either men or women as natural stewards of the environment, but to critically examine how evolving social relations shape people's roles and influence within resource management frameworks. Any effort to build community-based conservation models must intentionally engage with and incorporate these varying perspectives into the planning process (Shah & Shah, 1995).

If women passively benefit from decisions made by predominantly male community leaders risks reinforcing existing gender hierarchies. Moreover, it may place an unequal share of the burdens and costs of conservation on women (Sarin, 1995; Shah & Shah, 1995; Westermann et al., 2005). To ensure fairness and meaningful participation, it is vital to recognize women as independent members of community institutions rather than a part of the household unit. Relying on the household as the basis for membership often results in inequitable outcomes, particularly for women and younger family members in extended or joint families. Since membership in these bodies is linked to benefit entitlements, only way to align with the constitutional principle of gender equality is to offer direct and equal membership to all adult men and women (Shah & Shah, 1995).

The exclusion of women from direct membership and active decision-making roles in community forest management negatively impacts their well-being, that of their families, the effectiveness of local institutions, and the prospects for women's empowerment. In male-dominated forest protection groups, the rules and regulations often fail to reflect women's specific needs. For example, in several villages across eastern and western India, women have been prohibited from collecting even fallen twigs, a resource they previously relied on. Consequently, they are now compelled to travel farther, spending more time and risking social embarrassment for accessing resources in neighboring areas (Agarwal, 1998, 2000; Nightingale, 2006). Although some men acknowledge the value of involving women in conservation efforts, this recognition rarely translates into action.

Excluding women from formal decision-making spaces not only entrenches gender inequalities but also weakens women's influence both within the household and in the wider community (Agarwal, 1998; Nightingale, 2006; Westermann et al., 2005). Several factors beyond the official rules contribute to women's limited participation, even when policies appear inclusive:

- **Time and labor constraints**: Women often bear heavier workloads than men, juggling domestic responsibilities with field labor. Meetings are frequently scheduled during times when they are occupied with chores, leaving little room for attending or engaging meaningfully in discussions (Krishna, 2005; Tripathi et al., 2020).

- **Gender bias in official processes**: Government Forest officers usually consult men, overlooking women's input on key issues like species selection or micro-planning. Women have expressed frustration at being second-guessed or having their views dismissed when in conflict with men's (Agarwal, 2009; Nightingale, 2006; Sarin, 1995; Shah & Shah, 1995).
- **Social and cultural barriers**: Norms of female seclusion, expectations around modesty and public behavior, and perceptions of women's inferiority all limit their ability to engage in public forums. Meetings are commonly viewed as male spaces, with men's voices assumed to represent entire households (Agarwal, 1992, 2009; Shiva, 1992).
- **Lack of a critical mass**: Women are often hesitant to attend meetings where they are underrepresented. They feel more empowered to speak up and advocate for their interests when supported by a sizable group (Agarwal, 1998, 2000).
- **Marginalization in deliberations**: Even when women do attend meetings, their contributions are frequently ignored, leading many to disengage altogether (Agarwal, 2009).

7 Conclusion

Interpretations of the past rarely remain confined to historical reflection. They often shape future visions, whether implicitly or explicitly. In India, scholars and activists have drawn on historical narratives to advocate for the revival or replication of traditional practices and institutions. However, as this review illustrates, such an approach can be fraught with challenges, particularly because traditional conceptions of family, community, and institutional frameworks were not rooted in principles of social or gender equality.

Contemporary environmental institutions, despite emerging in a more modern context, have largely failed to challenge the patriarchal norms of the past. Consequently, they risk not only perpetuating historic gender imbalances but also fostering new forms of inequality related to access to resources, institutional power, and social relations.

Neglecting a gender lens in the design and operation of these institutions undermines opportunities for truly effective collective action. This oversight ignores the potential of women as key agents of cooperation, many of whom have long histories of mutual aid and collaboration within community-based networks, especially in rural settings of the Global South.

To make environmental movements and institutions more gender-inclusive, women must engage both from within these structures and independently, building their sources of strength and advocacy. Importantly, gender integration into environmental discourse requires deliberate negotiation, it is not an automatic outcome of sustainability initiatives. Women's historically subordinate status predates the current ecological crisis, and environmental reforms alone cannot guarantee gender equity. Assuming an inherent connection between environmental protection and women's empowerment risks oversimplifying the complexity of both struggles. Lasting change will depend on parallel efforts to transform deeper structures of inequality, namely, those surrounding labor, property ownership, and political representation.

References

Agarwal, B. (1992). The gender and environment debate: Lessons from India. *Feminist Studies, 18*(1), 119–158.

Agarwal, B. (1998). Environmental management, equity and ecofeminism: Debating India's experience. *Journal of Peasant Studies, 25*(4), 55–95.

Agarwal, B. (2000). Conceptualising environmental collective action: Why gender matters. *Cambridge Journal of Economics, 24*(3), 283–310.

Agarwal, B. (2009). Gender and forest conservation: The impact of women's participation in community forest governance. *Ecological Economics, 68*(11), 2785–2799.

Aswani, S., Lemahieu, A., & Sauer, W. H. (2018). Global trends of local ecological knowledge and future implications. *PLoS ONE, 13*(4), Article e0195440.

Barooah, M., & Pathak, A. (2009). Indigenous knowledge and practices of Thengal Kachari women in sustainable management of Bari system of farming. *Indian Journal of Traditional Knowledge, 8*(1), 35–40.

Basu, A. (1990). Indigenous feminism, tribal radicalism and grass roots mobilization in India. *Dialectical Anthropology*, 193–209.

Behera, H. C. (2021). Traditional agriculture, culture and the indigenous knowledge (IK) among the Kondhs in Odisha, India. *Journal of Huma Ecology, 73*(1–3), 44–55.

Carrin, M. (2022). Santal indigenous knowledge, cultural heritage, and the politics of representation. *Modern Asian Studies, 56*(5), 1438–1463.

Chhetri, G., Pandey, A., Gaira, K. S., & Joshi, R. (2024). Traditional knowledge, ethnic foods, and their conservation of indigenous Lepcha community of Darjeeling and Sikkim Himalaya. In S. Chakravarty, I. Vineeta, J. A. Bhatt, M. Kumar, & G. Shukla (Eds.), *Sustainable management and conservation of environmental resources in India* (pp. 413–434). Apple Academic Press.

Das, M., Das, A., Seikh, S., & Pandey, R. (2022). Nexus between indigenous ecological knowledge and ecosystem services: A socio-ecological analysis for sustainable ecosystem management. *Environmental Science and Pollution Research, 29*(41), 61561–61578.

Das, S. (2024). Women's experiences and sustainable adaptation: A socio-ecological study of climate change in the Himalayas. *Climate Change, 177*(4), 59.

Dhall, S. (2024). Women and the local knowledge system. In S. Somayaji, G. Somayaji, & J. P. Coelho (Eds.), *The plural social sphere: Insights from contemporary Indian Society*. Taylor & Francis.

Doubleday, K. F. (2020). Tigers and "good Indian wives": Feminist political ecology exposing the gender-based violence of human–wildlife conflict in Rajasthan, India. *Annals of the American Association of Geographers, 110*(5), 1521–1539.

Ekka, K. T., & Giangthandunliu, P. A. (2024). Theorizing adivasi/tribal feminism: Decoding voices from Chotanagpur and the Northeast Region of India. *Journal of International Women's Studies, 26*(2), 1–10.

Elias, M., Grosse, A., & Campbell, N. (2020). Unpacking 'gender'in joint forest management: Lessons from two Indian states. *Geoforum, 111*, 218–228.

Ellena, R., & Nongkynrih, K. A. (2017). Changing gender roles and relations in food provisioning among matrilineal Khasi and patrilineal Chakhesang Indigenous rural People of North-East India. *Maternal & Child Nutrition, 13*, Article e12560.

Hansepi, L., & Laisram, R. (2022). Karbi women and environmental conservation. *The Grassroots Journal of Natural Resources, 5*(3), 1–17.
Janaki, M., Pandit, R., & Sharma, R. K. (2021). The role of traditional belief systems in conserving biological diversity in the Eastern Himalaya Eco-region of India. *Human Dimensions of Wildlife, 26*(1), 13–30.
Jasmine, B., Singh, Y., Onial, M., & Mathur, V. B. (2016). Traditional knowledge systems in India for biodiversity conservation. *Indian Journal of Traditional Knowledge, 15*(2), 304–312.
Jayaram, N. (2012). Identity, community, and conflict: A survey of issues and analyses. *Economic and Political Weekly, 47*(38), 44–61.
Kelkar, M. (2007). Local knowledge and natural resource management: A gender perspective. *Indian Journal of Gender Studies, 14*(2), 295–306.
Krishna, S. (2005). Gendered price of rice in North-Eastern India. *Economic and Political Weekly, 40*(25), 2555–2562.
Krishnakumar, J., Fox, J., & Anitha, V. (2012). Non-timber forest products: Livelihoods and conservation. *Economic and Political Weekly, 47*(52), 132–139.
Lahiri-Dutt, K. (2001). From gin girls to scavengers: Women in Raniganj collieries. *Economic and Political Weekly, 36*(44), 4213–4221.
Leach, M., & Green, C. (1997). Gender and environmental history: From representation of women and nature to gender analysis of ecology and politics. *Environmental History, 3*(3), 343–370.
Lele, S., Wilshusen, P., Brockington, D., Seidler, R., & Bawa, K. (2010). Beyond exclusion: alternative approaches to biodiversity conservation in the developing tropics. *Current Opinion in Environmental Sustainability, 2*(1–2), 94–100.
Mathew, A. J., et al. (2024). *Preserving traditions: Unraveling the traditional architechture of Tudavar* (pp. 1–44). Doctoral dissertation, St. Teresa's College (autonomous) Ernakulam.
Mukhopadhyay, D. (2008). Indigenous knowledge and sustainable natural resource management in the Indian desert. In C. Lee & T. Schaaf (Eds.), *The future of drylands: International scientific conference on desertification and drylands research Tunis, Tunisia, 19–21 June 2006* (pp. 161–170). Springer.
Munshi, I. (2001). Women and forest: A study of the Warlis of western India. *Gender, Technology and Development, 5*(2), 177–198.
Nightingale, A. (2006). The nature of gender: Work, gender, and environment. *Environment and Planning D: Society and Space, 24*(2), 165–185.
Nosheen, M., Mumtaz, T., & Amjad, B. (2024). Gendered dynamics and economic influences in transboundary mountain ecology: A case study of the Himalayan Region. *J Asian Dev Stud, 13*(4), 349–367.
Oakley, E., & Momsen, J. H. (2005). Gender and agrobiodiversity: A case study from Bangladesh. *Geographical Journal, 171*(3), 195–208.
O'Neill, A. R., Badola, H. K., Dhyani, P. P., & Rana, S. K. (2017). Integrating ethnobiological knowledge into biodiversity conservation in the Eastern Himalayas. *Journal of Ethnobiology and Ethnomedicine, 13*, 1–14.
Pilgrim, S., Smith, D., & Pretty, J. (2007). A cross-regional assessment of the factors affecting ecoliteracy: Implications for policy and practice. *Ecological Applications, 17*(6), 1742–1751.
Priyadarshini, P., & Abhilash, P. C. (2019). Promoting tribal communities and indigenous knowledge as potential solutions for the sustainable development of India. *Env Dev, 32*, Article 100459.
Rajamohan, R. (2007). Ethnic knowledge and biodiversity conservation. In P. Singh (Ed.), *Biodiversity, conservation, and systematics* (p. 127). Scientific Publishers.
Rao, V. L. N., & Ramana, G. V. (2007). Indigenous knowledge, conservation and management of natural resources among primitive tribal groups of Andhra Pradesh. *Anthropol Spec, 3*, 129–134.
Resurrecci6n, B. P., & Elmhirst, R. (2008). *Gender and natural resource management*. Earthscan.
Resurrecci6n, B. P. (2017). Gender and environment in the Global South: From 'women, environment, and development' to feminist political ecology. In S. MacGregor (Ed.), *Routledge handbook of gender and environment* (pp. 71–85). Routledge.

Rist, L., Shaanker, R. U., Milner-Gulland, E. J., & Ghazoul, J. (2010). The use of traditional ecological knowledge in forest management: An example from India. *Ecology and Society, 15*(1), 3.
Rocheleau, D., Thomas-Slayter, B., & Wangari, E. (2013). Gender and environment: A feminist political ecology perspective. In D. Rocheleau, B. Thomas-Slayter, & E. Wangari (Eds.), *Feminist political ecology: Global issues and local experience* (pp. 3–23). Routledge.
Ruiz-Mallén, I., & Corbera, E. (2013). Community-based conservation and traditional ecological knowledge: Implications for social-ecological resilience. *Ecology and Society, 18*(4), 12.
Sarin, M. (1995). Regenerating India's forests: Reconciling gender equity with joint forest management. *IDS Bulletin, 26*(1), 83–91.
Sarker, T. (2017). Biodiversity: The role of women in North East India. *Quest Journals Journal of Research in Humanities and Social Science, 5*(7), 71–74.
Sawakar, N. (2017). Gender dimension towards right over traditional knowledge by Adivasi women of Melghat Tiger Reserve, India. *IOSR Journal of Humanities and Social Science, 22*(04), 58–67.
Sebestianus Lakra, S. J. (2019). Sustainable resource management through indigenous knowledge and practices —A case of food security among the Baiga Tribe in India. *European Journal of Sustainable Development, 8*(4), 233–233.
Shah, M. K., & Shah, P. (1995). Gender, environment and livelihood security: An alternative viewpoint from India. *IDS Bulletin, 26*(1), 75–82.
Sharma, M. (2012). Dalits and Indian environmental politics. *Economic Political Weekly, 47*(23), 46–52.
Shiva, V. (1992). Women's indigenous knowledge and biodiversity conservation, India. *International Centre Quarterly, 19*(1/2), 205–214.
Si, A., Agnihotri, S., & Madegowda, C. (2024). Age and gender-related variation in plant and animal naming ability in the Soliga/Solega community of Southern Karnataka, India. *Human Ecology, 52*(5), 1035–1047.
Singh, H. B., Yaipharembi, N., Huidrom, E., & Devi, C. A. (2022). Traditional knowledge, beliefs, and practices associated with ethnic people of Manipur, North East India in conservation of biodiversity. In S. C. Rai & P. K. Mishra (Eds.), *Traditional ecological knowledge of resource management in Asia* (pp. 61–75). Springer International Publishing.
Singh, A. B., & Teron, R. (2017). Ethnic food habits of the Angami Nagas of Nagaland state, India. *International Food Research Journal, 24*(3), 1061–1066.
Singh, N. C. (2023). Sustainable natural resource management through traditional ecological knowledge: A perspective on the role of Apatani tribal women, Arunachal Pradesh. In S. C. Rai & P. K. Mishra (Eds.), *Traditional ecological knowledge of resource management in Asia* (pp. 239–259). Springer International Publishing.
Singh, R. K., Pretty, J., & Pilgrim, S. (2010). Traditional knowledge and biocultural diversity: Learning from tribal communities for sustainable development in northeast India. *Journal of Environmental Planning and Management, 53*(4), 511–533.
Singh, R. K., Rallen, O., & Padung, E. (2013). Elderly Adi Women of Arunachal Pradesh: "Living encyclopedias" and cultural refugia in biodiversity conservation of the Eastern Himalaya, India. *Env Manag, 52*, 712–735.
Singh, R. K., Kumar, A., Singh, A., & Singhal, P. (2020). Evidence that cultural food practices of Adi women in Arunachal Pradesh, India, improve social-ecological resilience: Insights for Sustainable Development Goals. *Ecological Processes, 9*(1), 1–19.
Sugden, F., & De Silva, S. (2014). A framework to understand gender and structural vulnerability to climate change in the Ganges River Basin: Lessons from Bangladesh, India, and Nepal. *International Water Management Institute (IWMI), 50.* https://doi.org/10.5337/2014.230
Sunder, M. (2007). The invention of traditional knowledge. *Law & Contemp Probs, 70*(2), 97–124.
Tiwari, C., & Sunny, A. (2024). To the ones who listen to the mountains: Evaluating the role of traditional ecological knowledge in conservation efforts in the Himalayas. In A. Borthakur, & P. Singh (Eds.), *Learning 'from' and 'with' the locals: Traditional knowledge systems for environmental sustainability in the Himalayas* (pp. 3–22). Springer Nature Switzerland.

Tiwari, C., & Rani, M. (2025). Where the wild things are: Identifying the factors behind and mitigation measures for the biodiversity conservation crisis in the Himalayas. In A. Borthakur, & P. Singh (Eds.), *People and mountain environments: Interconnectedness for sustainable development in the Himalayas*. Springer Nature.

Tripathi, A., Yadav, R. A., & Panda, J. K. (2020). Gender and ecology: Understanding the role of women in biodiversity conservation in India. In A. K. Singh, S. K. Jain, S. Nargunde, A. K. Sharma, R. Sharma, & R. Gupta (Eds.), *Woman empowerment & sustainable development* (IC-WESD-2020) (pp. 9–15).

Tyagi, N., & Das, S. (2020). Standing up for forest: A case study on Baiga women's mobilization in community governed forests in Central India. *Ecological Economics, 178*, Article 106812.

Tynsong, H., Dkhar, M., & Tiwari, B. K. (2020). Traditional ecological knowledge of tribal communities of North East India. *Biodiversitas J Biol Divers, 21*(7), 3209–3224.

Venkatasubramanian, K., & Ramnarain, S. (2018). Gender and adaptation to climate change: Perspectives from a pastoral community in Gujarat. *India. Dev Change, 49*(6), 1580–1604.

Westermann, O., Ashby, J., & Pretty, J. (2005). Gender and social capital: The importance of gender differences for the maturity and effectiveness of natural resource management groups. *World Development, 33*(11), 1783–1799.

Zweifel, H. (1997). The gendered nature of biodiversity conservation. *NWSA Journal, 9*(3), 107–123.

References for Figure 2

Angami tribe Monya, R. (2022). Traditional Dress and festivals. *Wikimedia Commons*. Retrived on March 10, 2025. https://commons.wikimedia.org/wiki/File:Angami_Tribe_NAGALAND_INDIA_20.jpg

Apatani tribe Barooah, S. (2004). Apatani lady of Ziro in Arunachal Pradesh, India with beaded jewellery. *Wikimedia Commons*. Retrived on March 10, 2025. https://commons.wikimedia.org/wiki/File:Apatani_Lady.jpg)

Baiga tribe Sandy & Vyjay. (2023). Baiga Tribal Women of Madhya Pradesh. Wikimedia Commons. Retrived on March 10, 2025. https://commons.wikimedia.org/wiki/File:Baiga_Tribal_Women_of_Madhya_Pradesh,_they_are_known_for_their_unique_tattoos_that_they_make_all_over_their_bodies.jpg

Bhil tribe Burdak, L. (2006). Self created image of Bhil tribe girls in Jhabua district. *Wikimedia Commons*. Retrived on March 10, 2025. https://commons.wikimedia.org/wiki/File:Bhil_tribe_girls_in_Jhabua.jpg

Bhotia tribe Kansihk, G. (2021). Rongpa teen girl from the lower tibet himalayas of uttarakhand, India. *Wikimedia Commons*. Retrived on March 10, 2025. https://commons.wikimedia.org/wiki/File:Rongpa_teen_girl_from_the_lower_tibet_himalayas_of_uttarakhand,_India.jpg)

Bishnoi tribe Zenit. (2009). T raditionally dressed Bishnoi Woman. *Wikimedia Commons*. Retrived on March 10, 2025. https://commons.wikimedia.org/wiki/File:Bishnoi_Woman.JPG)

Chenchu tribe Sakti1234. (2024). Chenchu Tribal Women: Preserving Culture Through Dance and Whistle Sounds. *Wikimedia Commons*. Retrived on March 10, 2025. https://commons.wikimedia.org/wiki/File:Chenchus12.jpg)

Gaddi tribe Hill, J. (2004). Gaddi woman in traditional dress with children and dog, Manali, HP, India. *Wikimedia Commons*. Retrived on March 10, 2025. https://commons.wikimedia.org/wiki/File:Gaddi_woman_in_traditional_dress_with_children_and_dog.jpg)

Khasi tribe Burke, D. (2010). Two Khasi girls in traditional dress at the Shad Suk Mynsiem dance, Shillong, Meghalaya, India. *Wikimedia Commons*. Retrived on March 10, 2025. https://commons.wikimedia.org/wiki/File:Khasi_Girls.jpg

Kondh Tribe PICQ, Samsara. (2005). An ethnic Adivasi woman from the Kutia Khond tribal group in Orissa, India. *Wikimedia Commons*. Retrived on March 10, 2025. https://commons.wikimedia.org/wiki/File:Kutia_kondh_woman_3.jpg

Lepcha community Kandell, A. S. (1965). Lepcha Women in Traditional Clothing in Singhik, Sikkim. *Picryl*. Retrived on March 10, 2025. https://picryl.com/media/lepcha-women-in-traditional-clothing-in-singhik-sikkim

Map of India Rajesh, K. (2011). Map of India—States and union territories marked. *Wikimedia Commons*. Retrived on March 10, 2025. https://commons.wikimedia.org/wiki/File:India-map-en.svg)

Mising tribe Devi, B. (2021). Life style of Mising tribe in Assam. *Wikimedia Commons*. Retrived on March 10, 2025. https://commons.wikimedia.org/wiki/File:Life_style_of_Mising_tribe_in_Assam.jpg)

Naga tribe Chothe, H. P. (2020). People dancing on the field. *Pexels*. Retrived on March 10, 2025. https://www.pexels.com/photo/people-dancing-on-the-field-5763155/

Rabari community Gupta, S. G. (2014). Rabari girls from Dwarka. *Wikimedia Commons*. Retrived on March 10, 2025. https://commons.m.wikimedia.org/wiki/File:Rabari_girls_from_Dwarka.jpg)

Reang tribe Tripura, B. (2022). Tripura girls in their traditional attire. *Wikimedia Commons*. Retrived on March 10, 2025. https://commons.wikimedia.org/wiki/File:Tripura_girls_in_their_traditional_attire.jpg)

Santhal Tribe, Murmu, K. (2023). Baha Dance by Santhal tribe wearing their traditional attire on spring festival Baha. *Wikimedia Commons*. Retrieved on March 10, 2025. https://commons.wikimedia.org/wiki/File:Baha_Dance_by_Santhal_tribe.jpg

Solega tribe Johnsingh A. J. T., WWF, NCF. (2012). *Sholiga family*, BRT TR. *Wikimedia Commons*. Retrived on March 10, 2025. https://commons.wikimedia.org/wiki/File:Sholiga_family,_BRT_TR,_AJT_Johnsingh.jpg)

Toda tribe Lalwani, R. (2009). Toda Dancers. *Wikimedia Commons*. Retrieved on March 10, 2025. https://commons.wikimedia.org/wiki/File:Dhoda_(Toda)Dancers.jpg

Warli Art Kasambe, R. (2012). Warli wall paintings at SGNP Mumbai by Raju Kasambe. Paintings by unknown tribal artist. *Wikimedia Commons*. Retrived on 10/05/2025. https://commons.wikimedia.org/wiki/File:Warli_Paintings_SGNP_by_Raju_Kasambe_DSCF0200_%281%29_06.jpg

Protecting Water and Land: Indigenous Women and Mining Impacts

Noklenyangla

Abstract Protests against mining especially uranium mining are a global phenomenon, arising from concerns about potential environmental and health hazards associated with this industry. Worldwide, anti-uranium mining movements are active in many countries, reflecting widespread concerns about the risks associated with this industry. In India, there have been significant protests against uranium mining in areas like Jadugora and Nalgonda. Concerns about the contamination of water sources and the health of local communities are major drivers of these protests. About from the key issues of environmental contamination and health risks, the issue of indigenous rights has also found significant attention. Many uranium deposits are located on or near indigenous lands, raising concerns about the impact of mining on traditional ways of life and cultural heritage. Indigenous communities often lead protests against uranium mining to protect their lands and rights. The uranium mining in the West Khasi Hills of Meghalaya, India, has met with heavy objection whether in regard to indigenous land rights or environmental issues. The resistance to uranium mining in this region has been backed by women organisations too. Spility Lyngdoh Langrin with her sheer determination motivated the locals against the evils of uranium mining on the environment, making her the face of resistance to uranium mining in Meghalaya. Although, the issue of uranium mining in Meghalaya is still ongoing, in this paper, we will be discussing about the significant role played by the women in resisting uranium mining in this region with relation to environmental issues. This paper will also make an analysis of the other regions of the world where indigenous communities in general and women in particular have resisted the uranium mining.

Keywords Protests · Uranium · Mining · Environmental · West Khasi Hills · Meghalaya · Women · Indigenous · Resistance

Noklenyangla (✉)
Dimapur, Nagaland, India
e-mail: ayangti@gmail.com

N. Moran et al. (eds.), *The Participatory Role of Indigenous Women*,
https://doi.org/10.1007/978-3-032-24059-0_11

1 Introduction

Uranium is a vital element for several reasons, primarily due to its unique properties. Although uranium mining is a crucial component of the nuclear fuel chain, it has historically disproportionately impacted indigenous community worldwide. Despite global calls for increased nuclear energy production, the devastating environmental, health, and socio-cultural consequences of uranium extraction continue to threaten the sovereignty and well-being of indigenous peoples cannot be overlooked. This paper highlights the reasons for opposition against uranium mining by the indigenous community in general and women in particular leading the resistance movements, through their unique perspectives, deep ancestral connections to land and water, and the gendered impacts of extractive industries. Specifically, this paper discusses the contribution of indigenous women of Meghalaya against uranium mining in their land. Although, at the moment, there is no single international treaty wholly dedicated to uranium mining, there are several international agencies and organizations and there are various Acts in India that plays a key role in regulating the uranium extraction, all these are comprehensively examined in the paper. Over all, this paper discusses the challenges and successes of the indigenous communities in protecting their lands, communities, and future generations from the destructive force of uranium mining and the contributions of the indigenous women in championing this cause. The end section has laid down some policy recommendations in order to address the pertaining issue discussed in the paper.

2 Usage of Uranium: Why Is It Sought After?

Primarily uranium is mined for its use as a fuel in nuclear reactors to generate electricity. This is by far its most significant application today. Uranium is used in nuclear power generation; indeed the most crucial use of uranium is in nuclear power plants. The isotope Uranium-235 (U-235) is fissile, meaning it can be split to release a tremendous amount of energy through a process called nuclear fission.[1] This heat is then used to boil water, create steam, and drive turbines to generate electricity. Nuclear power is seen as a vital component in reducing reliance on fossil fuels and addressing climate change.

Furthermore, the demand for uranium can be comprehended by the need of it in the following fields and sectors[2]:

[1] Nuclear fission is a process in nuclear physics where the nucleus of a heavy atom splits into two or more smaller nuclei, releasing a tremendous amount of energy in the process.

[2] See, World Nuclear Association. (2025, May 16). What is Uranium? How does it work? Retrieved from https://world-nuclear.org/information-library/nuclear-fuel-cycle/introduction/what-is-urnaium-how-does-it-work.

(a) *Medical isotopes*: A smaller, but important, proportion of uranium is used to produce medical isotopes. These isotopes are essential in diagnostic imaging and cancer therapy.
(b) *Naval propulsion*: Nuclear reactors powered by uranium are also used for marine propulsion, especially in naval ships and submarines.
(c) *Nuclear weapons*: Historically, and to some extent still today, uranium is mined for its use in nuclear weapons. Highly enriched uranium (HEU), with a higher concentration of U-235, is used to create the fissile material for atomic bombs.
(d) *Depleted Uranium*: After the enrichment process, the remaining uranium (mostly U-238) is known as depleted uranium. Due to its high density, depleted uranium has applications in: Radiation shielding, its density makes it effective at blocking radiation. It is used as kinetic energy penetrators used in armor-piercing ammunition and counterweights in aircraft control surfaces and yacht keels due to its heavy nature and intensity.

Due to its high energy density, uranium offers an incredibly concentrated source of energy. A small amount of uranium can produce a massive amount of electricity, far exceeding what can be generated from fossil fuels. Uranium, while not as common as some other elements, is more abundant in the Earth's crust than gold, silver, or mercury, and similar in abundance to tin. This means there are sufficient reserves to meet demands for many decades. To its credit, uranium is considered as a source of clean energy when in operation as nuclear power plants, once built, do not emit greenhouse gases during operation, contributing to efforts to combat climate change. Uranium mining involves various techniques, including open-pit, underground, and in-situ recovery, depending on the depth and nature of the ore deposit. Due to its radioactive nature, stringent safety measures are in place to protect workers and the environment during mining and processing.[3]

3 Opposition Against Uranium Mining: The Contesting Arguments

Groups often oppose uranium mining due to a range of significant concerns, mainly revolving around environmental, health, and social impacts. These concerns are often amplified for indigenous communities whose traditional lands are frequently targeted for mining operations.[4] Here are some of the reasons for opposition:

- *Environmental Contamination*[5]

---

[3] See, World Nuclear Association. (2024, May 24). *Uranium mining overview*. Retrieved from https://world-nuclear.ord/information-library/nuclear-fuel-cycle/mining-of-uranium-mining-overview.

[4] See Graetz (2014).

[5] See U.S. Environmental Protection Agency. *Radioactive waste from uranium mining and milling*. Retrieved from https://www.epa.gov/radtown/radioactive-waste-uranium-mining-and-milling.

Radioactive waste: Uranium mining and milling produce large quantities of radioactive waste, including mine waste rock and mill tailings. These tailings contain radioactive decay products like thorium-230 and radium-226 and hazardous chemicals from extraction process. If not properly managed, these wastes can contaminate soil, surface, water, and groundwater for thousands of years.
Water Contamination: The mining process, particularly open-pit mining and in-situ leaching, can contaminate water supplies with uranium, heavy metals like arsenic and molybdenum, and other toxic chemicals. This poses a threat to drinking water, agricultural use, and aquatic ecosystems.
Radon Gas Release: Uranium ore naturally emits radon gas, a radioactive gas that can cause lung cancer. While proper ventilation in underground mines can mitigate risks to workers, uncovered waste dumps and tailings ponds can release radon into the atmosphere, potentially affecting surrounding communities.
Dust and Air pollution: Mining operations can generate radioactive dust, which can be dispersed by wind and contaminate populated areas.
Land Degradation: Open-pit mines can span vast areas, altering landscapes and requiring extensive land disturbances. Even after closure, proper reclamation is crucial to prevent ongoing environmental issues.

- *Health Risks*[6]

Radiation exposure: Workers and nearby communities can be exposed to radiation through inhalation of radon gas and radioactive dust, ingestion of contaminated water or food, and direct gamma radiation.
Cancer: Exposure to radon and other radioactive materials increase the risk of various cancers. Particularly lung cancer, as evidenced by studies of past uranium miners.
Kidney Damage: Uranium is chemically toxic to the kidneys. And ingestion can lead to kidney damage.
Birth Defects and other Illnesses: Some studies suggest links between uranium exposure and reproductive issues, birth defects, neurological disorders, and other chronic illness.

- *Social and Cultural Impacts, especially on Indigenous communities*[7]

Displacement and Disruption: Uranium mines are often located on or near the ancestral lands of indigenous communities. Mining can lead to forced displacement, disruption of traditional livelihoods, e.g., hunting, framing, water use, and irreversible damage to cultural sites.

[6] See Potential Human Health Effects of Uranium Mining, Processing, and Reclamation. *Uranium Mining in Virginia: Scientific, Technical, Environmental, Human Health and Safety, and Regulatory Aspects of Uranium Mining and Processing in Virginia.* National Library of Medicine. Retrieved from https://www.ncbi.nlm.nih.giv/books/NBK201047/.

[7] See Ref no.(iv).

Lack of Informed Consent: Historically, indigenous communities were often not adequately informed about the health risks associated with uranium mining and were not properly consulted or given genuine consent for operations on their lands.
Loss of Cultural Connection: For many indigenous cultures, land and water are intrinsically lined to their identity, and way of life.
Intergenerational Health Issues: The long term contamination and health problems from uranium mining have led to an Indigenous community, affecting multiple generations.

- *Economic Disparities*: While mining companies may promise economic development, some indigenous communities have reported receiving low royalty fees, experiencing disregard for safety standards, and ultimately facing poorer services despite mining payments.[8]
- *Environmental Justice*: The disproportionate impact of uranium mining on marginalized communities highlights issues of environmental injustice, where the burdens of resource extraction fall heavily on vulnerable populations.[9]
- *Long-term Liabilities*: Perpetual Management: the radioactive nature of uranium waste means that tailings and contaminated sites require monitoring and management for hundreds of thousands of years, placing a significant long-term burden on future generations and potentially public funds.
- *Abandoned Mines*: Many abandoned uranium mines, particularly from earlier periods, were not properly remediated and continue to pose environmental and health hazards.

Hence, opposition to uranium mining stems from an insightful concern for human health, environmental integrity, and the well-being and rights of affected communities, particularly those who have historically borne the brunt of its negative consequences.

3.1 Evolution of Uranium Protest Movement

Uranium protests are often linked to the wider anti-nuclear movement, which opposes both nuclear weapons and nuclear power. Over time, the anti-uranium movement has increasingly aligned with indigenous right campaigns, recognizing the disproportionate impact of mining on these communities and the importance of indigenous sovereignty. Protesters have utilized various tactics, including marches, petitions, legal challenges, direct action, and public awareness campaigns. The collaboration between environmental organizations, indigenous groups, and international activists has often been crucial to their success.[10]

---

[8] See Rohrer et al. (2001).

[9] See Graetz (2014).

[10] See Feasby et al. (2009).

While not all uranium mining protests have achieved their immediate goals of halting operations, they have played a fundamental role in raising public awareness, influencing policy, and preventing further expansion of the industry in many areas. The legacy of uranium mining, particularly its environmental and health costs continues to fuel these protests globally.[11]

The key regions of Uranium mining protest in India

India is not new to environmental related protests; it is a consistent feature of India's social and political landscape. These protests are driven by concerns over deforestation, large-scale development projects, industrial pollution, and climate change impacts. These protest movements are often led by or involves local communities, tribal groups, activists and NGOs who use various forms of protest, including peaceful demonstrations, awareness campaigns and challenges. Some of the recent protests include, Jaipur Forest Conservation Protest, Pune Riverfront Development Project, Ladakh Protest, Save Mollem Movement (Goa). Without a doubt, India has a rich history of environmental activism, with several landmark movements that have significantly influenced environmental policy and public awareness, some of the notable ones are, Bishnoi Movement (1730s), Chipko Movement (1970s), Silent Valley Movement (1970s-1980s), Appiko Movement (1983), Jungle Bachao Andolan (1980s), Tehri Dam Conflict and Narmada Bachao Andolan (1985 onwards).

Likewise, there have been several protests against uranium mining in some of the regions of India,

- Jharkhand: The Jadugoda region in Jharkhand has been a long-standing site of protests against uranium mining by Uranium Corporation of India Limited (UCIL). Local communities have reported severe health issues and environmental contamination.[12]
- Tummalapalle (Kadapa district), Andhra Pradesh: There have been ongoing protests against uranium mining here, with residents complaining of groundwater contamination and stunted crops.[13]
- Lambapur and Peddagattu (Nalgonda district), Telangana: Protests have occurred against uranium exploration in these areas, with fears of agricultural land losing fertility. More recently, in November 2024, there were protests in Kappatralla over the Centre's approval for uranium exploration.[14]
- Domiasiat, (West Khasi Hills district), Meghalaya: This region has seen strong opposition to uranium mining plans, with local organizations and student unions leading protests due to concerns about environmental and health impacts, as well as the impact on indigenous ways of life.[15]

---

[11] Ibid.

[12] See Sha and Patra (2003).

[13] See Reddem (2022).

[14] See Mahapatra and Dua (2003).

[15] See Tribal oppose uranium mining. *Down to Earth*. Retrieved from https://www.google.com/amp/s/www.downtoearth.org.in/amp/story/environment/tribals-oppose-uranium-mining-30028.

While the government and mining companies often state that all safety and environmental standards are maintained, the consistent pattern of protests and reported health issues from affected communities suggests a significant disconnect and ongoing public concern regarding the impacts of uranium mining in India.

4 Indigenous Communities in the World Against Uranium Mining

Carrying forward the arguments about the social and cultural impacts on indigenous communities highlighted in the preceding section, let us drive deeper into it. Indigenous communities worldwide have been at the forefront of opposing uranium mining due to its devastating and often long-lasting impacts on their health, environment, culture, and traditional ways of life. These communities frequently live on or near lands rich in uranium deposits, making them particularly vulnerable to the consequences of extraction.

Below are some prominent examples of indigenous communities and groups who are against uranium mining:

- *North America (United States and Canada)*

Navajo Nation and other South-western US tribes: The Navajo people, along with communities like the Laguna, Acoma, Zuni, Hopi, Ute Mountain Ute, Havasupai, and the Northern Arapaho, have a tragic history of uranium mining on their ancestral lands. They have suffered from high rates of cancer, birth defects, and other illnesses due to radiation exposure and contaminated water sources. Many of these tribes have actively campaigned against new mining proposals and for the cleanup of existing abandoned mines. The Navajo Nation, for example, has banned uranium mining on their land.

- *Oglala Sioux Tribe (Pine Ridge Indian Reservation, South Dakota):* This tribe has also faced contamination of water sources and health problems linked to uranium mining.[16]
- *Dene and Cree (Saskatchewan, Canada):* Many Dene and Cree communities in northern Saskatchewan, where Canada's current uranium mines are located, have been opposing the mines and further exploration for decades. They report radioactive dust, contaminated waterways, and harm to local fish, impacting their health and traditional practices.[17]

[16] See Nauman, T. Oglala Sioux Tribe appeal blocks water permits for uranium mining. *Buffalo's Fire*. Retrieved from https://www.buffalofire.com/oglala-sioux-tribe-appeal-blocks-water-permits-for-uranium-mining.

[17] See Toledano, Michael. Indigenous Canadians are Fighting the Uranium Mining Industry. *Vice*. Retrieved from https://www.vice.com/en/article/a-dene-alliance-formed-to-resist-uranium-and-tar-sands-mining-in-saskatchewan-892/.

- *Cree (Quebec, Canada):* The Cree in Quebec successfully stopped uranium mining plans on their territory through protests and cooperation with environmental organizations.[18]
- *Inuit (Nunavut, Canada):* The Inuit people in Nunavut successfully rejected a large-scale uranium mining project in 2016 after years of conflict.[19]
- *Australia (Mirrar Aboriginal people and other Traditional Owners):* Uranium mining in Australia land. While not all indigenous owners are opposed, many traditional owners are opposed to it. The Mirrar Aboriginal people, for instance, rejected the expansion of the Ranger Uranium mine and the Jabiluka Uranium Mine due to concerns about environmental impact and cultural heritage. Many Aboriginal communities have expressed concern and anger at insensitive behaviour by governments and mining companies.[20]

Key Reasons for Opposition

- *Health Impacts*: Uranium mining exposes communities to radioactive materials like radon gas, and can lead to various cancers (especially lung cancer), birth defects, respiratory illnesses, and other severe health problems.
- *Environmental Contamination*: Mining operations can contaminate surface and groundwater with radioactive waste and heavy metals, rendering waste unsafe for drinking, agriculture, and ceremonial purposes. This impacts entire ecosystems and the traditional food sources of indigenous communities.
- *Cultural and Spiritual Disruption*: Ancestral lands are often sacred and integral to indigenous cultural and spiritual practices. Uranium mining can destroy sacred sites, disrupt traditional ways of life, and sever the deep ties communities have to their land.
- *Lack of Informed Consent and Consultation*: Historically, indigenous communities were often not adequately informed about the dangers of uranium mining or meaningfully consulted about projects on their lands. This often leads to mistrust and a sense of injustice.
- *Intergenerational Legacy*: The impacts of uranium mining are often intergenerational, affecting health and livelihoods for decades, even centuries, after mines are closed.

The struggle against uranium mining by indigenous communities highlights critical issues of environmental justice, human rights, and self-determination. They continue to advocate for the protection of their lands, health, and cultural heritage from the dangers of this industry.

---

[18] See Cree Nation reaffirms its opposition to uranium mining in Eeyou Istchee. *Nation*. Retrieved from http://nationnews.ca/news/cree-nation-reaffirms-its-opposition-to-uranium-mining-in-eeyou-istchee/.

[19] See Inuit voices grow louder in fight over Nunavut mine expansion. *Aljazeera*. Retrieved from https://www.aljazeera.com/amp/news/2021/2/27/inuit-voices-grow-loider-in-fight-over-nunavut-mine-expansion-canada.

[20] See Jones (2024).

5 Women Against Uranium Mining

Apart from the indigenous communities, in the gender front, women globally have been at the forefront of resistance against uranium mining due to its profound and disproportionate impacts on their health, livelihoods, and communities. Their activism is often rooted in a deep understanding of the immediate environmental and social consequences, as well as long-term perspective on the intergenerational harm caused by radiation and contamination. Women often lead grassroots movements against uranium mining, organizing protests, raising awareness, advocating for policy changes, and engaging in legal battles. Their networks within communities are crucial for mobilizing resistance.[21] Women's personal experiences of illness, loss, and environmental degradation provide powerful narratives that highlight the human cost of uranium mining and galvanize action. There is also the element of intergenerational advocacy where many women activists are driven by a desire to protect future generations from the harms of uranium mining, ensuring that their children and grandchildren inherit a healthy environment. Inherently, it is a way of challenging patriarchal structures as women's anti-mining activism often challenges not only the mining industry but also existing patriarchal structures within their communities and broader society, as they assert their voices and leadership.[22]

Notable Examples and Organizations

- *Indigenous Women*: Indigenous women globally, particularly in areas like the Navajo Nation in the United States and tribal belts in India, have been at the forefront of the fight against uranium mining due to the disproportionate impact of their ancestral lands and communities.
- *Civil Society Women's Organisation (CSWO), Meghalaya, India*: This powerful women's group has actively opposed uranium mining plans in Meghalaya, highlighting the potential for widespread environmental and health destruction.
- *Women's Rights and Mining (WRM)*: This collaborative effort of NGOs, researchers, and government organizations aims to address gender concerns in the mining sector more broadly, including the impacts of uranium mining.
- *Spility Lyngdoh Langrin (Meghalaya, India)*: An iconic figure of resistance, her unwavering opposition to uranium prospecting and mining in her village despite pressure and offers of money inspired many others.[23]

---

[21] See Claussen, A. Uranium is also a feminist issue. Peace & Health Blog. *International Physicians for the Prevention OF Nuclear War*. Retrieved from https://peaceandhealthblog.com/2020/03/09/uranium/.

[22] See Topno, N.P. Impact of Uranium Mining on Women's Health in Tribal Belt. *South India Journal of Social Science,* Vol. XX No.1, pp. 132–144.

[23] Spility Lyngdoh Langrin (1925–2020) was a revered indigenous woman from the Khasi community in Meghalaya, India, who became an iconic figure in the decades-long movement against uranium mining in her homeland. Often referred to as the 'Iron Lady' or 'Matriarch of Domiasiat', her unwavering resistance made her a symbol of courage and a source of inspiration for activists across the region.

Women's activism against uranium mining is a crucial aspect of environmental and social justice movements worldwide, driven by the severe and often gender-specific impacts of this industry.

Additionally, Ecofeminism[24] offers a powerful lens through which to analyze and critique the practice of uranium mining, highlighting the interconnectedness of environmental destruction, patriarchal systems, and social injustice. Here's how Ecofeminism relates to uranium mining:

- *The Logic of Domination and its Consequences*: Uranium mining embodies the patriarchal 'logic of domination' that ecofeminists critique. It views the Earth and its resources as mere commodities to be extracted for profit and power, rather than as a living system to be respected and protected. The process of mining involves significant disturbance of land, water, and air, leading to pollution and ecosystem degradation.
- *Domination of Women and Marginalized Communities*: Ecofeminism emphasizes that this domination of nature often foes hand-in-hand with the oppression of women and other marginalized groups, particularly indigenous communities. Uranium mining disproportionately impacts these groups for several reasons.
- *Geographic Concentration*: Uranium deposits are often found on or near indigenous lands. Historically and presently, these communities have been targeted for resource extraction due to their perceived lack of political power and their territories being deemed 'wastelands; by dominant societies.
- *Health Impacts*: Women, especially those living in close proximity to mines, are exposed to higher levels of radiation and toxic chemicals through contaminated water, air, and soil. This exposure has severe health consequences.
- *Social and Cultural Disruption*: Mining operations disrupt traditional ways of life, subsistence practices (like framing and foraging), and cultural connections to the land, which often disproportionately affect women who are central to these activities. The social fabric of communities can also be strained due to shifts in gender roles and economic dependencies.
- *Stigma*: Women exposed to radiation may face social stigma, impacting their marriage prospects and overall well-being.
- *Lack of Information and Agency*: Often, communities, particularly women, are not adequately informed about the risks of uranium mining of included in decision-making processes, leading to a lack of agency in protecting their own health and environment.
- *Nuclear proliferation and militarism*: Uranium mining is directly linked to the nuclear industry, both for energy production and historically, for nuclear weapons. Ecofeminists have long highlighted the connection between environmental destruction, patriarchal power structures, and militarism. The pursuit of nuclear power and weapons represents the ultimate manifestation of a destructive, dominating ethos that disregards life and ecological balance.

---

[24] Ecofeminism is a political philosophy and social movement that brings together the concerns of feminism and environmentalism. It posits that there is a deep, interconnected relationship between oppression and exploitation of women and the degradation and domination of nature.

Women have been at the forefront of anti-nuclear movements globally, recognizing the profound threat that nuclear technology poses to life on Earth, from the mining stage to potential nuclear accidents or warfare.

Ecofeminist Activitism Against Uranium Mining

Throughout history, women-led movements have actively resisted uranium mining. Examples include:

- *Indigenous Women's Resistance*: Indigenous women across North America (example: Navajo Nation), Australia and other regions have been central to advocating mining on their ancestral lands, recognizing the intergenerational health impacts and the spiritual significance of their territories. They often bring a holistic understanding of the interconnectedness of land, community, and well-being.
- *Grassroots Movements*: In places like Nova Scotia, Canada, women's institutes played crucial roles in raising awareness and protesting against uranium prospecting in the 1970s, contributing to a moratorium on mining in the province.
- *Global Solidarity*: Ecofeminists connect local struggles against uranium mining to broader issues of environmental justice, colonialism, and global power imbalances.

An ecofeminist approach to uranium mining inherently uses an intersectional lens. It understands that the impacts are not uniform and are compounded by intersecting identities such as gender, race, class, and indigeneity. For instance, an indigenous woman from low-income community near a uranium mine faces multiple layers of oppression and vulnerability that a non-indigenous male in a privileged setting would not. To sum up, ecofeminism views uranium mining as a quintessential example of patriarchal and colonial systems exploiting both nature and vulnerable human populations. It calls for fundamental shift away from this extractive paradigm towards practices that prioritize care, interconnectedness, and the well-being of all life on earth, with a particular focus on empowering those most impacted by such destructive industries.[25]

6 Role of Indigenous Woman in Protecting the Environment

Indigenous women play an absolutely vital and multifaceted role in protecting the environment, often serving as the primary guardians of traditional ecological knowledge and the frontline defenders of their ancestral lands. Their contributions

[25] See Gallo-Cruz, S. Awaiting Spring and War: Insights from Ecofeminism. *DEP- Deportate, esuli, profughe* 49, Retrieved from https://www.unive.it/.

are deeply rooted in their cultural heritage, spiritual connection to nature, and the practical responsibilities.[26]

To have a closer view of the role played by indigenous women in protecting the environment, let us have a breakdown of their key roles:

- *Custodians of Traditional Ecological Knowledge (TEK)*: Indigenous women are often the primary inheritors and transmitters of TEK, which is a vast body of knowledge about local ecosystems, including medicinal plants, sustainable resource management, weather patterns, animal behavior, and agricultural practices, developed over centuries of living in harmony with the land. This knowledge is crucial for understanding and responding to environmental changes. They pass this knowledge down through generations, often through storytelling, daily practices, and ceremonies.[27]
- *Guardians of Biodiversity and Food Systems*: They play a crucial role in conserving biodiversity by cultivating native seeds, growing traditional foods, and caring for medicinal plants. Their practices often promote diverse and resilient food systems, which are essential for food security and adapting to climate change. They are often responsible for seed selection, preservation, and understanding which plants and animals are indicators of ecosystem health.[28]
- *Leaders in Sustainable Resource Management*: Indigenous women often lead community-based efforts in managing natural resources sustainably. This includes practices like water management, land use planning, and forest conservation. Their deep understanding of local dynamics allows them to develop effective and culturally appropriate strategies for environmental protection.[29]
- *Advocates and Activists for Land and Rights*: Indigenous women are at the forefront of movements to protect their ancestral lands from exploitation and degradation, such as deforestation, mining, and oil drilling. They often engage in advocacy and activism, sometimes at great personal risk, to challenge destructive practices and uphold their collective rights. Their activism is driven by a profound sense of duty to protect their land for future generations.[30]

---

[26] See The Role of Indigenous Women in Climate Resilience and Adaptation. *Indigenous Climate Hub*. Retrieved from https://indigenousclimatehub.ca/2024/10/the-role-of-indigenous-women-in-climate-resilience-and-adaptation/.

[27] See The Role of Indigenous Women in The Transmission of Traditional Knowledge. *Convention on Biological Diversity*. Retrieved from https://www.cbd.int/article/role-indigenous-women-transmission-traditional-knowledge-IDWIP2022.

[28] See Indigenous women researchers speak their knowledge in an academic voice. *FAO Feature Story*. Retrieved from https://www.fao.org/newsroom/story/indigenous-women-researchers-speak-their-knowledge-in-an-academic-voice/en.

[29] See International Women's Day 2023: Celebrating indigenous women leaders in inclusive conservation. Retrieved from https://iucn.org/blog/202303/intenrational-womens-day-2023-celebrating-indigenous-women-leaders-inclusive.

[30] See The Indigenous World 2024: Defending the Rights of Indigenous Women. *IWGIA*. Retrieved from https://iwgia.org/en/defending-the-rights-of-indigenous-women/5407-iw-2024-defending-rights-indigenous-women.html.

- *Innovators in Climate Resilience and Adaptation*: Drawing on their traditional knowledge and community connections, indigenous women are spearheading innovative methods to address climate change impacts. This includes developing climate-resilient food systems, implementing traditional adaptation strategies, and integrating their knowledge with contemporary environmental research to shape policy.[31]
- *Strengthening Community Resilience*: Through their leadership in community-based adaptation strategies, women's support networks, and intergenerational teaching programs, they help strengthen community resilience in the face of environmental challenges. They often emphasize collective and spiritual values into environmental decision-making.[32]
- *Cultural and Spiritual Connection*: Indigenous women often have a profound spiritual connection to Mother Earth, viewing ecosystems as sacred and interconnected. This spiritual reverence underpins their dedications to environmental protection and informs their sustainable practices.[33]

Despite their immense contributions, indigenous women often face significant barriers, including lack of land ownership, limited participation in decision-making processes, and disproportionate impacts from climate change. Recognizing and empowering their voices, leadership, and knowledge is essential for achieving a more just and sustainable future for all.

7 Protest Against Uranium Mining in Meghalaya: Role of Indigenous Women

Meghalaya is often called the '*Abode of Clouds*',[34] it is a picturesque state nestled in the north-eastern part of India. It is renowned for its stunning natural beauty, including lush green landscape, cascading waterfalls, and unique living root bridges. When we talk about the geography of Meghalaya, it is a mountains plateau, an extension of the Deccan plateau, with elevations ranging from 150 m to 1961 m. The central Khasi Hills have the highest elevations, followed by the Jaintia hills to the east. The Garo Hills in the west are relatively flatter. The state is also rich in biodiversity. The state's forests are considered among the richest botanical habitats in Asia, supporting a vast variety of flora and fauna. This includes rare species like the Red Panda, Hoolock Gibbon, and Coulded Leopard in its national park and wildlife sanctuaries. The state

---

31 See Ref no. xxvi.

32 Ibid.

33 See Indigenous Women and their connection to the Earth and Her Waters. *The Indigenous Foundation*. Retrieved from https://www.theindigenousfoundation.or/articles/indigenous-women-and-their-connections-to-earth-and-water.

34 The name 'Meghalaya' in Sanskrit translates to 'Abode of Clouds' and true to its name, Meghalaya receives abundant rainfall, making it one of the wettest places on Earth. Mawsynram and Cherrapunji hold world records for the highest rainfall.

hosts numerous rivers and caves and several rain-fed rivers flow through the state, creating deep gorges and stunning waterfalls. Meghalaya is also home to some of the longest and most complex cave systems in the world, with impressive stalactite and stalagmite formations.[35]

Meghalaya is unique for being home to some of the world's largest surviving matrilineal societies. Its society follows a matrilineal descent meaning, lineage and family names are traced through the mother's side. Children take their mother's surname. The prominent tribes practicing this system are the Khasi, Garo and Jaintia. Property and ancestral wealth are inherited by daughters, not sons. Among the Khasis, the youngest daughter, known as the *Ka Khadduh*, holds a particularly significant role. She inherits the ancestral property and is responsible for managing family resources, caring for elderly parents, and looking after her siblings' welfare and education. The Garos also follow a similar pattern, with the youngest daughter often inheriting property, though there can be some flexibility. The Jaintias are known for maintaining one of the strictest matrilineal systems, with children belonging exclusively to the mother's clan.[36]

Thus, we can say that women play a dominant role in family life, managing property, and often engaging in entrepreneurial activities. The birth of a girl is celebrated, and there is less social stigma around remarriage or children born out of wedlock compared to many parts of India. They often enjoy significant financial security and independence. Efforts are being made to preserve the unique cultural heritage of these communities. Education is seen as a crucial factor in helping the youth understand and appreciate the value of matrilineal customs. There are ongoing debates within the communities about how to adapt to modern challenges while preserving their core values.[37]

In short, Meghalaya's matrilineal societies offer a fascinating counterpoint to the predominantly patriarchal societies found elsewhere in India and the world, highlighting a unique cultural framework where the maternal line is central to identity, inheritance, and family structure.

7.1 *Indigenous Women of Meghalaya Leading the Protest Against Uranium Mining*

Indigenous women in Meghalaya, particularly from the Khasi community, have been at the forefront of protests against uranium mining for decades. Their resistance

[35] See Basic Facts of Meghalaya. Retrieved from https://web.archive.org/web/201201204955/, http://megipr.gov.in/basic_facts.html and Meghalaya Biodiversity Board. Retrieved from https://megbiodiversity.nic.in.

[36] See Singh (2024) and Meghalaya's Matrilineal Societies: The Khasis, Garos, and Jaintias. Democracy and Development in Northeast India. Retrieved from https://banotes.org/democracy-development-northeast/meghalaya-matrilineal-socities-khasis-garos-jaintias/.

[37] Ibid.

is driven by deep concerns over the severe environmental and health impacts of such operations, as well as the potential threat to their land, culture, and traditional way of life. The indigenous communities in Meghalaya depend heavily on their land for agriculture and other traditional livelihoods. Uranium mining projects often involve land acquisition, displacement, and disruption of these practices, threatening their economic and cultural survival. Besides, - there are fears that the influx of outsiders for mining operations would dilute their unique culture and social fabric. The protests also highlight a broader concern about the state's development, which they perceive as prioritizing national interests over the well-being and rights of indigenous populations.[38]

Figures like Spility Lyngdoh Langrin[39] have become symbols of this resistance. She famously refused large sums of money from the Uranium Corporation of India Limited (UCIL) to protect her land and community, inspiring many others. Her steadfastness in the face of pressure from powerful government and corporate entities made her a hero in Meghalaya and beyond. Her image was often featured on banners during protests, and her story continues to inspire human rights and environmental activists. Her struggle, even after passing continues to be a source of motivation. The protests by indigenous women in Meghalaya against uranium mining are a powerful example of local communities standing up against large-scale industrial projects that threaten their environment, health, and cultural identity. The protesters demand an immediate halt to all uranium exploration and mining activities, proper remediation of existing contaminated sites, and recognition of their rights to self-determination over their ancestral lands and resources.[40] And as of 2025, uranium mining has not commenced in Meghalaya in any significant capacity, primarily due to the strong and sustained protests from local indigenous communities and civil society organizations.[41]

The persistent and vocal opposition from the people of Meghalaya, particularly the indigenous women and community organizations, has been successful in preventing uranium mining in the state. Although, the potential for uranium deposits remains, the socio-political landscape at present makes any large-scale mining operations highly improbable.

---

[38] See Uranium Exploration in Meghalaya's Villages Continues Unabated Without Local Sanction and Despite Troubling Health and Environmental Concerns. *The Caravan.* Retrieved from https://caravanmagazine.in/vantage/uranium-exploration-meghalaya-sanction-health-enviro mental-concerns.

[39] See Ref no xxii.

[40] The Quint. (2021, August 19). *How a Village in Meghalaya Fought Against Uranium Mining.* The Quint. YouTube: https://youtu.be/wUCTncQSgAc?Si=boeoKz-YsHphaJ6s.

[41] See Meghalaya govt stands firm against uranium mining amidst BJP karyakarta's advocacy. *Northeast News.* Retrieved from https://nenews.in/meghalaya/meghalaya-govt-stands-frim-against-uranium-mining-amidst-bjp-karyakartas-adovacy/8160/.

8 International Laws and Regulations on Uranium Mining

It is worth noting that the protests against uranium mining might be socio-political in nature coupled with environmental issues. However, uranium mining, due to its inherent risks related to radioactivity and its strategic importance for nuclear energy and weapons, is subject to strict regulations globally and within individual nations. While there is not a single comprehensive international treaty solely dedicated to uranium mining, several international instruments and organizations play a crucial role in regulating it, primarily through the lens of nuclear safety, non-proliferation, and environmental protection:

- *Treaty on the Non-Proliferation of Nuclear Weapons (NPT), 1968*: This Treaty aims to prevent the spread of nuclear weapons and promote cooperation in the peaceful uses of nuclear energy. It allows states to extract uranium for peaceful purposes, provided it is subject to safeguards agreement with the IAEA. The export of uranium is generally lawful unless it is known to be for developing nuclear explosive devices.
- *Treaty on the Prohibition of Nuclear Weapons (TPNW), 2017*: While the TPNW does not directly address the legality of uranium mining, it implies that states party to it are permitted to export uranium for peaceful purposes, even to nuclear-armed states, as long as it is not for the development of nuclear explosive devices.
- *International Atomic Energy Agency (IAEA)*: The IAEA, established under the UN, plays a central role in promoting the safe, secure, and peaceful uses of nuclear technology.
- *IAEA Safeguards Agreements*: These agreements are legally binding and aim to verify that unclear material (including Uranium) is not derived from peaceful uses to nuclear weapons or other nuclear explosive devices. The IAEA establishes and promotes international safety standards for all aspects of the nuclear fuel cycle, unclosing uranium mining, milling, and waste management. While mandatory for activities undertaken with IAEA assistance, they serve as strong recommendations for other nuclear activities. These cover areas like radiation protection, radioactive waste management, and emergency preparedness.
- *Conventions*: The IAEA also facilities international conventions related to nuclear safety, such as the Convention on Nuclear Safety and conventions on Early Notification of a Nuclear Accident and Assistance in Case of a Nuclear Accident or Radiological Emergency.

8.1 International Commission on Radiological Protection (ICRP) and United Mining in India

In India, uranium mining and all activities related to atomic energy and strictly controlled by the Department of Atomic Energy (DAE) and its regulatory body, the Atomic Energy Regulatory Board (AERB). The primary legal framework is:

- *The Atomic Energy Act, 1962*: This is the foundational law governing all aspects of atomic energy in India, including the production, development, use and disposal of atomic energy.[42] Key provisions relevant to uranium mining include:
  (a) Exclusive Government Control: The Act reserves the subject "Atomic Energy" for the exclusive domain of the Central Government and its Public Sector Undertakings (PSUs). This means only government entities are generally permitted for exploration, mining, and processing uranium.
  (b) Prescribed Substances: Uranium is defined as a prescribed substance under the Act.
  (c) Reporting Discoveries: Any person discovering uranium or thorium in India must report it to the Central Government within three months.
  (d) Control over Mining and Concentration: The Central Government has extensive powers to control or even prohibit mining or concentration of substances containing uranium, with provisions for compensation in case of prohibition.
  (e) Disposal of Uranium: Disposal of minerals, concentrates, or other materials containing uranium is prohibited without the Central Government's prior written permission.
  (f) Acquisition Powers: The Central Government can compulsorily acquire any produced mineral, concentrate, or material containing uranium.
  (g) Safety Provisions: The Act provides for the control over radioactive substances and radiation-generating plants to prevent radiation hazards, ensure public safety, and facilitate safe disposal of radioactive wastes.

- *Atomic Energy (Working of Mines, Minerals and Handling of Prescribed Substances) Rules, 1984:* These rules, framed under the Atomic Energy Act, provide detailed regulations for the handling of 'prescribed substances', including uranium, in mining and milling operations.[43]
- *Atomic Energy (Radiation Protection) Rules, 2004*: These rules specify the requirements for radiation protection for workers and the public.[44]
- *Atomic Energy (Safe Disposal of Radioactive Waste) Rules, 1987*: These rules govern the safe management and disposal of radioactive waste generated from uranium mining and milling.[45]
- *Mines Act, 1952, and Metalliferous Mines Regulation, 1961*: While not specific to uranium, these acts and regulations apply to all mining activities in India, including

[42] See The Atomic Energy Act, 1962. *Atomic Energy Regulatory Board (AERB)*. Retrieved from https://www.aerb.gov.in.

[43] See Atomic Energy Mines Minerals Prescribed Substance Rules, 1984. *Atomic Energy Regulatory Board (AERB)*. Retrieved from https://www.aerb.gov.in.

[44] See Atomic Energy (Radiation Protection) Rules, 2004. *Atomic Energy Regulatory Board (AERB)*. Retrieved from https://www.aerb.gov.in.

[45] See Atomic Energy (Safe Disposal of Radioactive Waste) Rules, 1987. *Atomic Energy Regulatory Board (AERB)*. Retrieved from https://www.aerb.gov.in.

aspects of occupational health and safety in uranium mines, complementing the specific atomic energy regulations.[46]

- *Environmental Protection Act, 1986:* This Act governs the environmental protection aspects related to industrial activities, including uranium mining, in India.

Regulatory Bodies in India

- *Atomic Energy Regulatory Board (AERB)*: The AERB is responsible for laying down safety standards and framing rules and regulations for the establishment and operation of nuclear facilities and the use of radioactive sources in India, including uranium mining and milling. It also conducts inspections and ensures compliance.
- *Directorate General of Mines Safety (DGMS)*: The DGMS is responsible for the safety of workers in all mines, including uranium, with a focus on industrial aspects.

Uranium mining worldwide and in India is a heavily regulated industry, emphasizing safety, environmental protection, and non-proliferation of nuclear weapons. International guidelines and national laws work in tandem to ensure responsible practices.

9 Policy Recommendations

Mining operations often have disproportionately negative impacts on indigenous women, who are often marginalized in decision making process and bear the brunt of social, environmental, and cultural disruptions. Therefore, policy recommendations must address these unique vulnerabilities and empower indigenous women in all stages of mining projects. Below are some of the policy recommendations,

- Ensure all information about mining projects is provided in a timely, comprehensive, culturally appropriate and accessible manner, including in indigenous languages where applicable.
- Gender Based Impact Assessments (GBIAs) should comprehensively analyze how mining impacts indigenous women differently from men, considering social, cultural, economic, health, and environmental dimensions. This includes examining potential increases in gender-based violence, changes in traditional livelihoods, disruption of social structures, and impacts on women's health and well-being.
- Create accessible, culturally appropriate and gender sensitive grievance mechanisms that help indigenous women to report harms and seek redress without fear of reprisal. These mechanisms should be independent and transparent.

[46] See Metalliferous Mines Regulations, 1961. *Directorate General of Mines Safety*. Retrieved from https://www.dgms.gov.in.

- Provide legal aid and support services for indigenous women to pursue justice in cases of human rights violations related to mining and implement policies that protect indigenous women's rights to clean water, fertile land, and traditional resources, which are often directly impacted by mining.
- Recognize and incorporate indigenous women's traditional ecological knowledge in environmental management plans and reclamation efforts, also acknowledge their unique understanding of the land and its resources.
- Strengthen regulatory oversight and enforcement of environmental standards with particular attention to how environmental degradation disproportionately affects indigenous women's health and livelihoods.

By implementing these comprehensive policy recommendations, governments, civil societies and mining companies can work towards mitigating the negative impacts of mining on indigenous women and promoting their well-being, empowerment, and self-determination.

10 Conclusion

The narrative surrounding uranium mining is a stark depiction of the ongoing global conflict between resource extraction and the rights and well-being of the indigenous communities. As demand for nuclear energy grows, the lure of uranium deposits in traditionally held lands often clashes with deeply rooted concerns about environmental degradation, health crises, and the erosion of cultural heritage.

The fierce and sustained protests, particularly by indigenous women like Spility Lyngdoh Langrin in Meghalaya, are not merely acts of opposition but insightful assertions of self-determination and environmental stewardship. Their resistance highlights the distressful, long-term consequences of uranium mining, from radioactive contamination of water and soil to increased rates of cancer and birth defects, and the disruption of traditional livelihoods. While the economic benefits of mining are often publicized, the disproportionate burden of its negative impacts consistently falls upon those with the least political power, creating a legacy of injustice.

Ultimately, the future of uranium mining centres on a fundamental re-evaluation of development priorities. It demands greater transparency, stricter regulatory oversight, and crucially, the genuine recognition of indigenous land rights and their right to free prior, and informed consent. Until and unless, the voices of those who stand and lose the most are genuinely heard and respected, the protests against uranium mining will continue to serve as a powerful testament that progress cannot come at the cost of human health, cultural integrity, and a healthy and vibrant planet.

References

Feasby, D. G, Chambers, D. B., & Lowe, L. M. (2009). *The Uranium industry and the protest movements.* WM2009 Conference, March 1–5, Phoenix, AZ. Retrieved from https://archievedproceedings.econference.io

Graetz, G. (2014). Uranium mining and first peoples: The nuclear renaissance confronts historical legacies. *Journal of Cleaner Production, 84*, 339–347.

Jones, T. (2024). *Australia: Uranium mining to be banned at Indigenous site.* DW. Retrieved from https://www.google.com/amp/s/amp.dw.com/en/australia-uranium-mining-to-be-banned-at-indigenous-site/a-69785799

Mahapatra, R., & Dua, N. (2003). *ublic outcry greets uranium mining project in AP.* Down to Earth. Retrieved from https://www.downtoearth.org.in/environment/ublic-outcry-greets-uranium-mining-project-in-ap-13407

Reddem, A. (2022). *Allaying concerns over nuclear projects in Andra Pradesh.* The Hindu. Retrieved from https://www.thehindu.com/opinion/op-ed/it-is-time-to-allay-fears-over-nuclear-projects-in-andra-pradesh/articles65202101.ece/amp/

Rohrer, R. E., Mitchell, J., & Betak, J. (2001). The economic opportunities and challenges of Uranium mine cleanup in New Mexico. *Bureau of Business & Economic Research.* Retrieved from http://www.nmlegis.gov

Sha, S.P., & Patra, S.K. (2003). *Uranium mining in Jadugoda, Jharkhand, India.* EJAtlas. Retrieved from https://ejatlas.org/conflict/uranium-mining-in-jadugoda-jharkhand-india

Singh, B. (2024). Matrilineal society of Meghalaya: Historical roots and contemporary challenges. *South Asia Journal.* Retrieved from https://southasiajournal.net/matrilineal-society-of-meghalaya-historical-roots-and-contemporary-challenges

Eco-Custodians of the Himalayas: A Feminine Perspective on Sacred Natural Sites Conservation

Anil Sharma, Sanoj Kumar Patel, Anil Barla, Amit Kumar Tiwari, Sudhanshu Kumar, and Gopal Shankar Singh

Abstract Sacred natural sites in the Himalayas represent a deep amalgamation of spiritual faith, natural science, and local tradition. Indigenous women as guardians of these places are the focus of this chapter, where sacred groves, lakes, and sacred sites serve as bio-reserves of biodiversity and religious cultural landscapes. Based on UNESCO research and ethnographic case studies, the chapter examines how Indigenous conservation ethics, underpinned by spiritual beliefs, govern ecological balance. Indigenous women emerge as spiritual and environmental guardians, their ceremonial spaces during festival and ritual processes testifying to respect for natural resources. Uttarakhand and Ladakh seasonal water deity festivals show how gendered spiritual leadership creates ecological awareness and guarantees the preservation of water bodies in the era of climate crisis. The chapter also highlights the oral traditions, stories, and myths as agents for communicating conservation ethics, emphasizing the agency of Indigenous women as knowledge keepers. In the context of modernization and environmental degradation, their stories become a powerful resilient tool, leaving ecological awareness in collective memory. The chapter also explores the challenges of Indigenous women, i.e., encroaching development, policy overlooking, and shifting socio-economic forces. However, it highlights their agency in spreading sustainable practices rooted in the wisdom of ancestors. Through enlightening the convergence of gender, spirituality, and conservation, this chapter contends that Indigenous women are not only custodians of sacred natural sites but also designers of a resilient ecological ethos. Their role in preserving material and spiritual dimensions holds much to teach conservation practice, which in turn calls for a holistic, culture-oriented practice of nature conservation.

Keywords Biodiversity · Cultural heritage · Environmental stewardship · Gender and ecology · Indigenous knowledge

---

A. Sharma (✉) · S. K. Patel · A. Barla · A. K. Tiwari · S. Kumar · G. S. Singh
Institute of Environment & Sustainable Development, Banaras Hindu University, Varanasi, India
e-mail: sharmaanil328@gmail.com

N. Moran et al. (eds.), *The Participatory Role of Indigenous Women*,
https://doi.org/10.1007/978-3-032-24059-0_12

1 Introduction

The association between gender and sacred natural sites has arisen as a pivotal area of feminist studies. In contrast to sex, which pertains to biological differences as gender covers the socially constructed norms, behaviors, and roles associated with masculinity and femininity (Levy, 1992; MacGregor, 2017). Gender serves as both a framework for identity and an apparatus of societal regulation, elucidating the entrenched disparities between men and women (Verschuur et al., 2016). Concurrently, sacred natural sites (SNS) as spiritually significant landscapes or water bodies stewarded by Indigenous communities are found globally (excluding Antarctica) and possess diverse ecosystems, ranging from forests to rivers (Bhagwat & Rutte, 2006; Gopal et al., 2018; Wild & McLeod, 2008). Although, frequently modest in size, these sites are intricately interwoven into cultural and ecological frameworks, sustaining extensive spiritual and territorial affiliations (Odhiambo & Manuga, 2017; Studley, 2019). The spiritual importance of SNS is derived from the faith that they are inhabited by deities, which referred as nature spirits (Studley, 2019).

The stewards of these sacred sites engage in ceremonial practices aimed at establishing communication and acknowledging to these spiritual entities, thereby ensuring that the sites preserve their sacred potency and significance. These ceremonial practices serve as mechanisms of environmental governance, modulating human engagement with these domains (Byrne, 2012; Sinthumule et al., 2021). Beyond rituals, traditional conservation practices—such as taboos and customary laws—play a crucial role in safeguarding sacred sites (Gokhale & Pala, 2016). Unlike formal legal systems, these unwritten norms shape human behavior toward nature through cultural beliefs rather than government enforcement (Colding & Folke, 2001). For instance, activities like hunting, fishing, or gathering firewood within sacred sites are often prohibited, with violations believed to invite ancestral punishment—ranging from illness to natural calamities (Negi, 2017). Such traditions have proven effective in preserving sacred sites across diverse cultures.

The Himalayas are one of the most biologically and culturally varied regions in the globe. They are referred to as the "Third Pole" because of their enormous glacial reservoirs. It spans many national borders and is home to several rare and endangered species as well as a diverse spectrum of environments, from alpine meadows to subtropical forests (Kandel et al., 2021; Kulkarni & Karyakarte, 2014). In addition, the Himalayas are deeply symbolic for many Indigenous and religious communities, serving as the setting for centuries of ritual, pilgrimage, and traditional knowledge systems (Tiwari & Rani, 2025). Due to the rich biological diversity and cultural veneration of Himalayas, they are uniquely positioned as a landscape where nature and culture are not distinct but rather closely connected.

In this framework, sacred natural sites (SNS)—like mountain peaks, rivers, caverns, high-altitude lakes, and groves—play an important dual function. These sites serve as unofficial conservation zones because native flora and fauna are traditionally protected from exploitation (Negi, 2017). They are ingrained in rituals and spiritual beliefs that uphold reverence for the natural world through cultural traditions. Many of these holy places are preserved by deeply ingrained Indigenous worldviews that perceive them as the homes of ancestral spirits or deities rather than by official legislation (Oviedo & Jeanrenaud, 2007). Therefore, the Himalayan SNS are both cultural sanctuaries that promote ecological continuity via spiritual observance and biodiversity reserves.

It is impossible to comprehend the preservation of these sacred sites without recognizing the crucial role those Indigenous women play. In many communities in the Himalayas, women are both ecological stewards and spiritual practitioners. They frequently oversee the management of forest resources, perform rituals, transmit oral traditions, and uphold the moral standards that regulate interactions between humans and the natural environment (Ormsby & Bhagwat, 2010). Their contributions are shaped by their profound comprehension of local ecosystems and commitment to preserving community customs over several generations (Bhattarai, 1997). Despite their marginalization within mainstream environmental narratives, this dual function acts as indigenous women are significant contributors to the conservation of SNS (Negi, 2012). This chapter examines the role of indigenous women in the Himalayas as eco-stewards of the SNS. It highlights the integration of gender, ecology, and spirituality within conservation methodologies by leveraging anthropological, UNESCO-sponsored, and region-specific case studies. This chapter endeavours to furnish a comprehensive approach in which Indigenous women steer extensive socioeconomic and political transformations that influence their communities along with safeguarding biodiversity through culturally embedded traditions.

This chapter elucidates that the responsibilities of Indigenous women embody a distinctive amalgamation of environmental insight, cultural identity, and spiritual duty, which positioned them at the forefront of the sacred site conservation discourse. Their vital practices, rooted in indigenous epistemologies, actively influence and sustain conservation efforts, rather than merely participating in them. It is imperative to grasp their perspectives to formulate inclusive and culturally sensitive conservation strategies for the Himalayan region and beyond.

2 Sacred Natural Sites as Ecological and Cultural Refuge

Various SNS possess profound cultural and spiritual importance for Indigenous cultures located in the Himalayan region, which is renowned for its extensive altitudinal variation and biodiversity. These sites, encompassing pilgrimage routes, high-altitude lakes, sacred forests, and river confluences, which are integral to holy rituals and deeply embedded in ancient belief systems (Joshi & Bhardwaj, 2015). Furthermore, they play a crucial ecological role, often functioning as biodiversity rich regions, where customs and spiritual norms govern anthropogenic activity (UNESCO, 2008).

2.1 Definition and Characteristics of Sacred Natural Sites

SNS are areas of the Himalayas that are considered to be both culturally and spiritually significant. Rituals, taboos, and collective memory are used to safeguard these sites, which are thought to be the homes of deities, spirits, or ancestral entities (Table 1). For example, sacred groves are forest patches that have been kept in remembrance of regional spirits and deities. These groves are found across the mid-hill regions of Uttarakhand, Himachal Pradesh, and portions of Sikkim (Acharya & Ormsby, 2017; Anthwal et al., 2010; Malhotra et al., 2007; Negi, 2012). They are not only revered but also protected by unofficial bans on hunting, grazing, and tree-cutting. Consequently, many of these regions serve as gene banks for indigenous biodiversity and are home to rare plant species (Anthwal et al., 2006).

Similarly, holy lakes like Mansarovar (which borders the Tibetan Plateau and the Indian Himalayas), Gurudongmar (Sikkim), and Hemkund Sahib (Uttarakhand) are cherished by Buddhist, Hindu, Bon, and Indigenous peoples. As revered water sources, these lakes are frequently connected to religious pilgrimages and mythologies. Despite pressures from tourism and climate change, these lakes have maintained a considerable ecological integrity because to the spiritual ban against pollution and overexploitation (Shah et al., 2024).

Pilgrimage roads include more than simply religious routes; such as the Char Dham Yatra in Uttarakhand and the routes that lead to monasteries in Ladakh and Sikkim. These treks contain holy places like temples, trees, or springs and wind through sensitive mountain habitats (Chand & Mishra, 2024). Many ancient traditions, including ceremonial bathing, feeding wildlife, and gathering herbs with spiritual permissions, continue to foster a respectful relationship with the environment, even though over tourism has worsened ecological stress on some routes (Bain, 2017).

2.2 Spiritual and Indigenous Perspectives

Sacred natural sites are seen by Indigenous tribes in the Himalayas as living beings with agency and rights, rather than only as places of worship. For instance, Mount Khangchendzonga is regarded by the Lepchas of Sikkim as both an ecological protector and a spiritual god. To preserve balance between human settlements and natural forces, elders—including women—lead rituals (Arora, 2006). Environmental deterioration is viewed in these cultures not just as ecological harm but also as a spiritual imbalance, which emphasizes the necessity of conservation through spiritual methods (Bain, 2017; Borrini, 2010).

Table 1 Sacred groves in the Indian Himalayan Region

State	Sacred grove name/Region	Associated deity/Spirit	Tribe/Community	Cultural practices	Ecological features	Conservation status	References
Jammu & Kashmir	Gurez Valley Groves	Local spirits (nature gods), Ziarats (Sufi saints)	Gujjars, Paharis	Offerings, restricted access, ritual protection	Alpine meadows, juniper, wild medicinal plants	Declining due to militarization and climate change	Dad and Khan (2011), Khuroo et al. (2016)
Ladakh	Nubra Valley Shrines	Lha (Mountain spirits), Buddhist protective deities	Ladakhi Buddhists	Prayer flags, Lama rituals, sacred springs	Cold desert groves, high-altitude junipers	Relatively intact due to Buddhist conservation ethos	Wakgchok (2007), Singh, (2014)
Himachal Pradesh	Shakti Groves (e.g., Churdhar, Shikari Devi), Dev Van (forest of gods)	Devi (Durga, Kali, Shikari Devi)	Kinnauris, Gaddis, general Pahari communities	No tree felling, annual fairs, goat sacrifice in some groves	Oak-rhododendron, Himalayan cedar, biodiversity hotspots	Moderately conserved; tourism pressure rising	Malhotra et al. (2007), Bhandari et al. (2018)
Uttarakhand	Hariyali Devi, Mahasu Devta groves, Bugyal (sacred alpine meadow)	Hariyali Devi, Bhumiyal Devta, Mahasu Devta	Jaunsaris, Bhotiya, Garhwali	Sacred festivals (e.g., Harela), water sources worshipped	Sacred springs, dense oak-chir pine forests, rare herbs	Facing threats from road construction and pilgrimage expansion	Anthwal et al. (2010)
Sikkim	Kabi Lungchok, Khangchendzonga Sacred Landscape	Khangchendzonga (guardian deity), local Lepcha spirits; *Devithans* (goddess grove)	Lepchas, Bhutias	No hunting, no felling zones, community forest rules	Pristine temperate forests, glacial-fed rivers, endemic flora	Strong conservation through local laws and monastic influence	Arora (2006), Acharya and Ormsby (2017)

(continued)

Table 1 (continued)

State	Sacred grove name/ Region	Associated deity/ Spirit	Tribe/ Community	Cultural practices	Ecological features	Conservation status	References
Arunachal Pradesh	Donyi-Polo Groves (Siang, Tawang), Gumpa forest	Donyi (Sun), Polo (Moon), local forest spirits	Adi, Nyishi, Apatani, Monpa	Ritual restrictions, animal sacrifice, oral cosmology	Bamboo groves, evergreen forests, cloud forest vegetation	Threatened by shifting cultivation and development	Sarmah et al. (2009), Shimrah et al. (2012), Murtem et al. (2014), Chaudhry and Murtem (2015), Acharya and Ormsby (2017), Jha (2017)
Nagaland	Khonoma Sacred Grove	Ancestor spirits, local animistic deities	Angami Nagas	Forest use taboo, ancestral burial rites	Old-growth Forest with rare orchids, mammals, birds	Community-led protection model (Khonoma Green Village)	Singh and Teron (2015), Ao (2022), Tsopoe et al. (2025)
Manipur	Umang Lai (Sacred Forests of Meitei)	Lai (Forest Deities)	Meitei, Tangkhul, other hill tribes	Lai Haraoba festival, no cutting of trees, offerings	Mixed deciduous forests, sacred lakes and ponds	Legally protected, but urban encroachment is an issue	Khumbongmayum et al. (2004), Devi et al. (2005), Malhotra et al. (2007), Upadhyay et al. (2019)
Meghalaya	Mawphlang, Law Lyngdoh Ki Law Lynddoh (laws or rules of sacred groves or forest), Ki Law Kyntang (laws or rules of sacred groves or forest)	U Basa, U Ryngkew (forest spirits)	Khasi, Jaintia	Oral laws (Ka Hima), taboo zones, clan guardianship	Rich evergreen forests, high endemism, carbon sinks	Well-preserved due to cultural taboos; eco-tourism managed	Tiwari et al. (1998), Khumbongmayum et al. (2004), Ormsby (2013), Kharumnuid and Bharucha (2020), Diengdoh et al. (2021), Bhattacharjee (2024)

(continued)

Table 1 (continued)

State	Sacred grove name/ Region	Associated deity/ Spirit	Tribe/ Community	Cultural practices	Ecological features	Conservation status	References
Mizoram	Sialkal Sacred Grove	Nature spirits and ancestral souls	Mizo tribes	Ceremonial restrictions, clan-based forest zoning	Montane subtropical forest, bamboo species	Partially degraded due to shifting agriculture	Upadhyay et al. (2019), Lalzarzovi (2022)
Tripura	Chhaturmukhi Temple Grove	Manasa Devi, forest guardian spirits	Tripuri, Reang	Sacred groves near temples, local rituals	Sal, bamboo, medicinal shrubs	Fragmented but culturally protected	Upadhyay et al. (2019), (Gupta, 2020)
Assam (Hill districts)	Sacred Bamboo Groves (Karbi Anglong), *Than* (place, site, abode; sacred grove of ancestral deities), *Madaico* (sacred space, ritual gathering)	Karbi spirits, local deities	Karbi, Dimasa	Restricted cutting, ritual feasts	Bamboo-dominated groves, dense undergrowth	Moderately conserved, development pressures increasing	Baidya et al. (2020), Chanda (2021), Dhanaraju et al. (2022), Sharma and Gogoi (2006)

2.3 Ecological Significance as Biodiversity Hotspots

SNS frequently act as biodiversity micro-reserves. They often harbor undisturbed flora and fauna, including keystone species and rare medicinal plants, because of traditional constraints (Table 2). Sacred groves have more plant diversity and denser forest cover than adjacent non-sacred regions, according to research done in Uttarakhand's Garhwal region (Anthwal et al., 2006; Kandari et al., 2014). Despite their small size, holy lakes and wetlands in high-altitude zones are ecologically important because they provide habitat for migrating birds and unique alpine vegetation (O'Neill, 2019; Sharma et al., 2020).

2.4 UNESCO and Ethnographic Insights

UNESCO and IUCN (2008) have highlighted the significance of SNS worldwide, designating them as "no-go zones" for industrial development and suggesting their incorporation into conservation policies. UNESCO-supported research in the Himalayan region shows how these sites serve as locally and Indigenous-managed, culturally embedded conservation systems. Through oral traditions, community members—particularly women—serve as ecological monitors, ritual leaders, and carriers of conservation knowledge, as further shown by ethnographic study (Berkes, 2012; Nosheen et al., 2024).

For instance, women actively participate in seasonal rites associated with holy springs and forests among Bhotiya communities in Uttarakhand (Mishra et al., 2025). These ceremonies, which are synchronized with agricultural cycles, frequently include narrative performances, songs, and offerings that uphold ecological principles like sharing, self-control, and appreciation for the land (Manocha & Pasupuleti, 2024). The ecological stability and spiritual life of sacred sites depend heavily on this kind of gendered engagement.

Consequently, the Himalayan SNS serves as both a cultural and spiritual realm, as well as an ecological refuge. These sites not only protect vital ecosystems and biodiversity but also reinforce local identities and ethical values towards nature through traditional belief systems and conservation initiatives (Verschuuren, 2016). Acknowledging Indigenous perspectives and their intrinsic ecological perception is essential for understanding the dual function of Indigenous peoples. The focus is on how Indigenous communities, particularly women, play a crucial role in preserving these sacred landscapes. In various Himalayan communities, custodial responsibilities, environmental knowledge, and spiritual authority constitute the foundation of sustainable conservation practices (Kala, 2017).

Table 2 List of some sacred plants in Hindu mythology

Botanical name	Local name	Family	Beliefs/uses
Aegle marmelos (L.) Corrêa	Bail	Rutaceae	Sacred to Shiva, used in rituals and medicine
Artemisia nilagirica (C.B.Clarke) Pamp	Kunju	Asteraceae	Used in cleansing rituals
Artocarpus heterophyllus Lam	Kathal	Moraceae	Ritual food offering
Azadirachta indica A.Juss	Neem	Meliaceae	Purifying, medicinal, protective from evil
Blumea balsamifera (L.) DC	Bari ilaichi	Asteraceae	Used in rituals and traditional medicine
Betula utilis D.Don	Bhojpatra	Betulaceae	Sacred scripts written on bark
Calotropis procera (Aiton) W.T.Aiton	Aak	Apocynaceae	Associated with Shiva, used in rituals
Cedrus deodara (Roxb. ex D.Don) G.Don	Deodar	Pinaceae	Sacred Himalayan tree, used in temples
Cynodon dactylon (L.) Pers	Dhoob	Poaceae	Offered to Lord Ganesha
Datura stramonium L	Datura	Solanaceae	Offered to Shiva, has medicinal/toxic uses
Desmostachya bipinnata (L.) Stapf	Kush	Poaceae	Used in yajnas (Vedic rituals)
Elaeocarpus sphaericus (Gaertn.) Heer	Rudraksha	Eleocarpaceae	Sacred beads, associated with Shiva
Erythrina variegata L	Paribhadraka	Fabaceae	Used in rituals, culturally symbolic
Ficus benghalensis L	Bargad	Moraceae	Sacred tree, symbolizes immortality
Ficus religiosa L	Peepal	Moraceae	Sacred tree, connected to Buddha and Vishnu
Mangifera indica L	Mango	Anacardiaceae	Leaves used in puja and decoration
Musa × paradisiaca L	Banana	Musaceae	Used in rituals, leaf for offerings
Ocimum tenuiflorum L	Tulsi	Lamiaceae	Sacred, medicinal, worshipped daily
Phyllanthus emblica L	Aonla	Phyllanthaceae	Sacred, used in health tonics
Pinus roxburghii Sarg	Chir	Pinaceae	Resin used for incense, sacred in Himalayas
Prinsepia utilis Royle	Bhakel	Rosaceae	Used in local rituals
Prosopis cineraria (L.) Druce	Khachri	Fabaceae	Worshipped during festivals like Teej

(continued)

Table 2 (continued)

Botanical name	Local name	Family	Beliefs/uses
Prunus cerasoides Buch.-Ham. ex D.Don	Paiya	Rosaceae	Flower offerings
Punica granatum L	Daru	Lythraceae	Fruit offered in rituals
Santalum album L	Sandal	Santalaceae	Sacred paste used in worship
Saussurea obvallata (DC.) Sch.Bip	BramKamal	Asteraceae	Rare Himalayan flower, sacred in rituals
Terminalia arjuna (Roxb. ex DC.) Wight & Arn	Arjun	Combretaceae	Used in Ayurveda and rituals
Toona ciliata M.Roem	Tun	Meliaceae	Used in traditional ceremonies
Zanthoxylum acanthopodium DC	Timoor	Rutaceae	Sacred, used in local healing traditions
Zanthoxylum capense (Thunb.) Harv	Timoor	Rutaceae	Sacred, medicinal and ritual use
Quercus leucotrichophora A.Camus	Oak	Fagaceae	Sacred in Himalayan belief systems

3 Indigenous Women as Guardians of Conservation Ethics

3.1 Spiritual and Environmental Stewardship in the Himalayas

In many Indigenous communities in Himalaya, women are pivotal in maintaining equilibrium between ecological sustainability and spiritual practices (Fig. 1).

Beyond their everyday subsistence roles, they hold a vital position in the religious, ceremonial, and ecological tapestry of their cultures (Amritesh et al., 2025). Indigenous women have historically safeguarded sacred landscapes and community-centric conservation methodologies, from orchestrating seasonal festivities to transmitting ecological knowledge through oral traditions (Samakov & Berkes, 2017). Their active engagement in ceremonial practices that emphasize environmental stewardship constitutes a significant aspect of custodianship (Torri & Herrmann, 2011). Women are tasked with organizing and participating in ceremonies that pay homage to local deities, many of whom are associated with natural features such as rivers, lakes, forests, and mountains, particularly in Uttarakhand and Ladakh (Kaur & Balodi, 2016). For instance, Indigenous women's involvement is indispensable to the observance of the Nanda Devi Raj Jat Yatra, a lengthy pilgrimage venerating the mountain goddess Nanda Devi in Uttarakhand's Kumaon region (Patwal, 2016). Throughout this pilgrimage, which traverses high-altitude ecosystems recognized as ecologically and spiritually significant, they prepare offerings, lead ritualistic songs and uphold sacred traditions (Manocha & Pasupuleti, 2024).

Fig. 1 Pictorial representation of different sacred natural sites in Indian Himalayan states: **a** Sacred Lake in Uttarakhand, Himalaya; **b** Dream stones in Buddhism philosophy, Sandakphu, Darjeeling, Himalaya; **c** Buddhist monastery in Sikkim; **d** Goddess Durga, Himachal Pradesh, India Photo by: Poushali Dey

Similarly, women-led rituals that represent the rebirth of natural elements and community interactions with water sources are incorporated into water god festivals, such as Yuru Kabgyat and Losar in Ladakh (Angmo, 2024). During these celebrations, holy springs are cleaned, river deities are offered ceremonies, and prayers for seasonal balance. Women's ceremonial duties directly support the protection of local water supplies and their spiritual holiness, representing a type of gendered ecological consciousness (Butcher, 2015; Gagné, 2020). Indigenous women are important carriers of traditional ecological knowledge (TEK) in addition to serving as ritual leaders. Oral traditions, storytelling, songs, and proverbs are some of the ways that this information is transmitted, and they frequently incorporate ecological values (Juma, 2022).

Through these traditions, younger generations learn about the moral obligation to preserve non-human existence, the spiritual repercussions of excessive usage, and the proper seasons for harvesting (Gaur et al., 2024). According to older women in the Lepcha villages of Sikkim tell children stories about the Khangchendzonga mountain and its sacred plants, emphasizing the need of avoiding harming specific plants that are said to be the homes of guardian spirits (Arora, 2006). These practices exhibit those environmental principles are not only persist but also become ingrained in the culture.

Indigenous women directly influence sustainable management of natural resources as mentors, knowledge keepers, and ritual specialists. Their authority to continue community compliance with ecological rules at some extents are legitimized through moral and spiritual stories as opposed to governmental policies is where they are seen to have the greatest influence (Gaur et al., 2024). Ongoing protection of sacred natural sites and their associated biodiversity, their survival depended on these unofficial institutions, supplemented by ceremonies and myths (Berkes, 2012).

Other research finds that Indigenous women in the Himalayas have considerable power in the spiritual and environmental spheres is borne out by ethnographic data. They sometimes observe shifts in nature patterns ahead of human detection, like modifications of flowering calendars or changes in rainfall regimes, and respond by adjusting rituals or providing resources (Gaur et al., 2024). Because the information provided by this adaptive knowledge, inherited from centuries of experience and practice, is a case of gender-balanced and culturally embedded form of community-based conservation (Samal & Dhyani, 2006). Therefore, Indigenous women play a key role as guardians of conservation. Employing their environmental wisdom, storytelling, and ritual practice, they keep the spiritual and cultural institutions behind conservation intact in conjunction with the physical environment.

3.2 *Sacred Taboos and People-Centric Ecological Conservation: Case Studies*

From Renuka Lake (Himachal Pradesh) and Apatani Tribe (Arunachal Pradesh)

Renuka Lake, an SNS in the peaceful foot-hills of the Sirmaur district, Himachal Pradesh, is deeply ingrained in local myths and ecological management (Bhardwaj, 2022; Choden et al., 2022). The lake is consecrated as the symbol of Renuka Devi, the mother of Parashurama, a Hindu scholar is surrounded by the tracts of forestland and religious buildings. The community considers the lake to be a living god, particularly by women. Protective taboos dominate long enough that hunting, tree falling, and dumping trash in the catchment of the lake are prohibited due to this sacred nature (NITI Aayog, 2018.). The collection of firewood has long been discouraged and regulated by traditional norms in the vicinity of sacred site. These religious sanctity-founded informal norms have contributed greatly to maintaining the water quality and biodiversity of the lake, including migratory birds and native aquatic species (Gagné, 2020).

The Apatani tribe in Ziro Valley of Arunachal Pradesh provides a comparable illustration. The Apatani, who are renowned for their complex land-use systems, combine wet-rice farming with agroforestry in a way that strikes a compromise between ecological sensitivity and productivity (Singh, 2023). Strict customary taboos are in place to protect clan forests, water springs, and sacred trees (Donyi-Polo places of worship). In addition to being a social transgression, hunting, fishing, and tree-cutting in certain areas are seen as spiritual transgressions that might bring bad luck to the entire community (Chaudhry et al., 2011). Oral traditions, seasonal rituals, and community surveillance conducted by both youth and elder groups all serve to perpetuate these taboos.

Another such example is the Apatani people of Arunachal Pradesh's Ziro Valley. The Apatani, whose land-use system is highly complex, mix wet-rice cultivation with agroforestry in a manner which attains an interconnection between maintenance of ecological integrity and productivity (Singh, 2023). Stricter customary prohibitions are followed to conserve clan forests, water springs, and sacred trees (Donyi-Polo places of worship). In addition to being a social offense, hunting, fishing, and tree cutting in some of these places are deemed to be religious offenses that could cause bad luck to the whole community (Chaudhry et al., 2011). Oral traditions, seasonal rituals, and community surveillance conducted by both youth and elder groups all serve to perpetuate these taboos.

The regulated use of forest resources is one of the distinguishing characteristics of Apatani system (Barua et al., 2015). Private forestland is owned by each family, but harvesting is controlled by group decision-making. Due to spiritual beliefs that connect human health with the welfare of local wildlife, hunting wild animals during mating or breeding seasons is strictly prohibited (Chaudhry et al., 2011; Dollo et al., 2009). In spite of greater population pressures, these activities have served to protect the natural integrity of the region and protecting animals such as the endangered Mithun (*Bos frontalis*).

Both Renuka Lake and Apatani valley illustrate how environmentally embedded ethics i.e., taboos and ritualized norms, function as de facto conservation measures. The gender component of environmental stewardship is also highlighted by these cases. Women who participate in religious processions and ceremonies are unofficial caretakers of the lake and the forest that surrounds it in Renuka. Seed preservation, intercropping, and compliance with ritual protocols prescribing land and forest health are also the duty of Apatani women (Patnaik, 2020). These obligations have a tendency to be passed on by oral cultures, myth, and religious teaching passed from generation to generation, yielding a conservation ethic regardless of written law or external enforcement (Singh, 2023).

It is also notable that both case studies indicate towards the manner in which outside pressures such as road building in Ziro and tourism in Renuka are making these traditional systems more robust. Local resistance to development that degrades the environment persists, however, typically spearheaded by women's groups and village councils. They present an argument on grounds both of long-term ecological basis as well as of cultural rights. Apatani tribe's spiritual taboos and Renuka lake are the best demonstration of how religious beliefs and social norms structure Indigenous conservation practice. Besides highlighting the agency of the community, particularly indigenous women in preserving fragile ecosystems in the face of anthropogenic pressures, these localized practices have shown the low-impact, culturally connected ecological management.

3.3 Case Studies: Gendered Spiritual Leadership in Conservation

Water is not just a resource in the Indian Himalayan high-altitude areas; water is an ritualized and incarnated sacred material worshipped by myth and ritual (Kaur & Balodi, 2016). Festivals according to seasonal occasions for the water gods are major cultural events in Ladakh and Uttarakhand when inhabitants celebrate their gratitude towards local ecosystems (Gaur et al., 2024; Sharma et al., 2020). These activities are not only religious ceremonies but guarantee the cleanliness of water bodies in regions that are becoming more exposed to climatic uncertainty. Indigenous women are at the heart of these customs, playing important ceremonial and administrative duties that strengthen the ecological and spiritual aspects of water management (Gagné, 2020).

In Uttarakhand, the pre-monsoon season is frequently marked by the local spring-cleaning festivals and the Bhojpatra Mahotsav, which are held to honour and purify the water sources said to be inhabited by deities. It is the duty of women, especially older women, to plan the ceremonies, make the sacrifices, and lead the prayers and hymns (Sati, 2023). In addition to encouraging the physical upkeep of springs, these rituals also symbolize respect for natural cycles and the start of water-use limitations. Traditional ecological knowledge systems that support the sustainable utilization of regional resources are strongly aligned with these activities (Samal & Dhyani, 2006).

Similarly, women in Ladakh work alongside Buddhist monks to lead water purifying ceremonies during celebrations like Yuru Kabgyat and Losar (New Year). In these rituals, prayers are made to water spirits for protection and plenty, and sacred water sources, such glacier streams and springs, are ceremoniously cleansed (Gagné, 2020; Gaur et al., 2024). In a landscape where water resources are being affected by glacial melt and reduced snowfall, these activities both functionally and symbolically guarantee community involvement in water conservation (Angmo, 2024). Moreover, women's participation in these activities, often as the foremost planners and keepers of culture, illustrates the affirmed connection between religious leadership which is gendered and ecological resilience.

Aside from their ceremonial role, Himalayan Indigenous women are instrumental in locally driven conservation efforts, particularly in confronting environmental challenges such as pollution, forest depletion, and water scarcity (Torri et al., 2011). For instance, in the state of Uttarakhand, women's organizations such as the Mahila Mangal Dals have risen to protect holy forests and water sources through devotion and activism. To counter encroachment, development schemes, and illegal felling of timber that poses a threat to sacred sites, these tribes make use of traditional rights and religious folklores (Agarwal, 2001). Oral tradition and story-telling are very important tools in most of these conservation efforts because they allow for the transmission of environmental values from generation to generation (Juma, 2022). Women are essential to the preservation and storytelling of regional folklore that highlights the value of nature and cautions against environmental violations. In Himachal Pradesh and Sikkim, for example, mythology regarding guardian spirits punishing individuals who harm sacred trees or pollute sacred rivers still affects cultural attitudes. These stories represent but also provide regulating frameworks that shape local environmental governance (Arora, 2006).

The indigenous women become spiritual leaders and ecological mentors as a consequence of these exercises. A culturally based and locally adapted gendered conservation is formed through their participation in oral knowledge systems, communal involvement, and celebration of seasonal festivals (Mishra et al., 2025; Samal & Dhyani, 2006). While such leadership is not founded on explicit scientific jargon, it is no less powerful in inspiring durable environmental transformation and sustainable resource utilization. Conservation planning in the modern world would need to recognize the significance of this type of gendered spiritual leadership, particularly as individuals struggle with increasing environmental issues (Manocha & Pasupuleti, 2024; Nosheen et al., 2024). These Ladakh and Uttarakhand case studies suggest Indigenous women are not passive witnesses to nature, whose ecological knowledge and cultural wisdom can guide more inclusive and effective conservation.

3.4 *Challenges Faced by Indigenous Women in Conservation*

Encroaching Development, Socio-Economic Change, and Ecological Threats

The protection and upkeep of SNS has long been a major responsibility of Indigenous women throughout the Himalayan region. However, there are a number of factors that are posing a growing danger to the cultural and ecological systems they support. Indigenous knowledge systems have been displaced and sacred sites have been eroded as a result of climate change, growing development projects, changing economic priorities, and deteriorating traditional institutions (Gaur et al., 2024). These constraints impact Indigenous women's autonomy, means of subsistence, and traditional obligations. They are not just natural in nature, but they are also profoundly social (Singh, 2023).

Encroaching development poses one of the biggest risks. Roads, hydropower projects, mining, tourism, and other infrastructure growth frequently happen without enough community involvement (NITI Aayog, 2018; Bhardwaj, 2022). Despite their long-standing protection under traditional systems, sacred groves, pilgrimage routes, and forested regions are occasionally removed or limited under modern land-use rules (Verschuuren, 2016). Hydropower projects in Uttarakhand and Himachal Pradesh, for instance, have caused sacred rivers and forests to become submerged or fragmented, compromising the biological processes and ceremonial customs associated with those spaces (Baker, 2021; Bostancı & Yıldırım, 2025).

Neglecting policies exacerbates these problems even more. Despite the fact that many SNS serve as de facto protection zones, official environmental or cultural heritage regulations seldom ever acknowledge them. State-level conservation and forest projects frequently overlook the importance of Indigenous ecological knowledge or gender-specific responsibilities, which marginalizes women's voices in decision-making (Agarwal, 2001; Mishra et al., 2025). Women's ecological contributions and spiritual leadership are often dismissed as non-scientific or informal, which limits their capacity to shape policy even in areas where they are the main resource managers (Bain, 2017; Nosheen et al., 2024). Additionally, conventional conservation duties have been challenged by socioeconomic developments (Rotarangi & Russell, 2009). Changes in education, market integration, and migration are changing how communities engage with their surroundings (Esmail et al., 2023; Yu et al., 2024).

Intergenerational transmission of ecological knowledge may be weakened if younger generations travel for work or adopt lifestyles less connected to traditional activities (Lima Mota et al., 2023). Furthermore, unsustainable resource exploitation has occasionally resulted from the monetization of forest products and growing land pressure, undermining the spiritual values that formerly controlled usage (Ford et al., 2020). According to elderly women, sacred myths, taboos, and rituals are losing their power in the face of economic need. Another pressing issue is climate change. The natural rhythms and ceremonial calendars have been disrupted by glacial melting, unpredictable rainfalls, and changeable crop seasons (Cajete, 2020). For instance, in Ladakh, the schedule of water god festivals and agrarian rituals that used to be performed by women are being hindered due to decreasing snowfall and retardation of water courses (Angmo, 2016).

Such kinds of environmental shifts add an additional burden to women's ability to undertake their parental roles since sacred places are subjected to increased cultural as well as environmental degradation (Torri et al., 2011). Indigenous women have shown a lot of strength under adverse conditions. Women's collectives have mobilized across the Himalayan republics in defense of sacred forests, rehabilitating endangered customs, and resisting exploitative development (Kala, 2017). For example, Mahila Mandals have mobilized in Uttarakhand to resist deforestation around sacred sites and to reseed indigenous species by appealing to spiritual and ecological justifications. By cultural programs and community schools, Sikkim's women elders assist in recording oral traditions along with imparting lessons to generations about sacred places (Arora, 2006).

Their activism is typically a combination of cultural heritage and conservation, which is grassroots in nature and adjusts to change but remains resilient to outside pressure (Mishra et al., 2025). Indigenous women create culturally responsive, inclusive, and place-based alternative models of sustainability based on customary law, spiritual practice, and solidarity (Kandari et al., 2014). By placing technical competence ahead of embodied knowledge, these projects challenge current conservation practice and encourage more holistic approaches that identify Indigenous women as valuable collaborators (Joshi & Bhardwaj, 2015). On this basis, despite all the setbacks they work under in delivering their conservation roles, they are still intact agents (Negi, 2012). In their involvement with SNS, their work exhibits paradigmatic support for spiritual and environmental harmony. In the Himalayas and other

Indigenous lands, culturally rooted and participatory conservation benefits from the recognition and appreciation of their endeavours (Anthwal et al., 2010).

3.5 Indigenous Women as Agents of Resilience and Advocacy

Community-Based Conservation and the Reimagining of Ecological Futures

Indigenous women in the Himalayan region continue to implement effective community-based conservation strategies in response to mounting ecological and social constraints (Rotarangi & Russell, 2009). These women continue to be in the vanguard of attempts to preserve SNS and traditional ecological knowledge (TEK) systems in spite of significant obstacles, such as environmental deterioration, relocation brought on by development, and socioeconomic changes (Ford et al., 2020). Their answers blend advocacy, cultural practices, and local knowledge, providing crucial insights for more sustainable and inclusive conservation paradigms.

3.6 Indigenous Women as Representatives of Resilience and Advocacy

Reimagine the Fate of Ecological Conservation Through the Gendered Lens

Himalayan Indigenous women continue to develop helpful community-based conservation practices to adapt to increasing ecological and societal constraints (Rotarangi & Russell, 2009). These women remain at the forefront of the effort to conserve SNS and TEK systems in the midst of many obstacles, such as environmental degradation, development-induced displacement, and socioeconomic transformation (Ford et al., 2020). Their strategies integrate advocacy, cultural action, and local knowledge and have lessons for more sustainable and equitable conservation efforts.

Sacred forests, pilgrimage places, and high-altitude lakes are not formally protected in most Himalayan cultures but remain on the basis of community consciousness (O' Neill, 2019; Shah et al., 2024). Women's organizations frequently spearhead this unofficial protection, relying on collective memory, spiritual principles, and ancestors' wisdom to preserve natural harmony. For instance, in Uttarakhand, women from communities that depend on forests have established Mahila Mandals (women's collectives) and Van Panchayats (village forest councils) to coordinate replanting campaigns, keep an eye on the condition of the forests, and bargain with authorities over land use matters. These programs, which are based on the unspoken values of ecological restraint and reciprocity, frequently centre on culturally significant sites (Agarwal, 2001).

In the Bageshwar district of Uttarakhand, women stepped in to prevent unlawful timber harvesting close to a holy forest connected to a local deity. They petitioned district officials to stop the activity, called for community meetings, and invoked traditional taboos against damaging deity forests (Negi, 2017). Their strategy, which reflected how Indigenous women adapt traditional responsibilities to contemporary conservation difficulties, included religious conviction, customary authority, and strategic interaction with governmental systems (Joshi & Bhardwaj, 2015; Nosheen et al., 2024). Climate change has reduced the amount of glacier water availability in Ladakh; thus, women have mobilized around water management through traditional festivals and rituals (Butcher, 2015). Water sources are cleansed and sanctified during the yearly Losar festivities, serving as an opportunity for female elders to teach younger members about the ecological and spiritual functions of sacred springs (Gagné, 2020). A conservation ethic based on common sense and daily actions is strengthened by these endeavours (Angmo, 2024). Women are still in the forefront of attempts to modify ceremonial calendars, control water consumption, and preserve important ecological sites when glacier melt and shifting weather patterns endanger the water systems and the rituals that support them (Sati, 2023).

Despite the fact that official conservation frameworks frequently ignore the contributions of Indigenous women, their strategies are a reflection of traditional adaptive management approaches (Bostancı & Yıldırım, 2025). Environmental consciousness is ingrained in cultural identity through community-based enforcement of ecological rules, seasonal rituals, and oral storytelling (Juma, 2022). Through these methods, conservation is guaranteed to be viewed as a natural progression of cultural life rather than as an external imposition. Moreover, advocacy jobs outside of local communities has become more prevalent among Indigenous women. Women's organizations in Himachal Pradesh based on the forest have participated at the state level meetings to protest around commercialization in and around sacred groves and demand recognition of community-led conservation activities in forest policy (Arora, 2006).

These instances illustrate their ability to integrate traditional knowledge with contemporary campaigning, and depict the resourcefulness of Indigenous women. Their activities demonstrate how locally grounded, culturally derived knowledge systems tend to create effective stewardship, which is in contrast to top-down conservation tactics. Besides, they focus on the implementation of gender-sensitive conservation measures that recognize women as ecological leaders and active decision-makers rather than merely beneficiaries (Chaudhry et al., 2011; Singh, 2023). Therefore, during socio-environmental crises, Indigenous women across the Himalayas are important innovators and strong individuals (Rotarangi & Russell, 2009). Their collective contributions yield important knowledge towards transforming culturally appropriate, participatory, and long-term ecologically ethical models of conservation (Cajete, 2020). Community-based, equal-access, and sustainable environment governance relies upon recognition and empowerment of their agency.

4 Implications for Modern Conservation Paradigms: The Indigenous Women's and Culture-Based Holistic Roles

Conservation is not regarded as a separate, institutionalized activity but as an integral part of everyday life in most Himalayan Indigenous communities (Bostancı & Yıldırım, 2025). Women are integrated within it, especially Indigenous women. They preserve sacred places, manage natural resources, organize ceremonies that reinforce environmental ethics, and keep seeds in order to conserve local biodiversity (Dollo et al., 2009; Sati, 2023). These are not isolated processes; while, they form part of highly embedded cultural settings that blend religiosity, livelihood, and the environment. This consolidated paradigm to nature stewardship deviates from mainstream paradigms of conservation leaning towards wildlife reserves, protected areas, and scientific monitoring (Tiwari & Rani, 2025). These methods are typically imposed through top-down legal frameworks with negligible local community participation (Agrawal & Gibson, 1999). On the other hand, Indigenous systems emphasize decentralized, community-owned knowledge in the form of decades of lived experience that transmitted through oral tradition, ritual, and day-to-day practice (Berkes, 2012).

4.1 Incorporating Indigenous Knowledge into Conservation Policy

Growing appreciation of the importance of indigenous ecological knowledge (IEK) in climate adaptation, biodiversity conservation, and sustainable management of natural resources is becoming increasingly important (Verschuuren, 2016). Yet, institutional conservation efforts rarely draw or overlook such knowledge, particularly when they underestimate the women's roles into consideration. For example, even though women's knowledge of wild foods, medicine, seasons, and long-term harvesting methods is productive and sustainable, it is hardly documented or incorporated into conservation policies (FAO, 2011; Shiva, 2016). Conservation efforts may become more flexible, culturally appropriate, and socially inclusive by incorporating IEK into practice and policy. In the Indian Himalayas, case studies demonstrate that indigenous women's participation in seed banks, water user organizations, and forest management committees enhances ecological results and community involvement (Nimachow et al., 2011; Rai, 2007).

4.2 Spirituality and Gender in Ecological Consciousness

Environmental ethics and spirituality are inextricably linked in many Indigenous cultures. Water bodies, pilgrimage routes, and sacred grooves are conserved for their ecological significance as well as because they are thought to be inhabited by ancestral spirits or deities (Verschuuren, 2016). Women are frequently the major bearers of the taboos and myths that govern moral relationships with nature, as well as ritual specialists and custodians of sacred sites (Tiwari et al., 1998). The concept of ecological awareness, which stresses care, restraint, and reciprocity with the natural world, is culturally ingrained and is influenced by this gendered spiritual leadership (Kala, 2017). There are observable consequences to such consciousness. Ecosystems in regions where religious taboos prevent water contamination or tree cutting are better protected than those in nearby areas with just secular regulations (Gadgil & Vartak, 1976).

4.3 Indigenous Women as Architects of Ecological Ethos

The intangible cultural components of conservation, including rituals, taboos, and oral histories that define local ecological ethics, as well as the concrete aspects, such soil fertility, seed diversity, and water management, are both influenced by indigenous women. In Ziro Valley (Arunachal Pradesh), women regulate water using bamboo and paddy, whereas in Himachal Pradesh, women preserve Renuka Lake by means of religious customs that forbid destruction inside its sacred boundaries (Singh, 2023; Singh et al., 2011). In addition to protecting local ecosystems, their efforts strengthen cultural resilience, communal identity, and spiritual continuity. According to their contributions, Indigenous women are the creators of ecological ethos, whose practices closely resemble those of global conservation but come from a different epistemological foundation (Samakov & Berkes, 2017).

4.4 Implications for Global Conservation Efforts

Growing awareness that technological solutions alone are inadequate is a result of the evolution of global conservation initiatives to address environmental justice, biodiversity loss, and the climate catastrophe (Baker, 2021; Cajete, 2020). The International Union for Conservation of Nature (IUCN), the Convention on Biological Diversity (CBD), and the Intergovernmental Panel on Climate Change (IPCC) have all recognized the importance of traditional ecological knowledge (TEK) and the necessity of inclusive, culturally sensitive approaches. Official support for community rights, gender equity in decision-making, and support for sub-regional conservation initiatives at the juncture of ecological science and indigenous practice are among the structural changes required to incorporate these approaches. A real example of this integration may be seen in the leadership of Indigenous women, who show that conservation frameworks that priorities gender views, culture, and spirituality are the most effective means of achieving sustainability (UNEP, 2021).

5 Conclusion and Recommendations

Sacred natural sites, including groves, lakes, rivers, and pilgrimage paths, are ecological refuges and cultural repositories which harbor rich biodiversity. It embodied as Indigenous spiritual values and community identities across the Himalayan region. The delicate balance of ecological resilience and cultural heritage is kept by indigenous women, who play important role in environmental protection and cultural survival. Their taboo systems, cyclical rituals, and practices all exhibit their gendered religious authority, which is key to upholding the sanctity of such sites. Examples from Himachal Pradesh, Uttarakhand, Arunachal Pradesh, and Ladakh show how women serve as practitioners, storytellers, seed preservers, and conservationists, preserving ecosystems and integrating ecological ethics into community life while strengthening generation-old knowledge systems.

Indigenous women have historically served as protectors of sacred natural sites, greatly aiding in the conservation of both biological and intangible cultural diversity through their daily activities and spiritual stewardship. Their traditional ways of life are being increasingly disrupted by infrastructure growth, socioeconomic changes, and climate change, endangering not only the landscapes they safeguard but also their cultural authority and local governance systems. The legitimacy and effectiveness of Indigenous ecological knowledge, especially its gendered aspects, which have frequently been left out of decision-making processes, must be formally acknowledged by conservation policies in order to address these issues. To protect sacred natural sites from encroachment and extractive development, governments should strengthen legal protections and incorporate them into national and subnational conservation frameworks. Furthermore, it is crucial to support community-led conservation efforts where Indigenous women are already involved, such as traditional festivals, seed banks, and oral history documentation, with financial, legal, and technical assistance.

State and national conservation bodies must include Indigenous women as decision-makers rather than just stakeholders to guarantee that local perspectives influence policy from conception to implementation. In order to facilitate implementation, organizations such as UNESCO, CBD, and IUCN should broaden their frameworks to include spiritual and cultural values in ecosystem protection. They should also cultivate partnerships with organizations led by Indigenous women. Conservation paradigms must urgently change from top-down, technocratic models to gender-inclusive, culturally-based, holistic approaches that acknowledge Indigenous women as active designers of ecological ethics. Their methods teach important lessons about community cohesion, sustainability, and resilience—values that contemporary conservation urgently needs. Since Indigenous women's knowledge, leadership, and spiritual authority are recognized and supported as essential to environmental stewardship, protecting sacred natural sites necessitates not only ecological awareness but also cultural respect and gender sensitivity. This ensures that their frequently marginalized voices are heard at the core of conservation policy and practice.

References

Acharya, A., & Ormsby, A. (2017). The cultural politics of sacred groves: a case study of: Devithans: In Sikkim, India. *Conservation and Society, 15*(2), 232–242.

Agarwal, B. (2001). Participatory exclusions, community forestry, and gender: An analysis for South Asia and a conceptual framework. *World Development, 29*(10), 1623–1648.

Agrawal, A., & Gibson, C. C. (1999). Enchantment and disenchantment: The role of community in natural resource conservation. *World Development, 27*(4), 629–649.

Amritesh, A. R., Ramachandran, V. S., Ormsby, A. A., & Viswanathan, P. K. (2025). Sociocultural heritage of India: A review of studies of the sacred groves of Kerala. *For Trees Livelihoods, 34*(3), 187–206.

Angmo S (2024) Exploring Buddhist tourism potential: The case of Leh, Ladakh in the Indian Himalayas. In Arnab Gantait, A., Singh, K., & Sharma, R. (eds.), Special interest tourism: Exploring dimensions, impacts, and future trends. Astitva Prakashan, India (pp. 95–110).

Anthwal, A., Gupta, N., Sharma, A., Anthwal, S., & Kim, K. H. (2010). Conserving biodiversity through traditional beliefs in sacred groves in Uttarakhand Himalaya, India. *Resources Conservation and Recycling, 54*(11), 962–971.

Anthwal, A., Sharma, R. C., Sharma, A., & Singh, N. (2006). Sacred groves: Traditional way of conserving biodiversity in the Garhwal Himalaya, Uttaranchal. *Journal of American Science, 2*(2), 35–43.

Ao, L. (2022). Sacred grove: A cultural heritage of Nagaland. In *The cultural heritage of Nagaland* (pp. 165–181). Routledge.

Arora, V. (2006). The forest of symbols embodied in the Tholung sacred landscape of North Sikkim, India. *Conservation and Society, 4*(1), 55–83.

Baidya, S., Thakur, B., Malunguja, G., & Devi, A. (2020). Changes of land use land cover and floristic composition in sacred forests of West Karbi Anglong district, Assam, India. *Tropical Plant Research, 7*(2), 296–308.

Bain, W. K. (2017). Conservation of environment through traditional knowledge and wisdom with special reference to beliefs and practices in tribal India: An overview. *Heritage: Journal of Multidisciplinary Studies in Archaeology, 5*, 224–243.

Baker, J. M. (2021). Injustice and justice: The double movement of small hydro development in Himachal Pradesh. In *Advancing environmental justice for marginalized communities in* India (pp. 22–37). Routledge

Barua, K. K., & Slowik, J. (2015). *Traditional ecological knowledge and community-based sustainable natural resource management in the eastern Himalayas—A case study of the Apatani tribe.* Centre for Nature Conservation, Department of Conservation Biology, Georg-August University.

Berkes, F. (2012). *Sacred ecology* (3rd ed.). Routledge

Bhagwat, S. A., & Rutte, C. (2006). Sacred groves: Potential for biodiversity management. *Frontiers in Ecology and the Environment, 4*(10), 519–524.

Bhandari, H., Kaur, P., & Grover, A. R. (2018). An explorative study of Hindu goddess sites in Himachal Pradesh. *International Journal of Scientific Research and Review, 7*(6), 450–455.

Bhardwaj, R. L. (2022). Role of ecotourism in sustainable development of Himalayas: A case study of Himachal Pradesh. In *Rethinking Himalaya: Its Scope and Protection* (pp. 109–129).

Bhattacharjee, S. (2024). Indigenous nature conservation in Meghalaya: Environmental and religious dimensions of tribal land ownership among the Khasi community. In *Environmental humanities in India* (pp. 69–83). Springer Nature Singapore.

Bhattarai, N. K. (1997). Role of women in the use and conservation of biodiversity: Studies in the Nepal Himalaya. *Banko Janakari, 7*(2), 3–8.

Borrini, G. (2010). *Bio-cultural diversity conserved by indigenous peoples and local communities: Examples and analysis.* Companion document to IUCN/CEESP (p. 10).

Bostancı, S. H., & Yıldırım, S. (2025). Environmental justice in the Himalayan region. In *People and Mountain Environments: Interconnectedness for sustainable development in the Himalayas* (pp. 339–353). Springer Nature Switzerland.

Butcher, A. (2015). Ceremonial activities in development: The spiritual 'Problems' for Ladakh's secular encounter. *The Tibet Journal, 40*(2), 195–218.
Byrne, D. (2012). The enchanted Earth: Numinous sacred sites. In *Sacred natural sites* (pp. 79–87). Routledge.
Cajete, G. A. (2020). Indigenous science, climate change, and indigenous community building: A framework of foundational perspectives for indigenous community resilience and revitalization. *Sustainability, 12*(22), 9569.
Chand, S., & Mishra, L. (2024). Char dham yatra and pilgrimage tourism in Uttarakhand: Emerging opportunities, challenges and future trends. *International Journal of Innovations Science, Engineering and Management*, 114–119.
Chanda, K. (2021). Conservation through belief: A review of sacred groves in Karbi Anglong and Dima Hasao-only districts in Assam under Eastern Himalayan region. *Envis Bulletin Himalayan Ecology, 29*, 32.
Chaudhry, P., & Murtem, G. (2015). Role of sacred groves, value education and spirituality in conserving biodiversity with special reference to Arunachal Pradesh state of India. *International Journal of Society Systems Science, 7*(2), 151–180.
Chaudhry, P., Dollo, M., Bagra, K., & Yakang, B. (2011). Traditional biodiversity conservation and natural resource management system of some tribes of Arunachal Pradesh, India. *Interdisciplinary Environmental Review, 12*(4), 338–348.
Choden, Y., Sharma, M. P., Pandey, G., Gupta, S. K., & Dema, K. D. (2022). Ecological health assessment of Renuka lake, Himachal Pradesh, India. *Nature Environment and Pollution Technology, 21*(1), 167–174.
Colding, J., & Folke, C. (2001). Social taboos: "invisible" systems of local resource management and biological conservation. *Ecological Applications, 11*(2), 584–600.
Dad, J. M., & Khan, A. B. (2011). Threatened medicinal plants of Gurez valley, Kashmir Himalayas: Distribution pattern and current conservation status. *International Journal of Biodiversity Science, Ecosystem Services and Management, 7*(1), 20–26.
Devi Khumbongmayum, A., Khan, M. L., & Tripathi, R. S. (2005). Sacred groves of Manipur, northeast India: Biodiversity value, status and strategies for their conservation. *Biodiversity and Conservation, 14*(7), 1541–1582.
Dhanaraju, V., & Bijeta, R. K. (2022). Cultural ecology of sacred groves in Karbi Anglong District of Assam, India. *Indian Journal of Natural Sciences, 13*(73), 46860–46866.
Diengdoh, B. H., & Jayakumar, S. (2021). A Comparative study of local attitudes and perceptions of sacred groves between village communities in East Khasi Hills, Meghalaya, India. *International Journal of Ecology and Environmental Sciences, 47*(3), 197–208.
Dollo, M., Samal, P. K., Sundriyal, R. C., & Kumar, K. (2009). Environmentally sustainable traditional natural resource management and conservation in Ziro Valley, Arunachal Himalaya, India. *Journal of American Science, 5*(5), 41–52.
Esmail, N., McPherson, J. M., Abulu, L., Amend, T., Amit, R., Bhatia, S., Bikaba, D., Brichieri-Colombi, T. A., Brown, J., Buschman, V., & Fabinyi, M. (2023). What's on the horizon for community-based conservation? Emerging threats and opportunities. *Trends in Ecology and Evolution, 38*(7), 666–680.
FAO. (2011). *The role of women in biodiversity conservation*. Food and Agriculture Organization.
Ford, J. D., King, N., Galappaththi, E. K., Pearce, T., McDowell, G., & Harper, S. L. (2020). The resilience of indigenous peoples to environmental change. *One Earth, 2*(6), 532–543.
Gadgil, M., & Vartak, V. D. (1976). The sacred groves of Western Ghats in India. *Economic Botany, 30*(2), 152–160.
Gagné, K. (2020). The materiality of ethics: Perspectives on water and reciprocity in a Himalayan Anthropocene. *Wiley Interdisciplinary Reviews: Water, 7*(4), Article e1444.
Gaur, M. K., Goyala, R. K., Choudharyb, V., Kanwara, M. S., Singhc, N., Chichagharea, A. R., & Raghuvanshid, M. S. (2024). Traditional ecological knowledge and natural resource management in the Cold Desert of Leh-Ladakh, India. *Ecofeminism and Climate Change, 5*(2), 81–88.

Gokhale, Y., & Pala, N. A. (2016). Developing conservation management strategies for biodiversity rich sacred natural sites of Uttarakhand, India. *Asian Biotechnology and Development Review, 18*(3).

Gopal, D., von der Lippe, M., & Kowarik, I. (2018). Sacred sites as habitats of culturally important plant species in an Indian megacity. *Urban for Urban Green, 32*, 113–122.

Gupta, S. S. (2020). *Relation between nature and tribal culture in Tripura.* Tribal Research and Cultural Institute, Agartala, Tripura.

Jha, K. K. (2017). Heritage non-timber forest product use and management by indigenous community in Northeastern Himalayan hotspot, Arunachal Pradesh, India. *Journal of Plant Chemistry Ecophysiology, 2*(1), 1012.

Joshi, K., & Bhardwaj, N. (2015). Women and natural resource management: A study of 'communities of practice 'prevailing in women farmers' community management of water and forests of lesser Himalayan region in India. *International Journal of Advanced Research, 3*, 363–374.

Juma, F. A. (2022). Recapturing the oral tradition of storytelling in spiritual conversations with older adults: An afro-indigenous approach. *Religions, 13*(6), 563.

Kala, C. P. (2017). Conservation of nature and natural resources through spirituality. *Applied Ecology and Environmental Sciences, 5*(2), 24–34.

Kandari, L. S., Bisht, V. K., Bhardwaj, M., & Thakur, A. K. (2014). Conservation and management of sacred groves, myths and beliefs of tribal communities: A case study from north-India. *Environmental Systems Research, 3*(1), 16.

Kandel, P., Chettri, N., Chaudhary, S., Sharma, P., & Uddin, K. (2021). Ecosystem services research trends in the water tower of Asia: A bibliometric analysis from the Hindu Kush Himalaya. *Ecological Indicators, 121*, Article 107152.

Kaur, J., & Balodi, K. N. (2016). Lakes of the Gods: Sacred high-altitude lakes of Uttarakhand, India. In *Asian sacred natural sites* (pp. 219–229). Routledge.

Kharumnuid, W., & Bharucha, E. (2020). Management practices for biodiversity conservation through community conserved areas in Meghalaya, India. *Indian Journal of Ecology, 47*(1), 226–234.

Khumbongmayum, A. D., Khan, M. L., & Tripathi, R. S. (2004). Sacred groves of Manipur–ideal centres for biodiversity conservation. *Current Science, 87*(4), 430–433.

Khuroo, A. A., Mehraj, G., Muzafar, I., Rashid, I., & Dar, G. H. (2020). Biodiversity conservation in Jammu and Kashmir State: Current status and future challenges. In G. Dar, & A. Khuroo (eds.), *Biodiversity of the Himalaya: Jammu and Kashmir State.* Topics in biodiversity and conservation (vol. 18). Springer. https://doi.org/10.1007/978-981-32-9174-4_41

Kulkarni, A. V., & Karyakarte, Y. (2014). Observed changes in Himalayan glaciers. Current Science, 237–244.

Lalzarzovi, S. T. (2022). Plant species diversity in a tropical semi-evergreen forest in Mizoram (northeastern India): Assessing the effectiveness of community conservation. *Journal of Threatened Taxa, 14*(5), 21055–21067.

Levy, C. (1992). Gender and the environment: The challenge of cross-cutting issues in development policy and planning. *Environment and Urbanization, 4*(1), 134–149.

Lima Mota, M. R., Lauer-Leite, I. D., & Novais, J. S. D. (2023). Intergenerational transmission of traditional ecological knowledge about medicinal plants in a riverine community of the Brazilian Amazon. *Polibotánica, 56*, 311–329.

MacGregor, S. (ed.). (2017). *Routledge handbook of gender and environment*. Taylor & Francis.

Malhotra, K. C., Gokhale, Y., Chatterjee, S., & Srivastava, S. (2007). *Sacred groves in India.* Aryan Books International.

Manocha, A., & Pasupuleti, R. S. (2024). Investigating the concept of religious pathways: A case of Nanda Devi Raj Jaat Yatra in Uttarakhand, India. In A. Baghel, & S. Parikh (eds.), *Advancing heritage innovations in India* (pp. 75–87).

Mishra, A., Sekar, K. C., & Pant, H. (2025). Women protecting forests in Uttarakhand, India. In T. Rogelja, & L. Kroese (eds.), *Women as stewards of forests.* (Tropical Forest Issues 63). Tropenbos International. xvi + 198 pp.

Murtem, G., & Chaudhry, P. (2014). Sacred groves of Arunachal Pradesh: Traditional way of biodiversity conservation in Eastern Himalaya of India. *Journal of Biodiversity Management and Forestry, 3*(2), 14, 2.

Negi, C. S. (2012). Culture and biodiversity conservation: Case studies from Uttarakhand, Central Himalaya. *Indian Journal of Traditional Knowledge, 11*(2), 273–278.

Negi, C. S. (2017). The sacred natural sites, the social taboo system and the scope of developing some of the sites as biodiversity heritage sites, Uttarakhand, Central Himalaya. *Ethnobotany of India, Apple Academic Press, 4*, 445–477.

Nimachow, G., Joshi, R. C., & Dai, O. (2011). Role of Indigenous knowledge system in conservation of forest resources—A case study of the Aka tribes of Arunachal Pradesh. *Indian Journal of Traditional Knowledge, 10*(2), 276–280.

NITI Aayog. (2018). *Sustainable tourism in the Indian Himalayan region*. Report of Working Group II, NITI Aayog, Government of India.

Nosheen, M., Mumtaz, T., & Amjad, B. (2024). Gendered dynamics and economic influences in transboundary mountain ecology: A case study of the Himalayan region. *Journal of Asian Development Studies, 13*(4), 349–367.

Odhiambo, B. D. O., & Manuga, M. (2017). Tshatshingo pothole: A sacred Vha-Venda place with cultural barriers to tourism development in South Africa. *African Journal of Hospitality, Tourism and Leisure, 6*, 1–12.

O'Neill, A. R. (2019). Evaluating high-altitude Ramsar wetlands in the Eastern Himalayas. *Global Ecology and Conservation, 20*, Article e00715.

Ormsby, A. (2013). Analysis of local attitudes toward the sacred groves of Meghalaya and Karnataka, India. *Conser Soc, 11*(2), 187–197.

Ormsby, A. A., & Bhagwat, S. A. (2010). Sacred forests of India: A strong tradition of community-based natural resource management. *Environmental Conservation, 37*(3), 320–326.

Oviedo, G., & Jeanrenaud, S. (2007). Protecting sacred natural sites of indigenous and traditional peoples. *Protected Areas Spirit*, 77–99.

Patnaik, S. K. (2020). Multifunctional agriculture as a holistic tool in addressing sustainability issues among mountain communities of Ziro valley, Arunachal Pradesh. *Transactions, 42*(2), 189.

Patwal, A. S. (2016). Pilgrimage satisfaction of religious event-Nanda Devi Raj Jaat 2014 Uttarakhand. *PUSA Journal of Hospitality and Applied Sciences, 2*, 57–72.

Rai, S. C. (2007). Traditional ecological knowledge and community-based natural resource management in northeast India. *Journal of Mountain Science, 4*(3), 248–258.

Rotarangi, S., & Russell, D. (2009). *Social-ecological resilience thinking: Can indigenous culture guide environmental management*.

Samakov, A., & Berkes, F. (2017). Spiritual commons: Sacred sites as core of community-conserved areas in Kyrgyzstan. *International Journal of Commons, 11*(1).

Samal, P. K., & Dhyani, P. P. (2006). Gender in the management of indigenous knowledge: Reflections from Indian Central Himalaya. *Current Science* 104–108.

Sarmah, R., Arunachalam, A., & Upadhayay, N. (2009). Sacred values and biodiversity conservation around Namdapha National Park in Arunachal Pradesh. In *Bio-cultural diversity and sustainable development in North East India: Status, vision and challenges* (p. 158). Read worthy Publications (P) Ltd.

Sati, V. P. (2023). Diversity in fairs and festivals. In *Natural and cultural diversity in the Himalaya* (pp. 85–99). Springer Nature Switzerland.

Shah, S., Sen, S., & Sahoo, D. (2024). State of Indian Northwestern Himalayan lakes under human and climate impacts: A review. *Ecological Indicators, 160*, Article 111858.

Sharma, U. K., & Gogoi, J. (2006). In P. C. Trivedi (Ed.), *Ethno-medico-botany of some sacred plants of Dhemaji District of Assam* (pp. 100–105). International Publishing House.

Sharma, S., Gosavi, V. E., Kanwal, K. S., Agnihotri, V., Chandra Sekar, K., Arya, S. C., & Rawal, R. S. (2020). *High altitude wetlands in the Indian Himalaya: Conservation and management*. G. B. Pant National Institute of Himalayan Environment. Kosi - Katarmal, Almora.

Shimrah, T., Bharali, S., Rao, K. S., & Saxena, K. G. (2012). Cultural Landscapes: The basis for linking biodiversity conservation with the sustainable development in West Kameng, Arunachal Pradesh. In *Cultural landscape the basis for linking biodiversity conservation with the sustainable development* (pp. 105–147), UNESCO, NIE.

Shiva, V. (2016). *Staying alive: Women, ecology, and development*. North Atlantic Books.

Singh, A. (2014). *Buddhist traditional ethics: A source of sustainable biodiversity* (pp. 13–14). SAARC/CC/Research Grant.

Singh, A. B., & Teron, R. (2015). Diversity of wild edible plants used by the Angami-Nagas in Kohima District of Nagaland, India. *Pleione, 9*(2), 311–324.

Singh, H., Agnihotri, P., Pande, P. C., & Husain, T. (2011). Biodiversity conservation through a traditional beliefs system in Indian Himalaya: A case study from Nakuleshwar sacred grove. *The Environmentalist, 31*(3), 246–253.

Singh, N. C. (2023). Sustainable natural resource management through traditional ecological knowledge: A perspective on the role of Apatani tribal women, Arunachal Pradesh. *Traditional ecological knowledge of resource management in Asia* (pp. 239–259). Springer International Publishing.

Sinthumule, N. I. (2021). An analysis of communities' attitudes towards wetlands and implications for sustainability. *Global Ecology and Conservation, 27*, Article e01604.

Studley, J. (2019). *Indigenous sacred natural sites and spiritual governance: The legal case for juristic personhood*. Routledge.

Tiwari, B. K., Barik, S. K., & Tripathi, R. S. (1998). Biodiversity value, status, and strategies for conservation of sacred groves of Meghalaya, India. *Ecosystem Health, 4*(1), 20–32.

Tiwari, B. K., Barik, S. K., & Tripathi, R. S. (2010). *Sacred forests of Meghalaya: Biological and cultural diversity*. Regional Centre, National Afforestation and Eco-development Board, NEHU, Shillong.

Tiwari, C., & Rani, M. (2025). Where the wild things are: Identifying the factors behind and mitigation measures for biodiversity conservation crisis in Himalayas. In A. Borthakur & P. Singh (Eds.), *People and mountain environments: Interconnectedness for sustainable development in the Himalayas* (pp. 247–286). Springer.

Torri, M. C., & Herrmann, T. M. (2011). Spiritual beliefs and ecological traditions in indigenous communities in India: Enhancing community-based biodiversity conservation. *Nature and Culture, 6*(2), 168–191.

Tsopoe, M., Meetei, K. B., Baruah, R. C., Kalita, S. S., & Giri, K. (2025). Study on tree diversity, community utilization categories and soil carbon stock in alder-based Jhum farming and community forest of Khonoma Village, Nagaland. *Vegetos*, 1–13.

UNESCO. (2008). *Sacred natural sites: Guidelines for protected area managers*. IUCN.

Upadhyay, K. K., Japang, B., Singh, N. S., & Tripathi, S. K. (2019). Status and socio-ecological dimensions of sacred groves in Northeast India. *Journal of Applied Natural Science, 11*(3), 590.

Verschuur, C., Gu´erin, I., Gu´etat-Bernard, H. (eds.). (2016). *Under development: Gender*. Springer.

Verschuuren, B. (2016). How the cultural, spiritual and philosophical underpinnings of sacred natural sites can make conservation in Asia more effective and sustainable. In *Asian sacred natural sites: Philosophy and practice in protected areas and conservation* (pp. 297–309).

Wakgchok, S. (2007). Sacred landscapes in the Nubra valley. *Rivista Degli Studi Orientali, LXXX*(supplemento 2), 1000–1013.

Wild, R., & Mcleod, C. (2008). *Sacred natural sites: Guidelines for protected area managers*. IUCN.

Yu, Y., Appiah, D., Zulu, B., & Adu-Poku, K. A. (2024). Integrating rural development, education, and management: Challenges and strategies. *Sustainability, 16*(15), 6474.

Beyond Essentialism: Rethinking the Women-Nature Nexus in an Age of Globalization

K. N. Ajoykumar

Abstract Although conservation and environmental discourses frequently reproduce gendered biases, gender equity is widely acknowledged as being crucial to attaining sustainable development. Essentialist perspectives, which portray women as inherently closer to nature, have long influenced environmental ethics and policy, particularly within climate change debates. However, social ecofeminist critiques challenge these assumptions, highlighting the fact that social, political, and economic structures—rather than biological fate—determine women's environmental roles. This study adopts a multifaceted approach to examine how globalization both enables and constrains women's environmental involvement. It explores how market-driven economies, conservation interventions, and climate-induced displacement restructure women's access to and control over natural resources. With particular attention to the Himalayan context, the analysis highlights women's agency in negotiating shifting ecological and social realities. Women are not merely passive custodians of nature but active political actors whose strategies of adaptation and resistance challenge deterministic narratives of gender and environment. The study highlights the need for policy interventions that goes beyond essentialism to arrive at socially rooted and context-specific understandings of the gender environment relationship. By situating women's role it proceeds to place the role of women within the broader dynamics of globalization, environmental governance, and cultural change, thereby developing a social ecofeminist perspective, in which diversity, agency, and structural inequality are prioritized in creating a constantly evolving nexus between women and nature.

Keywords Climate change · Women-nature nexus · Conservation · Eco feminism · Political ecology · Globalization · Gender bias · Social ecofeminism · Essentialism · Himalayan women

---

K. N. Ajoykumar (✉)
Department of Plant Science, Kannur University, Kannur, India
e-mail: knajoykumar@gmail.com

N. Moran et al. (eds.), *The Participatory Role of Indigenous Women*,
https://doi.org/10.1007/978-3-032-24059-0_13

1 Introduction- Gender, Environment, and Changing Realities

Over the past few decades, the relationship between gender, ecology and development has become a central theme in environmental science. At the core of this discourse lies a familiar assumption—that women possess a special closeness to nature. This view, rooted in roles such as caregiving, subsistence farming, and water or forest management, has shaped environmental movements and found its strongest expression in strands of ecofeminist thought. (Shiva, 1988; Mies & Shiva, 1993; Elmhirst & Resurreccion, 2021). While this concept has shaped global environmental discussion, it has also tended to hide the socio-political, economic, and historical contexts that structure women's actual engagement with ecological systems. However, the idea that women possess an innate bond with nature feels increasingly inadequate in today's globalized world. As climate change accelerates, markets expand, and societies undergo rapid transformation, these essentialist views risk oversimplifying women's complex and shifting relationships with the environment. When women are framed merely as "natural" stewards or victims of environmental harm, their political agency, economic roles, and diverse lived experiences are often overlooked. (Elmhirst & Resurreccion, 2021; Nightingale, 2021; Sultana, 2022).

Ecofeminism, emerging in the late twentieth century, sought to connect critiques of environmental degradation with critiques of patriarchal domination, presenting these as mutually reinforcing systems of exploitation (Gaard, 1993; Merchant, 2020). Early ecofeminist narratives—especially those rooted in cultural and spiritual traditions—often relied on essentialist assumptions that women's closeness to nature was biologically determined (Agarwal, 1992; Salleh, 1997). While these ideas helped to focus the gendered dimensions of environmental struggles, they have since been critiqued for romanticising women's roles and for neglecting how differences of class, caste, ethnicity, and political economy mediate women's relationships with the environment (Arora-Jonsson, 2014; Sanchez-Garcia et al., 2025).

An alternative framework emerged through feminist political ecology, which argues that gendered environmental relationships are shaped not by innate qualities but by historically specific power structures, systems of resource access, and socio-cultural contexts (Rocheleau et al., 1996; Nightingale, 2021). This approach shifts the analytical focus from an assumed universal female–nature connection to the differentiated realities of women's lives, thereby highlighting the political nature of environmental involvement.

The accelerating forces of globalisation have further transformed these dynamics. Market integration, the commodification of natural resources, changing patterns of land use, and intensified conservation regimes have reshaped rural livelihoods and altered gendered divisions of ecological labour (Baviskar, 2020; Sultana, 2022). In the Himalayas—an ecologically fragile and culturally diverse region—such transformations are intensified by climate change, large-scale infrastructural expansion, out-migration, and tourism. Women in these mountainous areas are both disproportionately exposed to environmental risks and central actors in adaptation, conservation, and livelihood diversification. It has been recognized that the effects of climate change will be harshest in tropical countries in the South and will affect the poor the most. As the global market economy and the development process started to make inroads into traditional communities, women's economic position became increasing insecure (Afshar 1985; Sen & Grown, 1987; Mies, 1986). This insight has led some to claim that women are more vulnerable to the effects of climate change for a number of reasons (Arora Jonsson, 2011).

This review examines the changing women–nature nexus under the combined pressures of climate change, neoliberal development, and socio-political restructuring, with the Himalayas as the focal region. Drawing from ecofeminist thought, feminist political ecology and post colonial perspectives, it situates women's environmental roles within wider systems of power, access, and knowledge. The article focus on the limitations of essentialist methods and advances a post-essentialist, justice-oriented method. Such an approach recognises that women's environmental agency is mediated by caste, class, ethnicity, and other axes of social differentiation, and that it must be placed within the broader political economy of resource governance (Crenshaw, 1991; Elmhirst & Resurreccion, 2020).

The review trace the historical and theoretical evolution of women–environment relationship, explore empirical evidence from the Himalayan context, analyse grassroots activism and policy interventions, and assess how these engagements challenge reductive portrayals of women as either 'natural stewards' or passive victims. Instead, it positions women as diverse, situated, and politically engaged actors actively shaping ecological futures.

2 Women–Environment Relationships in Historical and Theoretical Perspective

2.1 *Historical Context*

The relationship between women and the environment has been historically mediated by political ideologies, economic transformations, and socio-cultural norms. Across civilisations, women have been alternately celebrated as "guardians" of nature and relegated to domestic labour, with both narratives tending to simplify the complexity of women's ecological engagement. Conceptualising the relationship between women and nature as an essential one thus helped to define a global women's environmental movement and challenged the hegemony of (male) scientific knowledge as the privileged source of information about environmental change (Nightingale, 2006).

Colonialism, capitalism, and patriarchal governance structures have simultaneously relied on women's environmental labour while limiting their access to resources and decision-making (Agarwal, 1992; Harcourt & Nelson, 2015; Rocheleau et al., 1996; Sultana, 2022). In many Indigenous and premodern societies, women's environmental roles were legitimised through spiritual and ecological worldviews. Nature was often personified in feminine forms—such as the Hindu goddess Ganga or the Greek deity Demeter—casting women as intermediaries between human society and the natural world (Ruether, 2005; Shiva, 1989).

While cultural adoration could confer symbolic respect, it rarely translated into social equality. Women often remained excluded from formal governance and land ownership, revealing that symbolic "Mother Earth" narratives could coexist with entrenched hierarchies (Arora-Jonsson, 2014; Plumwood, 1993). Ecofeminism has contributed a great deal both to activist struggle and to theorising links between women's oppression and the domination of nature over the last two decades. In some versions it has engaged with all four forms of exploitation encompassed in race, class, gender and nature. At the same time, ecofeminism has been stereotyped in some quarters both as theoretically weak and as doubtfully liberated, and also as exclusively linked to cultural feminism Internal debate on the issue of political ecology (which has involved social ecology, deep ecology and ecofeminism) is also the question of liberatory coherence and of the relationship between the radical movements and critiques of oppression each of these internal green positions is aligned with.(Plumwood, 1993) Contemporary post-essentialist views emphasizes that such roles are historically conditioned, shaped by governance systems, labour divisions, and intersecting power dynamics rather than innate biological affinity (Nightingale, 2021).

2.2 *The Legacy of Ecofeminism and Essentialism*

Ecofeminism emerged in the 1970s at the intersection of second-wave feminism and environmentalism, advancing the argument that the domination of women and the degradation of nature are interconnected outcomes of patriarchal and capitalist systems (d'Eaubonne, 1974; Merchant, 1980; Shiva, 1988). Ecofeminism was initially more rooted in the humanities—in history philosophy and religion (Merchant, 1980; Plumwood, 1993; Ruether, 1995; Ojeda et al., 2022) and was often closely articulated with Indigenous cosmologies, politics, and literature. Cultural ecofeminism emphasised spiritual and symbolic ties between women and nature, drawing on goddess traditions and earth-based spiritualities (Griffin, 1978; Starhawk, 1982). Social ecofeminism, in contrast, placed environmental struggles in material realities—land rights, labour, and political economy—critiquing the exploitative nexus between capitalism and environmental degradation (Mies & Shiva, 1993; Salleh, 1997). Another significant strand emphasized how cultural and political regimes of production and reproduction generate distinct ecologies (Merchant, 1989). Yet another important constellation of ecofeminists influenced by Marx criticized the pervasiveness of patriarchy within Marxism and the push for "modernization" and hyper consumption promoted within some Marxist traditions (Luxemburg,

2015, Mies, 1986, Bennholdt & Mies, 1999). Marxist-influenced eco-socialist feminists, including Evelyn Reed (1975), Alexandra Kollontai (1920), and Tithi Bhattacharya (2017), extended this critique by emphasising how capitalism depends on the appropriation of women's unpaid reproductive labour and the commodification of nature. Anuradha Ghandy's (2016) linked Ecological destruction to class exploitation in postcolonial contexts, highlighting how rural and Indigenous women bear the brunt of extractive economies, situating it alongside radical, anarchist, socialist, and post-modern approaches.

There has been a tendency to identify ecofeminism with an essentialist point of view. It is on the assumption that there is a biological unity between nature and women. Man is excluded from this. It essentialises woman and natural world. A false image of women is created with this according to the critics. There itself inequalities exist among women. That is discarded by this view (Meller, 1997). For some ecofeminists, women's affinity to nature comes from 'their physiological functions (birthing, menstrual cycles) or some deep element of their personalities (life-oriented, nourishing/caring values. In this way they "understand" nature, whereas men do not and cannot (Meller, 1996).

Cultural ecofeminism found popular resonance but was often critiqued for biological determinism and universalising women's experiences (Warren, 2000). Social ecofeminists countered that women's environmental agency arises from material conditions and political struggles, not from an essentialised connection to nature (Gaard, 2015; Mellor, 1997).

2.3 *Eco-socialist Feminist Perspectives*

Social ecology stresses environmental problems as social problems, arising from the domination of human by human, but has little sensitivity to the domination of non-human nature (Plumwood, 1993). Social ecofeminism aligned closely with Marxist and socialist feminist traditions, focusing on structural oppression and resource control. This perspective resonates with eco socialist critiques that capitalism treats both nature and women's reproductive labour as "free gifts" to the economy (Mellor, 1996). Reed's (1975) anthropological analysis placed women's subordination within the historical transition from communal to private property systems, while Kollontai (1920) argued for collective social responsibility over reproduction, childcare, and resource management.

Ecofeminists such as Shiva (1988) and Mies and Shiva (1993) analysed how colonial science and development projects undermined women's ecological knowledge systems, replacing them with monocultures and market-driven agriculture. These critiques remain relevant in contemporary debates on food sovereignty, seed politics, and biodiversity conservation.

2.4 *Social Ecofeminism and Political Economy-Capitalism, Gender, and Ecology*

Eco-socialist feminism brings together the critical insights of socialist feminism, Marxist ecology, and environmental justice, situating the gender–environment relationship within the structural dynamics of capitalism, imperialism, and ecological crisis. Unlike essentialist frameworks that portray women's environmental roles as a natural inheritance, eco-socialist feminism frames them as historically produced within exploitative systems of production and reproduction (Biro, 2011; Mellor, 1997). The approach builds on Marx's notion of the metabolic rift—the structural dislocation between human societies and ecological systems under capitalist production (Foster, 2000; Marx, 1867)—and expands it to include gendered labour, recognising that women's unpaid care work and ecological stewardship are systematically appropriated without remuneration. Unpaid domestic labour has continued to be a crucial site of feminist inquiry to uncover the oppression of women in capitalist patriarchal societies (Bhattacharya, 2017; Varela, 2019).

Eco-socialist feminists argue that the subordination of women and the commodification of nature are mutually reinforcing processes embedded in the political economy of capitalism. The reproductive labour performed by women—food production, seed preservation, child-rearing—remains economically invisible, even though it sustains both the labour force and ecological systems (Bhattacharya, 2017). This erasure is not incidental but constitutive of capitalist accumulation. Genuine liberation requires dismantling the privatised burden of reproduction, a point that eco-socialist feminists extend to include the collective stewardship of natural resources (Kollontai, 1920; Merchant, 2020, Mellor, 1997).

Extractivism is a part of capitalist exploitation which can be defined as a model of overexploitation of nature under which local communities and environments suffer the damage, but do not benefit from extractivist activities (Gomez-Barris, 2017; Veltmeyer & Bowles, 2014). Drawing from ecofeminist studies, current critiques of extractivism address masculinist productions of nature and how they represent a continuation of colonial violence (Mies, 1986; Shiva, 1988; Plumwood, 1993). Eco socialist feminists situate environmental degradation within the structural logics of capitalism, patriarchy, and imperialism, emphasising that the exploitation of nature and the oppression of women are not parallel phenomena but mutually reinforcing processes (Kovel, 2002; Salleh, 2017). They argue that capitalist systems rely on the appropriation of unpaid reproductive labour, disproportionately performed by women, as well as on the commodification and over-extraction of natural resources.

Most importantly, eco socialist feminists highlight the key differences between the lived realities of working-class or peasant women and those who occupy elite or political positions of power. They caution that certain strands of ecofeminism risk romanticising women's history and experiences, promoting a universalised category of "woman" that overlooks intersecting class, racial, and geopolitical inequalities (Pear, 2021). By centring the struggles of marginalised women—especially those engaged in subsistence farming, informal economies, and grassroots environmental defence—eco socialist feminism resists depoliticised narratives and aligns itself with broader anti-capitalist and anti-colonial movements (Bauhardt & Harcourt, 2019; Foster et al., 2023). Capitalism emerged from a patriarchal feudal society in which male private property inheritance demanded women's bodies and lives were subordinated to the needs of the family. Ecological feminists inspired by feminist Marxism have also questioned the neat separation between workers and communities within capitalism, while once again reminding us of the inseparability of production and social reproduction (Araya, 2022).

In the South Asian context, the studies on the intersection of caste, gender, and class provide a vital perspective for understanding how eco-socialist feminism must address not only capitalism but also feudal and caste-based modes of exploitation (Ghandy, 2016). Ghandy stated that any movement that fails to challenge these intertwined systems risks reproducing existing hierarchies, even within progressive environmental struggles. Such insights are particularly relevant in the Himalayan context, where women's environmental work is shaped by agrarian structures, caste-based exclusions, and state-led conservation policies that prioritise market and tourism revenues over local livelihoods (Agarwal, 1994; Rege, 1998; Sundar, 2001).

Eco-socialist feminism also draws inspiration from environmental pioneers such as Rachel Carson, whose Silent Spring (1962) exposed the ecological and human health consequences of chemical-intensive agriculture. While Carson did not explicitly identify as a socialist, her work revealed the structural links between industrial capitalism, ecological degradation, and public health—a link later developed by eco-socialist feminists to critique agro-industrial regimes that displace sustainable, women-led agricultural practices. Simone de Beauvoir's (1949) philosophical interrogation of women's 'otherness' is similarly instructive: just as women have been defined as "the Other" in patriarchal societies, nature has been constructed as a subordinate realm to be exploited, creating parallel logics of domination.

2.5 *Eco-socialist Feminism as Theory and Praxis*

Researchers (Gaard, 2015; Salleh, 2017) emphasises that eco-socialist feminism is not merely analytical but deeply praxis-oriented. It advocates for modes of production and reproduction grounded in collective ownership, agroecology, and democratic decision-making—rejecting both capitalist commodification and patriarchal control. Social reproduction theory extends this praxis by framing ecological sustainability as inseparable from the struggle to de commodify basic needs, including water, energy, and food systems (Bhattacharya's, 2017). Many scholars are of the view that eco socialism that is not ecofeminist at heart is not worth (Barca, 2019; Brownhill & Turner, 2019, 2020; Feder, 2019; Giacomini et al., 2018; Kovel, 2005).

In the Himalayan region, an eco-socialist feminist perspective offers a way to reinterpret women's environmental activism—such as forest protection, opposition to hydropower projects, and seed sovereignty movements—not as acts of 'natural' care but as conscious resistance to capitalist extraction and state dispossession. Women's struggles here can thus be understood as defending both social reproduction and ecological integrity against the twin logics of accumulation and patriarchal control. This framework also reveals the limits of technocratic climate adaptation strategies that ignore questions of ownership, land rights, and control over productive resources.

By integrating historical materialism, feminist critique, and ecological analysis, eco-socialist feminism provides a robust theoretical grounding for a post-essentialist understanding of the women–nature nexus. It recognises that achieving gender justice and ecological sustainability requires dismantling the economic structures that commodify both women's labour and the earth itself, and replacing them with systems rooted in equality, solidarity, and ecological care (Foster, 2000; Hartsock, 1998; O'Connor, 1998).

2.6 Against Essentialism: Postcolonial and Post Structural Interventions

Critics from feminist political ecology, postcolonial theory, and poststructuralism have consistently challenged ecofeminism's essentialist strands for portraying women as a homogeneous category defined by an ahistorical "closeness" to nature (Mohanty, 1988; Rocheleau et al., 1996; Spivak, 1988). They argue that such narratives erase intra-gender differences shaped by caste, class, ethnicity, race, and geography (Crenshaw, 1991; Sultana, 2024).

Essentialist framings depoliticise women's environmental labour, naturalising care and subsistence work while obscuring the structural inequalities that compel it (Foucault, 1978; Jackson, 1993). This naturalisation can be co-opted by neoliberal and "green growth" agendas, positioning women as ideal environmental stewards without challenging exploitative structures (Jadhav & Shrestha, 2024).

Post-essentialist approaches instead foreground power relations, political economy, and scale. They situate women's environmental actions within historical struggles over land, water, and forest governance, rejecting deterministic binaries of "nurturers" versus "victims" (Sundberg, 2004; Nightingale, 2006). Contemporary studies also integrates queer, trans, and non-binary perspectives, recognising the fluidity of gender subjectivities in environmental governance (Whyte, 2017; Stein, 2022).

3 Climate Change, Livelihoods, and Gendered Vulnerability

Global literature on gender and climate change shows that gender relations are integral to both climate change and the social transformations it activates (Denton, 2002; Dankelman, 2010; Arora-Jonsson, 2011; Pearse, 2017). Climate change in the

Himalayan region is not merely an environmental phenomenon but a socio-political process that reconfigures livelihoods, resource governance, and everyday survival. The Himalayas are warming at a rate significantly higher than the global average, with glacial retreat, altered monsoon patterns, and increased frequency of extreme events already evident (Bolch et al., 2019; IPCC, 2022). These changes interact with market integration, land tenure arrangements, and state development policies, producing differentiated vulnerabilities among local populations.

A post-essentialist approach resists framing "women" as a single, homogeneous category of climate victims. Instead, it situates vulnerability within the intersection of gender, caste, class, ethnicity, age, and migratory status (Sultana, 2022; Nightingale, 2021). For example, in Nepal's Koshi Basin, landless women from Dalit communities have limited adaptive capacity due to insecure tenure, lack of credit access, and exclusion from disaster preparedness planning (see reports of the International Centre for Integrated Mountain Development 2017). On the other hand, women from politically connected, land-owning households may use remittances or social networks to relocate, diversify livelihoods, or invest in adaptation technologies (Rasul et al., 2019). In many situations race was more important in shaping their experiences of inequality and thus the connections of race, class and gender need to be considered, according to Latino and African American women. (Nightingale, 2006; Hooks, 1990; Kobayashi & Peake, 1994; Moraga & Anzaldua, 1988).

3.1 *Stuctural Inequalities in Climate Adaptation*

Himalayan communities are exposed to a growing set of environmental hazards: glacial lake outburst floods (GLOFs), flash floods, landslides, prolonged droughts, and unpredictable precipitation (Pritchard, 2019; Tambe & Rawat, 2020a, 2020b). However, exposure does not directly translate into vulnerability; rather, vulnerability is socially mediated. Migrant-sending households in Uttarakhand, where men work seasonally in urban centres, often rely on women for agricultural work, fodder collection, and water management. This feminisation of labour can increase women's workload without corresponding gains in decision-making authority or resource control (Banerjee & Bell, 2023).

Adaptive strategies also vary widely. In Sikkim, women's seed-saving practices have been modified to incorporate hybrid varieties that meet both market demands and subsistence needs (Sherpa, 2020). In Ladakh, women's cooperatives have adapted irrigation schedules in response to glacial meltwater unpredictability, negotiating water rights with local councils to maintain equitable distribution (Angmo et al., 2021). These examples demonstrate that women's responses are neither automatic nor purely "instinctive," but are shaped by governance structures, access to knowledge, and political agency.

3.2 Inequalities and Post-disaster Recovery

Recovery pathways following disasters further reveal structural disparities. After the 2013 Kedarnath floods, upper-caste women were more likely to secure housing compensation and state rehabilitation support, aided by literacy and social connections (Mehta, 2016). Dalit women, in contrast, often remained excluded from aid networks. Similarly, during the 2021 landslides in remote Lepcha villages of North Sikkim, women-headed households faced barriers in accessing relief due to geographic isolation, language differences, and bureaucratic requirements (Tambe & Rawat, 2020a, 2020b).

These patterns highlight that climate change acts as a "threat multiplier" (UN Women, 2024), amplifying pre-existing inequalities rather than creating new ones in isolation. Addressing such inequalities requires moving beyond vulnerability mapping to an analysis of the power relations embedded in adaptation policy and practice (Djoudi et al., 2016; Nagoda & Nightingale, 2017).

3.3 Livelihood Diversification and Resilience

Livelihood diversification—through small-scale tourism, medicinal plant trade, or high-value horticulture—is often promoted as an adaptive measure in Himalayan development plans (ICIMOD, 2023). Yet, benefits are unevenly distributed. For instance, eco-tourism cooperatives in Himachal Pradesh have enhanced incomes for women from dominant caste groups with access to capital and land, while women from marginalised groups remain in seasonal wage labour with little say in cooperative governance (Rai et al., 2021).

Resilience, therefore, cannot be understood solely in terms of household asset portfolios; it must be assessed in relation to social justice, redistribution of resources, and equitable participation in decision-making. Post-essentialist frameworks emphasise that adaptation is a contested and negotiated process, embedded in power-laden institutions that often privilege certain actors while marginalising others (Elmhirst & Resurrecci6n, 2020; Carr & Thompson, 2014).

4 Women's Participatory Roles in Himalayan Conservation

The Himalayan region is not only an ecological hotspot but also a complex sociocultural landscape where conservation practices are deeply intertwined with local livelihoods and gender relations. Women have historically played pivotal roles as custodians of biodiversity, managing forests, water sources, and agro-ecosystems. Their everyday labour —from fuelwood and fodder collection to crop diversity maintenance—constitutes thefoundation of community-based resource management

(Agarwal, 2010; Lama & Buchy, 2002). Yet, these roles have evolved under the influence of climate change, migration, market integration, and state-led conservation policies. Far from being static, women's environmental participation reflects a constant process of adaptation, negotiation, and innovation.

In mountain agroecological systems, women's ecological work is inseparable from household survival. They tend kitchen gardens, safeguard traditional crop varieties, and manage water distribution for both domestic and agricultural purposes (Leach, 2007; Nazarea, 1999). In Uttarakhand, women's seed-saving networks maintain genetic diversity that underpins climate resilience, while in Bhutanese Himalayan villages, women-led irrigation groups allocate water during dry spells through consensus-based rules (Tsering et al., 2021).

Such practices embody detailed, place-specific ecological knowledge, often transmitted matrilineally. However, conservation and development programmes frequently overlook these knowledge systems, privileging externally driven technical solutions over locally grounded strategies (Agrawal & Redford, 2009; Kelkar, 2009).

4.1 Community-Based Conservation

Across the Himalayas, women have been active in forest user groups, watershed committees, and biodiversity management boards. In Nepal's community forestry programme—a globally recognised model—women's leadership has increased over the past decade, partly due to quota-based inclusion policies (Pokharel et al., 2020). Nonetheless, representation does not automatically translate into influence. Studies from Dolakha and Kaski districts show that while women hold official positions, major decisions about forest product sales and revenue allocation often remain male-dominated (Giri & Darnhofer, 2021).

Where women's voices are substantively integrated, conservation outcomes tend to be more equitable. In Kinnaur, Himachal Pradesh, Mahila Mandals (women's collectives) have initiated afforestation drives and negotiated grazing rights with panchayats, aligning conservation objectives with livelihood needs (Lahiri-Dutt, 2006; Nightingale, 2006). These examples challenge the reductive framing of women as passive beneficiaries, showing them instead as active shapers of conservation priorities.

4.2 State Power and the Politics of Conservation

State-led conservation in the Himalayas has often been implemented through exclusionary models—the "fortress conservation" approach—which prioritises biodiversity protection over local access to resources (Brockington & Wilkie, 2015; Ghosal & Kjosavik, 2015). The establishment of national parks and wildlife sanctuaries has, in many cases, restricted entry to forests traditionally used for fuel, fodder, and medicinal plants. Women, as primary household resource managers, are disproportionately affected by these restrictions (Lama & Buchy, 2002).

Large-scale infrastructure and hydropower projects compound these challenges by displacing entire communities. Compensation mechanisms, tied to formal land titles usually held by men, often exclude women from restitution packages, deepening economic marginalisation (Kelkar, 2009; Mehta & Srinivasan, 2000). The resulting loss of access to place-based ecological knowledge is not only a cultural erosion but also a conservation setback, as relocation sites seldom allow for the same biodiversity management practices.

4.3 Reframing Participation: From Representation to Transformation

While some conservation projects have introduced microcredit schemes, skill training, or women's eco-tourism ventures, these initiatives can be superficial if they do not address underlying inequalities such as land rights, mobility restrictions, or exclusion from political decision-making (Agarwal, 2010; Cornwall, 2003). Genuine gender-transformative conservation requires recognising women not as a marginal interest group but as political actors whose environmental agency is shaped by structural conditions.

Evidence from the Kangra Valley shows that when women's collectives are granted formal rights to forest products and a role in governance, there is a measurable improvement in forest regeneration, fuel security, and household nutrition (Rai et al., 2020). Such models demonstrate that conservation outcomes and gender justice are mutually reinforcing when women's participation is both substantive and structurally supported.

4.4 Himalayan Conservation and Anthropocentrism: An Eco Socialist Critique

Conservation in the Himalayan region has long been shaped by competing visions of environmental value. Anthropocentrism remains deeply embedded in many conservation strategies (Crist, 2012; Kopnina & Washington, 2016). In practice, this means that ecosystems and species are often conserved to the extent that they serve human welfare, whether in the form of ecosystem services, tourism revenues, or cultural symbolism. While anthropocentric logics may promote certain conservation outcomes, they can also perpetuate extractive or exclusionary practices that undermine ecological integrity and social justice.

Conservation in the Himalayan region has often been guided by a program that places human needs and economic priorities at the centre of environmental decision-making. This anthropocentric orientation reflects a broader trend in global conservation discourse, where the value of ecosystems is frequently articulated in terms of their utility to humans—whether as sources of ecosystem services, tourist revenue, hydropower potential, or carbon sequestration credits. While such frameworks are often justified on pragmatic grounds, they can marginalise the intrinsic ecological and cultural values of Himalayan landscapes, and they frequently sideline the perspectives of local communities, particularly women, whose livelihoods are intimately tied to these environments (Rai et al., 2021).

From an ecosocialist perspective, anthropocentrism is deeply entangled with the logics of capitalist accumulation. Michael Lowy (2015) argues that under capitalism, nature is commodified—transformed into a series of resources to be extracted, managed, or traded. In the Himalayas, this commodification manifests in the push for large-scale hydropower projects, high-altitude tourism infrastructure, and biodiversity offset schemes, all of which recast ecosystems as economic assets rather than as commons stewarded by communities. (Buscher & Fletcher, 2019). John Bellamy Foster (2020) notes that this "metabolic rift"—the severing of human societies from their ecological foundations—is particularly acute in frontier regions like the Himalayas, where global capital seeks to integrate previously marginal landscapes into circuits of accumulation.

4.5 Capitalist Conservation Models

The economy" and "the environment" are not independent of each other. Capitalism is not an economic system; it is not a social system; it is a way of organizing nature (Moore, 2015). The central problem posed by the capitalization of nature can be overcome to the extent that the inner contradiction finds an outer outlet. The rising organic composition of capital tends towards socio-ecological disequilibrium, whose systemic expression is the rising capitalized composition of world-nature (Foster, 2000; Moore, 2015; O'Connor, 1998; Saito, 2017).

The conservation strategies while aimed at preserving biodiversity, the protectionist model often reflects a colonial legacy of controlling both land and people, reframing conservation as a technocratic project divorced from local governance traditions. In Sikkim, Arunachal Pradesh, and parts of Himachal Pradesh, the creation of national parks and biosphere reserves has involved restricting pastoral grazing routes, timber extraction, and medicinal plant collection activities that are central to the survival strategies of rural women and Indigenous communities (Chettri, 2024; Mehta, 2021).

From an eco socialist standpoint, such policies are not neutral environmental measures but expressions of class and power relations. Andreas Malm (2018) emphasizes that capitalist states often treat conservation as a tool to manage nature for strategic economic ends—whether for future tourism revenue, carbon credit markets, or as a reserve of resources for extraction at a later stage. In the Himalayas, conservation zones sometimes function as land banks for eco-tourism concessions or hydropower licensing, revealing the proximity between state-led environmentalism and capital accumulation.

4.6 Anthropocentrism and Green Developmentalism

Anthropocentric conservation is closely tied to the discourse of green developmentalism, which seeks to reconcile economic growth with environmental protection. In this context, Moore's (2015) notion of the 'Capitalocene' offers a critical viewpoint. Environmental crises are not simply the product of generic human activity (as in the Anthropocene framing) but are structured by capitalist relations of production. In the Himalayan context, "green" initiatives—such as eco-tourism resorts in Uttarakhand, carbon offset afforestation in Nepal, and market-based medicinal plant cultivation in Himachal are presented as win–win solutions. Yet these projects often impose monocultures, restrict community access to multi-use landscapes, and channel profits toward external investors rather than local households (Gurung et al., 2021; Sharma, 2025).

By considering conservation primarily in terms of ecosystem services to humans, such programmes risk reproducing anthropocentric hierarchies where nature's worth is contingent on its ability to serve human economic objectives. This perspective conceals the intrinsic value of non-human species and ecosystems, and it ignores the long histories of biocentric ethics embedded in Himalayan cosmologies, which see rivers, forests, and mountains as sentient entities deserving respect beyond their economic utility (Sherpa, 2020). Deep ecology, despite its resonance with various religious traditions and personal transformation philosophies, remains limited by its inability to engage meaningfully with critiques of human oppression. In this respect, an ecologically oriented feminism appears to offer the most promising theoretical foundation, one capable of integrating the emancipatory concerns of both feminist and ecological thought (Plumwood, 1993).

4.7 Gendered Implications of Anthropocentric Policies

The anthropocentric bias in conservation has gendered consequences. In many Himalayan communities, women bear primary responsibility for collecting fodder, fuelwood, and water. When conservation restrictions are imposed without consultation, these daily provisioning tasks become more labour-intensive, often requiring longer travel times and greater physical effort (Clement et al., 2019; Lama & Buchy, 2002). Furthermore, compensation schemes for resource access loss tend to be channelled through male landowners or local elites, marginalising women who lack formal land titles (Agarwal, 2010).

Eco socialist feminist scholars such as Ariel Salleh (2017) argue that such arrangements exemplify "meta-industrial labour"—work that is ecologically regenerative and socially necessary, yet systematically undervalued under capitalist systems. In the Himalayas, women's meta-industrial labour sustains not only household economies but also the ecological functions of agro-pastoral systems. Policies that ignore or undermine this labour are not merely socially unjust; they are ecologically self-defeating.

4.8 Anthropocentric Logics

Tourism has become one of the fastest-growing sectors in the Himalayan economy, promoted as a sustainable alternative to extractive industries (Lama, 2022; Nepal, 2020). However, dominant forms of tourism commodify landscapes, wildlife, and cultural practices for external consumption (Cater, 2021; Nepal, 2000; Singh & Sharma, 2021). This transformation redefines communities as service providers for a global leisure economy (Brockington et al., 2018), displacing traditional land uses and creating unevenly distributed economic benefits (Nepal & Chipeniuk, 2021).

Gendered impacts are evident in conservation-linked tourism ventures. (Gurung & DeCoursey, 2000; Lama & Buchy, 2002). Women's growing involvement in hospitality and handicraft production may enhance income but often reinforces precarious, low-paid labour without structural empowerment (Nepal & Thapa, 2009). Ecotourism often relies on women's unpaid or underpaid labour in homestays, handicrafts, and cultural performances, while excluding them from higher-value roles (Baral et al., 2023). Tourism can also marginalise subsistence priorities, increasing dependence on imported goods and reducing climate resilience (Bajracharya et al., 2020).

Sacred landscapes, once governed by customary ecological ethics, are reframed as "eco-cultural heritage" for visitors, justifying infrastructure development in fragile zones (Goodwin, 2021). This contributes to waste generation, path way erosion, and wildlife disturbance, challenging the sustainability claims of tourism-led conservation.

4.9 Political Economy of Resource Commodification

From an eco-socialist feminist perspective, the rapid spread of tourism, hydropower, and other large-scale projects in the Himalayas reflects global capitalist strategies of commodifying resources rather than neutral "modernisation." Such projects reshape both ecosystems and social relations, with gendered effects often absent in policy discourse. Tourism, while promoted as revenue-generating, channels profits to corporations and state agencies, leaving local women in low-paid, seasonal service roles and eroding subsistence livelihoods (Becken & Hay, 2012). Cultural commodification further entrenches stereotypes of women as caretakers or performers. Hydropower schemes, similarly justified as vital for national energy security, disrupt riverine ecologies, displace communities, and undermine traditional agricultural systems.

For women, displacement is not just a matter of losing land—it entails the loss of commons-based resources such as grazing areas, medicinal plants, and access to water sources that sustain household and community reproduction. Changing policies introduced mechanisms that make it easier for private companies to profit from hydropower projects leading to a large number of new players, many without any previous experience in the hydropower sector, signing contracts for building dam projects (Dharmadhikary, 2008). Eco-socialist feminists argue that this is an extension of the capitalist logic that treats rivers, forests, and mountains as sites of extraction rather than collective stewardship. In many ways, India's colonial legacy reinforced the pursuit of hydraulic capitalism through the large dam. The large dam in India is today less of a political conundrum. It is now widely accepted that it is part of a contested political, economic, and ecological terrain. As the above discussion indicates, large dams, though announced as neutral technological artifacts, have been deeply implicated in several processes integral to capitalist expansion, and in the manner in which it casts its imprint upon the natural world (Rohan D'Souza, 2008).

The cumulative effects of tourism, hydropower, and conservation-driven displacement deepen the metabolic rift in the Himalayan socio-ecological system. Women are often positioned as the shock absorbers of these disruptions, tasked with finding alternative livelihoods, adapting household provisioning, and sustaining community welfare in contexts of ecological scarcity. Bhattacharya's (2017) considerations of social reproduction helps understand how these gendered burdens are systematically externalised from development cost–benefit analyses, rendering them invisible in planning processes.

By situating these processes within the broader global capitalist economy, eco-socialist feminism exposes the structural drivers behind environmental change and gendered vulnerability. This perspective also challenges the "green growth" narrative that reinforces many tourism and renewable energy initiatives. Without redistributive justice, land rights reforms, and genuine participatory governance, such projects risk becoming another frontier of capitalist accumulation—deepening, rather than alleviating, gender and ecological inequalities (Salleh, 1997).

5 Towards Sustainable Conservation

A post-anthropocentric Himalayan conservation framework would recognise the intrinsic rights of non-human species, integrate community-led governance, ensure gender equity, and resist reducing the Himalayas to a "service provider" for distant markets (Sultana, 2022).

This aligns with ecosocialist visions that connect ecological sustainability with social justice (Foster, 2020; Lowy, 2015), emphasizing the restoration of gender-inclusive community forest councils, recognition of customary rights to pastures and forest products, promotion of low-impact locally owned tourism, and integration of Indigenous ecological knowledge into conservation science.

Such an approach reframes Himalayan ecosystems as co-constituted socio-ecological systems, where human well-being is interdependent with non-human flourishing, aligning feminist ecological ethics of care with eco socialist calls for dismantling the capitalist structures driving environmental decline.

6 Political Ecology of the Himalayan Region

The Himalayan region, straddling multiple national borders and encompassing diverse ecological zones, offers a striking example of how environmental change, political economy, and gendered labour are deeply intertwined. From a political ecology perspective, environmental governance here cannot be understood without tracing the historical accumulation of power, capital, and knowledge systems that shape both landscapes and livelihoods (Bryant & Bailey, 1997; Robbins, 2012).

The commodification of Himalayan land, water, and forests reflects the broader logic of extracting surplus from both nature and labour (Marx, 1867). State-led conservation regimes, hydropower expansion, and tourism industries often rely on the dispossession of communities from common resources, forcing their labour into market-oriented production. In many Himalayan states, this has meant enclosing forests under formal conservation while restricting subsistence uses, simultaneously expanding market opportunities in agriculture, horticulture, and ecotourism (Baviskar, 2020; Gururani, 2014).

State forestry departments, climate adaptation projects, and international conservation NGOs deploy technical assessments, zoning maps, and "community participation" protocols that not only regulate resource use but also shape the subjectivities of conservation (Agarwal, 2010). Women's involvement in forest user groups or water committees is often framed as evidence of empowerment, yet such participation is circumscribed by institutional rules, bureaucratic hierarchies, and donor-defined agendas (Elmhirst, 2011; Nightingale, 2011).

From a socialist feminist viewpoint these processes are inseparable from the gendered division of labour. The eco socialist writings of social reproduction and Marxist anthropology underline how women's unpaid subsistence work and care labour sustain not only households but also the economic systems benefiting from resource extraction (Kollontai, 1921; Reed, 1975; Analytical Monthly Review., 2008; Foster, 2015, 2023). In the Himalayas, women's time-intensive work in fodder collection, water management, and agricultural maintenance is essential to the functioning of both local economies and conservation regimes—yet it remains undervalued in policy metrics.

The capitalist and patriarchal systems co-produce exploitation in peripheral economies (Bhattacharya, 2017). In the Himalayan context, the gendered burden of maintaining ecological resources amid male out-migration mirrors broader patterns in which women's labour subsidises both state conservation and capitalist production without adequate recognition or redistribution.

Political ecology thus demands an intersectional and post-essentialist analysis: gendered environmental practices are not simply "traditional" inheritances but are actively produced through interactions between global markets, state policies, local institutions, and entrenched social hierarchies of caste, class, and ethnicity (Sultana, 2021; Onta & Resurreccion, 2023). This framing exposes the political nature of environmental struggles—struggles that are often flattened or depoliticised when recast solely as "community participation" or "women's empowerment" projects.

By locating Himalayan conservation within global circuits of capital and knowledge, and by drawing from socialist feminist insights, we see that women's ecological roles are embedded in contested terrains of power. This understanding sets the stage for examining grassroots movements—spaces where women, far from being passive beneficiaries, articulate political demands that challenge both environmental degradation and the structures that sustain it.

7 People's Movements and Ecological Struggles

Grassroots movements in the Southern part of the world and the impact of environmental factors on women are closely associated with a broader movement concerned with women's economic and social position within the 'development process' (Kabeer, 1994). Women's movements in the Himalayan region occupy a critical space in resisting environmental degradation, defending local livelihoods, and asserting claims to justice. These movements do not emerge in a political vacuum; they are shaped by the intersection of economic marginalisation, caste- and ethnicity-based exclusion, and uneven access to political representation. From anti-logging protests to anti-dam campaigns, women have mobilised both collective memory and contemporary legal frameworks to demand recognition of their rights to land, forests, and water. Their activism is often deeply place-based, drawing on lived experiences of ecological change and the historical memory of dispossession (Agarwal, 1992; Rai et al., 2020). The women's movement had drawn global attention as early as two decades earlier, with apex moments during two world conferences on women: in Mexico in 1975, followed by Nairobi in 1985 (Resurreccion, 2013).

One of the most emblematic examples is the Chipko movement in Uttarakhand, often depicted in popular discourse as a spontaneous, women-led defence of forests motivated by nurturing instincts. A post-essentialist reading complicates this narrative. Women's tree-hugging actions in the 1970s were strategic responses to commercial logging that had eroded local access to forest produce, intensified fuelwood scarcity, and undermined subsistence livelihoods. Far from being solely a cultural expression of care for nature, these protests were political acts against a state forestry regime that privileged industrial interests over local needs (Bose, 2020; Guha, 2001). In contemporary Uttarakhand, women's forest user groups continue to negotiate with state authorities over subsistence harvesting rights, balancing conservation goals with livelihood imperatives.

In Sikkim, grassroots activism has often been tied to the state's much-celebrated organic farming policy. While the policy has gained international acclaim, it has also increased women's unpaid labour burdens, especially in weeding and pest management (Lama, 2020). Indigenous Bhutia and Lepcha women sometimes enjoy higher political visibility than women of Nepali origin, but participation in cooperatives and panchayat structures remains stratified by caste and ethnicity (Subba, 2018). Anti-dam protests in Dzongu—a Lepcha reserve—have mobilised cultural identity alongside environmental arguments, strategically framing the river as sacred to gain leverage in negotiations with the state (Tambe & Rawat, 2020a, 2020b).

In Nepal, the Community Forestry Programme (CFP) is often cited as a participatory governance success story. However, despite formal quotas for women's representation, decision-making is frequently dominated by higher-caste, land-owning women, leaving marginalised groups with limited influence (Shrestha & Clement, 2019). Climate-driven male out-migration in districts like Dolakha and Solukhumbu has shifted agricultural and community management responsibilities onto women, without equivalent control over revenues or adaptation funding (Gurung et al., 2021). Grassroots campaigns have emerged to demand more equitable distribution of forest revenues and gender-sensitive adaptation measures.

7.1 De-politicising Tendencies in Environmental Discourse

Mellor observed that When green movement call upon people to think globally but act locally they often overlook the fact that it is women who live locally (Mellor, 1994) they have little choice but to think locally. While grassroots struggles are inherently political, dominant environmental narratives—including some strands of ecofeminism and state-led conservation rhetoric—often de-politicise women's activism. By framing women as "natural carers" of the environment, these narratives reduce complex political demands to moral appeals, erasing structural critiques of land tenure regimes, resource allocation, and economic inequality (Gaard, 2011; Harcourt & Nelson, 2015). This "victim–steward" binary portrays women either as passive sufferers of environmental change or as inherently virtuous custodians of nature, thereby limiting the political space for challenging underlying systems of exploitation.

The process of de-politicisation is reinforced by technocratic gender mainstreaming, which can reduce gender inclusion to mere numerical representation. In water-user committees, for example, token female membership may satisfy donor requirements while decision-making remains concentrated in local elites (Lama & Buchy, 2002). This not only obscures intra-community inequalities but also sidelines women's demands for redistributive justice.

From a post-essentialist standpoint, politicising environmental struggles requires recognising them as contests over power, rights, and governance, rather than as expressions of feminine moral authority. There are some differences in the environmental issues that face women in the North and the South. In non-industrial societies problems revolve around access to clean water and other resources as well as issues of poverty and health (Asian & Pacific Women's Resource Collection Network, 1989). This means situating women's activism within broader political economy contexts, acknowledging how state policies, corporate extraction, and global climate finance mechanisms shape local resource politics (Elmhirst, 2011; Resurreccion, 2017). In practice, many grassroots movements already enact this shift—combining identity-based mobilisation (e.g., as "mothers of the Himalayas") with explicit legal and economic claims.

8 From Policy Critique to Gender Justice

Policy frameworks addressing the intersection of gender and environmental sustainability have evolved over the past three decades, reflecting shifts in global development discourse, feminist scholars and environmental governance. International agreements such as the Beijing Platform for Action (1995), the Convention on the Elimination of All Forms of Discrimination Against Women (CEDAW), and the 2030 Agenda for Sustainable Development have formally recognised gender equality as a cross-cutting priority. In principle, Sustainable Development Goal 5 (SDG 5) to "achieve gender equality and empower all women" is not only a stand-alone target but also an enabling condition for achieving other goals, including climate action, poverty reduction, and sustainable resource management (UN, 2015; UN Women, 2018).

However, critical feminist scholars have pointed out that gender mainstreaming—the primary strategy promoted by these frameworks—often falls short of its transformative potential. As Cornwall and Rivas (2015) note, mainstreaming frequently degenerates into "tick-box" exercises, focusing on representation (numbers of women in projects or meetings) without addressing deeper structural inequalities in land tenure, political authority, and economic opportunity. In the Himalayan region, for example, women may be nominally included in water-user committees or forest management bodies, yet real decision-making power often remains concentrated among upper-caste male elites (Lama & Buchy, 2002).

8.1 Beyond Representation: Towards Structural Transformation

Women's oppression cannot be addressed in isolation from the capitalist structures that generate both environmental degradation and economic exploitation (Ghandy, 2016). Genuine emancipation, likewise, requires collective social transformation rather than piecemeal reforms that leave exploitative relations intact (Kollontai, 1920). Together, these perspectives expose the limits of technocratic gender mainstreaming, which prioritizes efficiency over justice and treats gender equity as a means to development outcomes rather than an end in itself. From an eco socialist feminist view, several scholars have emphasised that social reproduction—the unpaid and often invisible labour of sustaining human life—is inseparable from ecological sustainability (Bhattacharya, 2017, Mies & Shiva, 1993, Mellor, 1997, Salleh, 1997, Gaard, 2011) Policies that address women's environmental roles without confronting the undervaluation of care work risk reinforcing the very hierarchies they aim to dismantle. In the Himalayan context, the burden of unpaid ecological labour—such as water collection, fodder gathering, and soil restoration—intensifies under climate change, yet is rarely factored into climate adaptation budgets or remuneration schemes (Arora-Jonsson, 2014; Bose, 2020).

8.2 Social Dimensions in Policy Practice

Intersectionality (Crenshaw, 1989) demands that policies move beyond the category of "women" to address how gender interacts with caste, class, ethnicity, and migration status in shaping environmental access and vulnerability. Without this approach, interventions risk funnelling resources to already-privileged groups—as documented in Nepal's glacier melt adaptation funds, which disproportionately benefited politically connected households (Nagoda & Nightingale, 2017; Ojha et al., 2022).

Embedding intersectionality in policy means moving beyond representation-based metrics—such as simply counting how many women are present—to focusing on transformation-based outcomes that redistribute resource rights, workload, and decision-making authority (Rao et al., 2017). It also requires disaggregated data collection, participatory monitoring, and accountability mechanisms that enable marginalised voices to influence both policy design and implementation (Nellemann et al., 2011).

8.3 *Context-Specific Sustainability Agendas*

Another persistent weakness of top-down policy design is the wholesale importation of sustainability models from international development agendas into ecologically and culturally distinct contexts (Leach et al., 2010). This "one-size-fits-all" approach often overlooks the complex livelihood strategies, customary tenure systems, and cultural practices of mountain communities. In the Himalayas, lowland agricultural policies have been transplanted into fragile high-altitude ecosystems, with mixed or even harmful results (Gentle & Maraseni, 2012).

Participatory governance models offer an alternative by recognizing local knowledge holders—especially women from diverse social backgrounds—not as passive recipients but as equal partners (Reed et al., 2018). In Nepal's mid-hills, for instance, community forestry initiatives have been most effective when governance rules were negotiated to accommodate seasonal male out-migration and women's time constraints, thereby aligning conservation goals with livelihood realities (Gurung et al., 2013; Sunam & McCarthy, 2016).

8.4 *From Peripheral to Transformative Policy Approaches*

Gender-transformative environmental policy must dismantle the structural factors of inequality rather than simply "adding women" to existing systems (Maathai, 2003; Saigal & Srivastava, 2025). This means recognising and remunerating unpaid ecological labour as vital for both human survival and ecosystem health, while also securing women's rights to land, water, and forests through meaningful legal reform and enforcement (Hirway,2024). It further requires that all stages of environmental planning and funding allocation incorporate a multidimensional analysis of social inequalities (Pandya, 2023). Finally, such a policy framework depends on building strong alliances between grassroots women's movements, civil society, and progressive state actors to ensure institutional accountability.

Women are not born with fixed social roles but are socially constructed within material and ideological systems—systems that can be transformed. The biological and social sciences no longer admit the existence of unchangeably fixed entities that determine given characteristics, such as those ascribed to woman, the Jew, or the Negro (Beauvoir, 1949). In the environmental arena, such transformation depends on confronting the intertwined logics of patriarchy, capitalism, and ecological exploitation, rather than accommodating them through symbolic inclusion.

9 Towards the Future

The debates surrounding the women–nature nexus have travelled a long path—from early ecofeminist concepts of an "innate" bond between women and nature, to the more refined, politically aware perspectives of contemporary feminist political ecology and eco socialist feminism. This review has argued that essentialist framings, while historically significant in mobilising certain grassroots movements, risk flattening the diversity of women's experiences and depoliticising environmental struggles. In the Himalayan context, where socio-ecological fragility intersects with deep-seated structural inequalities, these simplifications can obscure the systemic forces—from land tenure regimes to neoliberal conservation policies—that shape women's ecological roles.

A post-essentialist approach does not deny the profound ecological knowledge and environmental stewardship found among women in mountain communities. Instead, it situates these practices within wider systems of power, production, and resistance. This means understanding Himalayan women not as timeless "guardians" of nature, but as historical actors whose strategies of survival and adaptation are shaped by changing climates, shifting labour regimes, market forces, and state interventions. It also means recognising that women's environmental agency is often exercised collectively—through Mahila Mandals, forest user groups, and resistance movements—and is bound up with struggles for rights, autonomy, and justice.

From the recognition that women must be understood as subjects of history rather than objects of destiny, (Beavoir 1949) to the realization that environmental politics is inseparable from health, justice, and corporate accountability, (Carson, 1962) to critiques of patriarchal and capitalist systems (Ghandy, 2016) in shaping oppression, it becomes clear that gender–environment analysis must remain firmly political. Early advocacy for collective provisioning, the social reproduction framework, (Kollontai, 1920), and Marxist-feminist anthropology (Mies, 1993, Leacock, 1983) further reinforce the need to link ecological sustainability with the transformation of social relations. As Wangari Maathai has pointed out, while the women of the South are struggling against visible problems such as desertification and soil erosion, 'at the global level we are fighting an invisible enemy' (Maathai & Jones, 1983). Future researchers must deepen these connections between ecological justice and social reproduction, drawing more explicitly from eco socialist feminism to analyse the shared logics of environmental degradation and gender oppression. In the Himalayan region, this means moving beyond tokenistic gender-mainstreaming in development plans, towards genuine gender-transformative policies that address structural inequalities in land rights, resource access, and decision-making power. It also means amplifying the knowledge of women across caste, ethnic, and class lines, integrating their strategies into climate adaptation, conservation, and livelihood diversification.

Policy frameworks must acknowledge that women's ecological agency is inseparable from broader struggles for economic and political rights. Activism and research should resist the de politicisation of environmental struggles, keeping the focus on power, inequality, and justice rather than on abstracted ideals of harmony with nature. As the pressures of climate change and globalisation intensify, the stakes for both gender equality and ecological sustainability will only grow sharper. Addressing these twin crises requires dismantling the social and economic systems that exploit both women's labour and the earth's resources—a task that, as eco socialist feminists remind us, is inseparable from the fight for a just and sustainable future.

References

Afshar, H. (Ed.). (1985). *Women, work and ideology in the Third World*. Tavistock Publications.

Agarwal, B. (1992) The gender and environment debate: Lessons from India. *Feminist Studies, 18*(1), 119–158. Available at: https://doi.org/10.2307/3178217

Agarwal, B. (1994). *A field of one's own: Gender and land rights in South Asia*. Cambridge University Press.

Agarwal, B. (1998). Environmental management, equity and ecofeminism: Debating India's experience. *The Journal of Peasant Studies, 25*(4), 55–95. Available at: https://doi.org/10.1080/03066159808438679

Agarwal, B. (2010). *Gender and green governance: The political economy of women's presence within and beyond community forestry*. Oxford University Press.

Agrawal, A., & Redford, K. (2009). Conservation and displacement: An overview. *Conservation and Society, 7*(1), 1–10.

Analytical Monthly Review. (2008, August 18). Southeastern Himalayan slopes: The frontline of revolutionary political ecology. *Monthly Review Online*. https://mronline.org/2008/08/18/southeastern-himalayan-slopes-the-frontline-of-revolutionary-political-ecology/

Angmo, P., Angmo, D., & Namgail, D. (2021). Water crisis: A big constraints for agricultural practices in high himalaya of Ladakh and affordable strategies to mitigate the crisis. *Agriculture Practices, 134*.

Araya, B. A. (2022). The capitalist unconscious: The social reproduction of labour and the limits of consciousness-raising. In T. Bhattacharya (Ed.), *Contradictions of capital and care* (pp. 101–121). Pluto Press.

Arora-Jonsson, S. (2011). Virtue and vulnerability: Discourses on women, gender and climate change. *Global Environmental Change, 21*(2), 744–751. Available at: https://doi.org/10.1016/j.gloenvcha.2011.01.005

Arora-Jonsson, S. (2014). *Gender, development and environmental governance: Theorizing connections*. Routledge.

Asian and Pacific Women's Resource Collection Network. (1989). *Asian and pacific women's resource and action series: Health*. Asian and Pacific Development Centre.

Bajracharya, B., et al. (2020). Matches, mismatches and priorities of pathways from a climate-resilient development perspective in the mountains of Nepal. *Environmental Science & Policy, 125*, 135–145.

Banerjee, S. B., & Bell, E. (2023). Ecologies of business: Global capitalism and climate crisis. *Organization Studies, 44*(5), 697–718. Available at: https://doi.org/10.1177/01708406221092061

Baral, N., et al. (2023). Women's participation and gender equality in sustainable ecotourism. *Tourism Management Perspectives, 43*, Article 100779.

Barca, S. (2019). The labor(s) of degrowth. *Capitalism Nature Socialism, 30*(2), 207–216.

Bauhardt, C., & Harcourt, W. (2019). Ecosocialist-feminist struggle, a degrowth-feminist contribution. In E. D'Alisa & F. Demaria (Eds.), *Degrowth and feminisms: A radical-ecological economics*. Routledge.

Baviskar, A. (2020). The politics of environment in India: An introduction. *Seminar, 727*, 1–6.
Beauvoir, S. (1949). *The second sex* (H. M. Parshley, Trans.). Vintage Books.
Becken, S., & Hay, J. E. (2012). *Climate change and tourism: From policy to practice* (xv, 279 pp.). Routledge.
Bennholdt-Thomsen, V., & Mies, M. (1999). *The subsistence perspective: Beyond the globalised economy*. Zed Books.
Bhattacharya, T. (2017). *Social reproduction theory: remapping class*. Recentering Oppression.
Biro, A. (Ed.). (2011). *Critical ecologies: The Frankfurt school and contemporary environmental crises*. University of Toronto Press.
Bolch, T., Mishra, A., Mukherji, A., & Shrestha, A. B. (Eds.). (2019). *The Hindu Kush Himalaya Assessment: Mountains, climate change, sustainability and people*. Springer Nature.
Bose, P. (2020). Caste, gender and forest governance: Lessons from Uttarakhand, India. *Mountain Research and Development, 40*(2). https://doi.org/10.1659/MRD-JOURNAL-D-19-00071.1
Brockington, D., & Wilkie, D. (2015). Protected areas and poverty. *Philosophical Transactions of the Royal Society B, 370*(1681), 20140271. Available at: https://doi.org/10.1098/rstb.2014.0271
Brockington, D., Duffy, R., & Igoe, J. (2018). *Nature unbound: Conservation, capitalism and the future of protected areas*. Earthscan.
Brownhill, L., & Turner, T. E. (2019). Ecofeminism at the heart of ecosocialism. *Capitalism Nature Socialism, 30*(1), 1–10. https://doi.org/10.1080/10455752.2019.1570650
Brownhill, L., & Turner, T. E. (2020). Ecofeminist ways, ecosocialist means: Life in the post-capitalist future. *Capitalism, Nature, Socialism, 31*(1), 1–14. https://doi.org/10.1080/10455752.2019.1710362
Bryant, R. L., & Bailey, S. (1997). *Third world political ecology*. Routledge.
Buscher, B., & Fletcher, R. (2019). Towards convivial conservation. *Conservation and Society, 17*(3), 283–296.
Buscher, B., Sullivan, S., Neves, K., Igoe, J., & Brockington, D. (2012). Towards a synthesized critique of neoliberal conservation. *Capitalism Nature Socialism, 23*, 4–30. https://doi.org/10.1080/10455752.2012.674149
Buscher, B., & Fletcher, R. (2020). *The conservation revolution: Radical ideas for saving nature beyond the Anthropocene*. Verso
Carson, R. (1962). *Silent spring*. Houghton Mifflin.
Carson, R. (2000). *Silent spring*. Penguin Classics.
Carr, E. R., & Thompson, M. C. (2014). Gender and climate change adaptation in agrarian settings: Current thinking, new directions, and research frontiers. *Geography Compass, 8*(3), 182–197.
Cater, E. (2021). The commodification of culture in global tourism: Balancing authenticity, sustainability, and ethical practices amidst disruption. *Tourism Management Perspectives, 38*, Article 100779.
Chettri, P. (2024). Gender and development discourses in Northeast India: A feminist political ecology approach. *Environmental Sociology, 10*(1), 1–13.
Chipkomovement. (n.d.). In *Wikipedia*. Retrieved August 18, 2025, from https://en.wikipedia.org/wiki/Chipko_movement
Clement, F., Pradhan, P., & Van Koppen, B. (2019). Community institutions in water governance for sustainable livelihoods. *Practical Action Publishing*. https://doi.org/10.3362/9781780449961
Cornwall, A. (2003). Whose voices? Whose choices? Reflections on gender and participatory development. *World Development, 31*(8), 1325–1342.
Cornwall, A., & Rivas, A.-M. (2015). From gender equality and 'women's empowerment' to global justice: Reclaiming a transformative agenda for gender and development. *Third World Quarterly, 36*(2), 396–415.
Crenshaw, K. (1989). Demarginalizing the intersection of race and sex: A Black feminist critique of antidiscrimination doctrine. *University of Chicago Legal Forum, 1989*(1), 139–167.
Crenshaw, K. (1991). Mapping the margins: Intersectionality, identity politics, and violence against women of color. *Stanford Law Review, 43*(6), 1241–1299. Available at: https://doi.org/10.2307/1229039

Crist, E. (2012). Half the earth for people (or More)? addressing ethical questions in conservation. *Biological Conservation, 152*, 251–260.
d'Eaubonne, F. (1974). *Le féminisme ou la mort*. P. Horay.
Dankelman, I. (Ed.). (2010). Gender and climate change: An introduction (pp. 21–29). Earthscan/Routledge.
Denton, F. (2002). Climate change vulnerability, impacts, and adaptation: Why does gender matter? *Gender and Development, 10*(2), 10–20.
Dharmadhikary, S. (2008). Mountains of concrete: Dam building in the Himalayas. esocialsciences.com, Working Papers.
Djoudi, H., Locatelli, B., Vaast, C., Asher, K., Brockhaus, M., & Sijapati, B. B. (2016). Beyond dichotomies: Gender and intersecting inequalities in climate change studies. *Ambio, 45*(Suppl. 3), 248–262.
Elmhirst, R. (2011). Introducing new feminist political ecologies. *Geoforum, 42*(2), 129–132. Available at: https://doi.org/10.1016/j.geoforum.2011.01.006
Elmhirst, R., & Resurrecci6n, B. P. (2021). *Negotiating gender expertise in environment and development: Voices from feminist political ecology*. Routledge.
Feder, H. (2019). Never waste a good crisis: An interview with Mary Mellor. *Capitalism Nature Socialism, 30*(4), 44–54.
Foster, J. B. (2000). *Marx's ecology: Materialism and nature*. Monthly Review Press.
Foster, J. B. (2015). Marxism and ecology. *Monthly Review, 67*(7). https://monthlyreview.org/2015/12/01/marxism-and-ecology/
Foster, J. B. (2020). *The return of nature: Socialism and ecology*. Monthly Review Press.
Foster, J. B., & Maskey, M. (2023). Marxian ecology and sustainable human development. *Monthly Review, 75*(7). https://monthlyreview.org/2023/12/01/marxian-ecology-and-sus tainable-human-development/
Foucault, M. (1978). *The history of sexuality:* (Vol. I). Pantheon.
Gaard, G. (1993). *Ecofeminism: Women, animals, nature*. Temple University Press.
Gaard, G. (2011). Ecofeminism revisited: Rejecting essentialism and re-placing species in a material feminist environmentalism. *Feminist Formations, 23*(2), 26–53.
Gaard, G. (2015). 'Ecofeminism and climate change', *Women's Studies International. Forum, 49*, 20–33.
Gentle, P., & Maraseni, T. N. (2012). Climate change, poverty and livelihoods: Adaptation practices by rural mountain communities in Nepal. *Environmental Science & Policy, 21*, 24–34. https://doi.org/10.1016/j.envsci.2012.03.007
Ghandy, A. (2016). *Philosophical trends in the feminist movement*. Foreign Languages Press.
Ghosal, S., & Kjosavik, D. J. (2015). Living with REDD+: The human rights-based approach to forest conservation in India. *Forest Policy and Economics, 61*, 1–9. Available at: https://doi.org/10.1016/j.forpol.2015.07.007
Giacomini, T., Turner, T. E., Isla, A., & Brownhill, L. (2018). Ecofeminism against capitalism and for the commons. *Capitalism Nature Socialism, 29*(1), 1–6.
Giri, K., & Darnhofer, I. (2021). Feminist political ecology and sustainable agriculture: Gendered knowledge and practices in rural systems. *Journal of Rural Studies, 82*, 75–84.
Gomez-Barris, M. (2017). *The extractive zone: Social ecologies and decolonial perspectives*. Duke University Press.
Goodwin, H. (2021). Sustainable tourism governance: A study of the impact of culture. *Journal of Governance and Regulation, 13*(2), 475–490.
Griffin, S. (1978). *Woman and nature: The roaring inside her*. Harper & Row.
Guha, R. (2001). The Chipko movement: A people's history. In *The Chipko movement: A people's history* (pp. 1–15). Permanent Black.
Gurung, A., Bista, R., Karki, R., Shrestha, S., Uprety, D., OH, S.-E. (2013). Community-based forest management and its role in improving forest conditions in Nepal. *Small-scale Forestry, 12*(3), 377–388. https://doi.org/10.1007/s11842-012-9217-z

Gurung, C. P., & DeCoursey, M. (2000). Too much too fast: Lessons from Nepal's lost kingdom of Mustang. In *Tourism and development in mountain regions* (pp. 167–184). CABI Publishing.
Gurung, L. J., Miller, K. K., Venn, S., & Bryan, B. A. (2021). Contributions of non-timber forest products to people in mountain ecosystems and impacts of recent climate change. *Ecosystems and People, 17*(1), 447–463. https://doi.org/10.1080/26395916.2021.1957021
Gururani, S. (2014). Geographies that make resistance": Remapping the politics of gender and place in Uttarakhand, India. *Himalaya, the Journal of the Association for Nepal and Himalayan Studies, 34*(1), 12.
Harcourt, W., & Nelson, I. L. (2015). *Practising feminist political ecologies: Moving beyond the 'green economy.'* Zed Books.
Hartsock, N. (1998). *The feminist standpoint revisited and other essays.* Westview Press.
Hooks, B. (1990). *Yearning: Race, gender and cultural politics* (South End Press, Boston).
Hirway, I. (Ed.). (2024). *India social development report 2023.* Oxford University Press.
ICIMOD. (2019). *The Hindu Kush Himalaya assessment: Mountains, climate change, sustainability and people.* Springer.
Intergovernmental Panel on Climate Change. (2022). Climate change 2022: Impacts, adaptation and vulnerability. Contribution of Working Group II to the Sixth Assessment Report of the Intergovernmental Panel on Climate Change. Cambridge University Press.
International Centre for Integrated Mountain Development. (2017). The Koshi River Basin: Status report on climate change, water resources, and vulnerability. ICIMOD.
Jackson, C. (1993). Doing what comes naturally? Women and environment in development. *World Development, 21*(12), 1947–1963.
Jadhav, R., & Shrestha, N. (2024). Beyond participation: Gendered labor and greenwashing in South Asian climate adaptation projects. *Development and Change, 55*(2), 112–135.
Kabeer, N. (1994). *Reversed realities: Gender hierarchies in development thought.* Verso.
Kelkar, G. (2009). The feminization of agriculture in Asia: Implications for women's agency and productivity. *UNIFEM Discussion Paper Series, 1*, 1–17.
Kobayashi, A., & Peake, L. (1994). Unnatural discourse, 'race' and gender in geography" gender. *Place and Culture, 1*, 225–243.
Kollontai, A. (1920). Communism and the family. In C. Mouat & G. Bowlby (Eds. And Trans.), *Women and revolution: A discussion of the unhappy marriage between marxism and feminism* (pp. 49–58). Zed Books.
Kollontai, A. (1921). The social basis of the woman question. In *Selected writings of Alexandra Kollontai.* Allison and Busby, 1977.
Kopnina, H., & Washington, H. (2016). Discussing why population growth is still ignored or denied. *Chinese Journal of Population Resources and Environment, 14*(2), 133–143.
Kovel, J. (2005). The ecofeminist ground of ecosocialism. *Capitalism Nature Socialism, 16*(2), 1–8.
Kovel, J. (2002). *The enemy of nature: The end of capitalism or the end of the world?* (2nd ed.). Zed Books.
Lahiri-Dutt, K., & Macintyre, M. (2006). *Women miners in developing countries: Pit women and others.* Ashgate Publishing Ltd.
Lama, A., & Buchy, M. (2002). Gender, class, caste and participation: The case of community forestry in Nepal. *Indian Journal of Gender Studies, 9*(1), 27–41.
Lama, K. (2020). Gendered implications of organic farming in Sikkim. *Economic & Political Weekly, 55*(50), 44–51.
Lama, R. (2022). Exploring the thrills: Adventure tourism in the Indian Himalayan region. In *Mountain tourism and ecological impacts: Himalayan region and beyond* (pp. 74–84). Springer.
Leach, M. (2007) *Pathways to sustainability: Gender and social–ecological resilience.* STEPS Centre.
Leach, M., Scoones, I., & Stirling, A. (2010). Dynamic sustainabilities: Technology, environment, social justice. Earthscan.
Leacock, E. (1983). Interpreting the origins of gender inequality: Conceptual and historical problems. *Dialectical Anthropology, 7*(4), 263–284.

Leon Araya A. (2023). The fruits of labour or the fruits of nature? Towards a political ecology of labour in Central America. In B. Bustos, S. Engel di Mauro, G. García, F. Milanez, D. Ojeda (Eds.),*Routledge handbook of Latin America and the environment*. Routledge.

Lowy, M. (2015). *Ecosocialism: A radical alternative to capitalist catastrophe*. Haymarket Books.

Luxemburg, R. (2015). *The accumulation of capital*. Routledge.

Malm, A. (2018). *The progress of this storm: nature and society in a warming world*. Verso.

Marx, K. (1867). Capital: a critique of political economy (vol. 1).

Maathai, W. W., & Jones, M. (1983). *Greening the desert: Women in Kenya reclaim land*. University of Nairobi.

Maathai, W. (2003). *The Green Belt Movement: Sharing the approach and the experience*. Lantern Books.

Mehta, M. (2016). Caste, gender, and disaster recovery in Uttarakhand. *Indian Journal of Gender Studies, 23*(3), 453–472.

Mehta, M. (2021). Reframing discourses on the Pahadi woman," in *Critical approaches to gender in mountain ecosystems* (IUCN). (Addresses gendered narratives and policy impacts in Uttarakhand/Himalaya; relevant to conservation restrictions.)

Mehta, L., & Srinivasan, B. (2000). Balancing gains and pains: A perspective paper on gender and large dams. Thematic Review 1.1: Social Impacts (Contributing Paper to the World Commission on Dams). World Commission on Dams Secretariat, Cape Town.

Mellor, M. (1994). Breaking the boundaries: Towards a feminist green socialism. *Environmental Values, 3*(3).

Mellor, M. (1996). The politics of women and nature: Affinity, contingency or material relation? *Journal of Political Ideologies, 1*(2).

Mellor, M. (1997). *Feminism and ecology: An introduction*. Polity Press.

Merchant C. (1989). *Ecological revolutions: Nature, gender, and science in New England*: Univ. North Carolina.

Merchant, C. (1980). *The death of nature: Women, ecology, and the scientific revolution*. Harper & Row.

Merchant, C. (2020). *Earthcare: Women and the environment*. Routledge.

Mies, M. (1986). *Patriarchy and accumulation on a world scale: Women in the international division of labour*. Zed Books.

Mies, M., & Shiva, V. (2015). *Ecofeminism*. Zed Books.

Mies, M., & Shiva, V. (1993). *Ecofeminism*. Fernwood Publications

Mohanty, C. T. (1988). Under Western eyes: Feminist scholarship and colonial discourses. *Feminist Review, 30*, 61–88.

Moore, J. W. (2015). *Capitalism in the web of life: Ecology and the accumulation of capital*. Verso.

Moraga, C., & Anzaldua, G. (1988). *This bridge called my back: writings by radical women of color*. Women of Color Press.

Nagoda, S., & Nightingale, A. J. (2017). Participation and power in climate change adaptation policies: Vulnerability in food security programs in Nepal. *World Development, 100*, 85–93.

Nazarea, V. D. (1999). *Ethnoecology: Situated knowledge/located lives*. University of Arizona Press.

Nellemann, C., Verma, R., & Hislop, L. (2011). *Women at the frontline of climate change*. UNEP.

Nepal, S. K. (2000). Tourism in protected areas: The Nepalese Himalaya. *Annals of Tourism Research, 27*(3), 661–681. https://doi.org/10.1016/S0160-7383(99)00105-X

Nepal, S. K. (2020). An exploring sustainable tourism development in Nepal. *Asian Journal of Management Cases, 17*(1), 1–15.

Nepal, S. K., & Chipeniuk, R. (2021). Tourism-led commodification of place and rural transformation in the Himalayas. *Land Use Policy, 104*, Article 105345.

Nepal, S. K., & Thapa, B. (2009). Tourism and sustainable community development in Nepal. *Tourism Management, 30*(1), 1–12.

Nightingale, A. J. (2006). The nature of gender: Work, gender and environment. *Environment and Planning D: Society and Space*, *24*(2), 165–185. Available at: https://doi.org/10.1068/d01k

Nightingale, A. J. (2011). Bounding difference: Intersectionality and the material production of gender, caste, class and environment in Nepal. *Geoforum, 42*(2), 153–162.
Nightingale, A. J. (2021) Commoning for inclusion? Competing ontologies and the politics of environmental governance. *Environment and Planning E: Nature and Space*, *4*(2), 402–424.
O'Connor, J. R. (1998). Natural causes: Essays in ecological Marxism. Guilford Press.
Ojeda, D., Nirmal, P., Rocheleau, D., & Emel, J. (2022). Feminist ecologies. *Annual Review of Environment and Resources, 47*(1), 149–171. https://doi.org/10.1146/annurev-environ-112320-092246
Ojha, H. R., Shrestha, K. K., & Subedi, Y. R. (2022). Adaptive governance for climate resilience in Nepal's rural landscapes. *Climate and Development, 14*(8), 719–731.
Onta, N., & Resurrección, B. P. (2023). Gendered climate adaptation in the Eastern Himalayas. *Climate and Development, 15*(2), 143–155.
Pandya, R. (2023). An intersectional approach to neoliberal environmentality: Women's engagement with ecotourism at Corbett Tiger Reserve, India. *Environment and Planning E: Nature and Space, 6*(1), 355–372.
Pear, J. (2021). *Ecofeminism and class struggle: Rethinking the universal woman.* Earth Politics Press.
Pearse, R. (2017). Gender and climate change. *Wires Climate Change, 8*, Article e451. https://doi.org/10.1002/wcc.451
Plumwood, V. (1993). *Feminism and the mastery of nature.* Routledge.
Pritchard, H. D. (2019). Asia's shrinking glaciers protect large populations from drought stress. *Nature, 569*, 649–654. https://doi.org/10.1038/s41586-019-1240-1
Pokharel, B. et al. (2020). Gender perspectives in community forestry and natural resource management in Nepal. *Forest Policy and Economics, 115*, 102147.
Rai, S. C. et al. (2020). Forest-based ecosystem services and their contribution to household nutrition and fuel security in Kangra Valley, Himachal Pradesh, India. *Ecological Economics, 176*, 106707.
Rai, R. K. et al. (2021). Household air pollution, health, and climate change: Implications for women in South Asia. *Environmental Research Letters, 16*(4), 045002. Available at: https://doi.org/10.1088/1748-9326/abe74b
Rao, N., Lawson, E. T., Raditloaneng, W. N., Solomon, D., & Angula, M. N. (2017). Gendered vulnerabilities to climate change: Insights from the semi-arid regions of Africa and Asia. *Climate and Development, 11*(1), 14–26.
Reed, E. (1975). *Woman's evolution: From matriarchal clan to patriarchal family.* Pathfinder Press.
Reed, M. S., Vella, S., Challies, E., De Vente, J., Frewer, L., Hohenwallner-Ries, D., Huber, T., Neumann, R. K., Oughton, E. A., Sidoli del Ceno, J., & Van Delden, H. (2018). A theory of participation: What makes stakeholder and public engagement in environmental management work? *Restoration Ecology, 26*, S7–S17.
Rege, S. (1998). Dalit women talk differently: A critique of 'difference' and towards a Dalit feminist standpoint. *Economic and Political Weekly, 33*(44), WS39–WS46.
Resurrección, B. P. (2017). Gender and environment from "women, environment and development" to feminist political ecology. In H. MacGregor & B. P. Resurrección (Eds.), *The Routledge handbook of gender and environment* (pp. 71–85). Routledge.
Resurreccion, B. P., & Elmhirst, R. (2020). *Negotiating gender expertise in environment and development: Voices from feminist political ecology* (1st ed.). Routledge. https://doi.org/10.4324/9781351175180
Resurreccion, B. P. (2013). Persistent women and environment linkages in climate change and sustainable development agendas. *Women's Studies International Forum, 40*, 33–43. Available at: https://doi.org/10.1016/j.wsif.2013.03.011
Robbins, P. (2012). *Political ecology: A critical introduction* (2nd ed.). Wiley-Blackwell.
Rocheleau, D., Thomas-Slayter, B., & Wangari, E. (1996). *Feminist political ecology: Global issues and local experiences.* UK Routledge.

Rohan D'Souza. (2008). Framing India's hydraulic crisis: The politics of the modern large dam monthly review 2008 60(03) (July-August)
Ruether, R. R. (1995). *New woman, new earth: Sexist ideologies and human liberation.* Beacon Press.
Ruether, R. R. (2005). *Integrating ecofeminism, globalisation and world religions.* Rowman & Littlefield.
Saigal, N., Srivastava, S. (2025). Urgent Imperatives: advancing gender equality in climate action. *Journal of Human Rights and Social Work.* timesofindia.indiatimes.com
Saito, K. (2017). *Karl Marx's ecosocialism: Capital, nature, and the unfinished critique of political economy.* Monthly Review Press.
Salleh, A. (1997). *Ecofeminism as politics: Nature, Marx and the postmodern.* Zed Books.
Salleh, A. (2017). *Ecofeminism as politics: Nature, Marx and the Postmodern* (2nd ed.). Zed Books.
Sanchez-García, P. A., Galaz, V., Rocha, J. C., & Barbour, F. (2025, June). Climate change, nature degradation, and financial stability: A review of domino-effects between finance, climate, and the biosphere. *Ecology and Society, 30*(2), Article 36. https://doi.org/10.5751/ES-16130-300236
Sen, G., & Grown, C. (1987). Development, crises, and alternative visions: Third World women's perspectives. DAWN (Development Alternatives with Women for a New Era).
Sharma, A. (2025). Bulletin of nexus. *Bulletin of Nexus, 2*(2).
Sherpa, P. (2020). Seed systems and women's adaptation strategies in Eastern Himalayas. *Mountain Research and Development, 40*(3), 233–243.
Shiva, V. (1988). *Staying alive: Women, ecology and development.* Zed Books.
Shiva, V. (1989). *Ecology and the politics of survival.* Sage.
Shrestha, K., & Clement, F. (2019). Unpacking gendered realities in community forestry: Evidence from Nepal. *Forest Policy and Economics, 108*, Article 101905. https://doi.org/10.1016/j.forpol.2019.05.029
Singh, A., & Sharma, R. (2021). Impact of tourism on vegetation structure and soil health in the Indian Himalayan region. *Biological Forum—An International Journal, 13*(2), 19–24.
Spivak, G. C. (1988). Can the subaltern speak? In C. Nelson & L. Grossberg (Eds.), *Marxism and the interpretation of culture* (pp. 271–313). University of Illinois Press.
Starhawk. (1982). *Dreaming the dark: Magic, sex & politics.* Beacon Press.
Stein, S. (2022). A colonial history of the future: Rethinking feminist ecology and sustainability. *Environmental Humanities, 14*(1), 1–23.
Subba, T. B. (2018). *Ethnicity, state and development: A case study of Sikkim.* Oxford University Press.
Sultana, F. (2024). Confronting Climate Coloniality: Decolonizing Pathways for Climate Justice. https://doi.org/10.4324/9781003465973
Sultana, F. (2021). Critical climate justice. *The Geographical Journal, 188*(1), pp. 118–124. Available at: https://doi.org/10.1111/geoj.12417
Sultana, F. (2022). *Water politics: Governance, justice and the right to water.* Routledge.
Sunam, R. K., & McCarthy, J. F. (2016). Reconsidering the links between poverty, international labour migration, and agrarian change: Critical insights from Nepal. *Journal of Peasant Studies, 43*(1), 39–63.
Sundar, N. (2001). Is devolution democratisation? *World Development, 29*(12), 2007–2023.
Sundberg, J. (2004). Identities in the making: Conservation, gender and race in the Maya Biosphere Reserve, Guatemala. *Gender, Place & Culture, 11*(1), 43–66.
Tsering, D. et al. (2021). Gender roles and climate change adaptation in Himalayan communities. *Environmental Development*, *39*, 100609.
Tambe, S., & Rawat, G. S. (2020a). Building sustainability in the Eastern Himalaya: Linking evidence to action. *Environment, Development and Sustainability, 22*(6), 5887–5903.
Tambe, S., & Rawat, G. S. (2020b). Ecology, economics and equity of the pastoral systems in the Khangchendzonga National Park, Sikkim Himalaya. *India. Ambio, 38*(2), 95–100.
Women, U. N. (2024). *Gender and climate resilience: Policy guidance for transformative action.* UN Women.

United Nations. (2015). *Transforming our world: The 2030 Agenda for Sustainable Development*. United Nations.

UN Women. (2015). *Progress of the world's women 2015–2016: Transforming economies, realizing rights*. United Nations.

UN Women. (2018). *Turning promises into action: Gender equality in the 2030 Agenda for Sustainable Development*. UN Women.

Veltmeyer, H., & Bowles, P. (2014). Extractivist resistance: The case of the Enbridge oil pipeline project in Northern British Columbia. *Extr. Ind. Soc., 1*, 159–168.

Varela, P. (2019). Gender and class: An interview with Tithi Bhattacharya. *Journal of Labor and Society, 22*(2), 399–406. https://doi.org/10.1111/lands.12399

Warren, K. J. (2000). *Ecofeminist philosophy: A Western perspective on what it is and why it matters*. Rowman & Littlefield.

Whyte, K. (2017). Indigenous climate change studies: Indigenizing futures, decolonizing the Anthropocene. *English Language Notes, 57*(1), 153–162.

Displacement, Gender, and Hydropower: Reassessing Development in the North-Western Himalaya

Isha Manjhkhola, Vikas Bharati, Radhey Shyam Sharma, and Vandana Mishra

Abstract Hydropower development in the north-western Himalaya is celebrated as a cornerstone of India's renewable energy expansion, yet it unfolds within one of the world's most fragile ecological and socially complex regions. Large dams such as Tehri, Pong, Bhakra, Nathpa Jhakri, Chamera, and Baglihar have reshaped river systems, fragmented habitats, and displaced thousands of people. The costs of these projects are borne unevenly: women, along with Scheduled Castes and Tribes, face the sharpest disruptions to livelihoods, health, and social security. Despite the existence of national policies on resettlement and gender empowerment, the lived realities of women—whose survival strategies are embedded in subsistence economies, kinship networks, and ecological knowledge—remain largely absent from hydropower planning and scholarship. This study reassesses hydropower-led displacement in the Himalaya through a gendered lens, guided by two questions: (i) how does displacement reconfigure women's access to resources, livelihoods, health, and agency? and (ii) what pathways can foster more inclusive and sustainable development? A mixed-methods approach is employed: qualitative review of literature, policy frameworks, and field-based evidence is paired with a quantitative case analysis of census data (1971–2011) from dam-affected and control sub-districts. Indicators include female population growth, literacy, workforce participation, and Scheduled Caste/Tribe representation. Findings show that displacement undermines subsistence economies, intensifies women's unpaid labour, and excludes them from decision-making. Census patterns reveal declining female population

I. Manjhkhola · V. Bharati · R. S. Sharma · V. Mishra (✉)
Bioresources and Environmental Biotechnology Laboratory, DepartmentofEnvironmentalStudies, University of Delhi, Delhi, India
e-mail: vandanamishra77@gmail.com

R. S. Sharma
Delhi School of Climate Change and Sustainability, Institute of Eminence, University of Delhi, Delhi, India

V. Mishra
Centre for Interdisciplinary Studies on Mountain and Hill Environment (CISMHE), University of Delhi, Delhi, India

DU-DDA Biodiversity Parks Programme, CEMDE, University of Delhi, Delhi, India

N. Moran et al. (eds.), *The Participatory Role of Indigenous Women*,
https://doi.org/10.1007/978-3-032-24059-0_14

growth in Tehri and Pong, rising SC/ST shares in affected areas, and persistent gender gaps in workforce participation despite literacy gains. Yet women also exhibit resilience through agricultural adaptation, barter economies, and ecological stewardship. Consequently, sustainable hydropower demands site-specific EIA norms, gender-disaggregated monitoring, participatory rehabilitation, and benefit-sharing frameworks. By situating women as both vulnerable and resilient, it advances an intersectional framework for rethinking development in the fragile mountain contexts of the north-western Himalaya.

Keywords Dam-induced displacement · Gendered vulnerabilities · Women · Rehabilitation and resettlement · North-western Himalaya · Land acquisition · Social sustainability

1 Introduction

Hydropower development in the Himalaya embodies a paradox: while celebrated as a cornerstone of India's renewable energy transition, it unfolds within one of the world's most fragile mountain systems, already stressed by climate change, seismic instability, and rapid socio-economic transformation (Huber, 2019; Grumbine & Pandit, 2013). The Indian Himalayan Region (IHR), with its steep river gradients and high discharge, offers immense hydropower potential (Sahu, 2020). Yet in the north-western Himalaya, dams are more than engineering infrastructure; they are agents of profound ecological and social transformation, altering river regimes, fragmenting habitats, and reshaping livelihoods, cultural practices, and social resilience (Erlewein, 2013; Kumar & Katoch, 2015). These disruptions fall unevenly, with particular consequences for women whose everyday labour in sustaining forests, water, and subsistence economies remains systematically marginalized in development planning.

Globally, large dams are recognized as a leading driver of development-induced displacement, with India and China accounting for the majority of those affected (WCD, 2000; Mishra et al., 2024b). In India, iconic Himalayan projects such as Tehri, Pong, Bhakra Nangal, Chamera, Nathpa Jhakri, and Baglihar have uprooted hundreds of thousands (Thakkar, 2000). Although legal safeguards such as the Right to Fair Compensation and Transparency in Land Acquisition, Rehabilitation and Resettlement Act (2013) exist, assessments often underestimate impacts by neglecting indirect losses from altered hydrology, landslides, catchment interventions, and downstream ecological changes (Cernea, 1996; McCully, 1996).

The burdens of displacement are unevenly distributed. Marginalized groups—particularly Scheduled Castes (SCs), Scheduled Tribes (STs), and economically disadvantaged households—experience heightened vulnerability, while women within these communities face compounded disadvantages rooted in patriarchal norms, gendered divisions of labour, and limited access to assets, education, and decision-making (Bisht, 2009; Gururani, 2002). In the IHR, women's subsistence-based livelihoods depend heavily on forests, water, and non-timber products—resources often rendered inaccessible after displacement. Male out-migration in search of wage labour further intensifies women's agricultural, domestic, and care-giving burdens while eroding traditional kinship support networks. These intersecting pressures diminish women's adaptive capacity and threaten fragile gains in agency and autonomy (Chopra & Ghosh, 2000; Pandey, 2021).

Research on dams in the Himalaya has largely centred on engineering feasibility, environmental impacts, or aggregate socio-economic costs (Erlewein, 2013; Grumbine & Pandit, 2013). While scholars have called for gender-sensitive development approaches (Agarwal, 1994; Mehta & Srinivasan, 2000), systematic analyses of hydropower's gendered impacts remain scarce. Few studies interrogate how displacement and resource reconfiguration reshape women's access to commons, health and well-being, safety, and participation in governance; fewer still examine women's agency in sustaining resilience amid these transformations (Gururani, 2002; Mishra et al., 2024a).

This study addresses these gaps by reassessing development through the lens of displacement, gender, and hydropower in the north-western Himalaya. Drawing on scholarship, policy documents, and field-based evidence, it examines women's vulnerabilities, adaptive strategies, and contributions to sustainability. It further incorporates a quantitative case analysis of census indicators (1971–2011) to compare female population dynamics, literacy, and workforce participation in dam-affected versus control sub-districts. By integrating qualitative insights with demographic data, the chapter advances an intersectional understanding of displacement and underscores the need for inclusive, gender-responsive policy frameworks in Himalayan hydropower development.

2 Gendered Impacts of Dam-Induced Displacement

Rural as well as semi-urban communities in the Western Himalaya have experienced profound economic and socio-cultural turbulence as a result of development-induced displacement (Parasuraman, 1993; Judge, 1997; Diduck et al., 2013). Although involuntary displacement affects both men and women, its impacts are often more severe for women because they are mediated by long-standing socio-historical processes that institutionalise a gendered division of labour, further compounded by biological roles and reproductive responsibilities (Pedraza, 1991; Aathana, 2012; Melketo et al., 2023). Traditionally, women have shouldered the responsibility of managing household provisioning and land-based resources, while men have been more involved in wage-earning and other labour-intensive activities. Consequently, resettlement and rehabilitation programmes frequently operate on male-centric assumptions, revealing a structural insensitivity to women's specific needs, responsibilities, and constraints.

In prevailing policy frameworks, Project Affected People (PAP) are typically treated as a homogeneous category, with minimal recognition of intra-community differences. India's *National Resettlement and Rehabilitation Policy* (2007) stipulates that "land may be allotted in the joint name of husband and wife"; however, because resettlement is a state subject, the design and implementation of policies vary considerably across states. This decentralisation often fails to capture the distinct ways in which women experience displacement and resettlement, leaving gaps between statutory intent and lived realities. As a result, existing guidelines concerning PAPs fall short in addressing gender-specific impacts in any meaningful or systematic way.

Gender becomes a crucial analytical lens in the context of development and globalisation for several reasons: biological differences between the sexes, the intersection of labour divisions with race, caste, and class, the unequal distribution of power and property, and the differentiated ways in which men and women interact with and depend upon natural resources. These factors shape not only the degree of exposure to environmental and social change but also the adaptive capacities available to men and women in responding to displacement (Agarwal, 1994). Numerous studies globally have underscored the importance of centring women's experiences in displacement analyses (Colson, 1971, 1999; Indra, 1999; Mehta & Srinivasan, 2000; Scudder, 1993; Bisht, 2009; Koenig, 2019).

This section therefore focuses on the challenges faced by women displaced by hydropower projects in the Western Himalaya. It synthesises insights from published literature, surveys, and reports concerning prominent dams in Uttarakhand, Himachal Pradesh, and Jammu and Kashmir, highlighting the multiple disadvantages and heightened risks faced by women, and underscoring the urgent need for more inclusive, gender-sensitive resettlement and rehabilitation policies.

2.1 Displacement and Housing Insecurity

Hydropower-related resettlement in the Himalaya is ostensibly designed to help families and communities rebuild their institutions, value systems, and social bonds while offsetting the economic and socio-cultural disruptions of displacement. In principle, compensation packages are intended to ensure fair rehabilitation. For example, the Tehri Dam Displacement Policy, formalised in 1995 by the Tehri Hydro Development Corporation (THDC), promised land, cash, and housing to the displaced (Hanumantha Rao Committee [HRC], 1997; THDC Report, 1998). While comprehensive on paper, its implementation was undermined by bureaucratic delays, corruption, and the exclusion of families lacking formal land titles, leaving many unable to access the promised benefits.

The scale of displacement was immense: approximately 100,000 people were directly uprooted by the Tehri project, with a further 80,000 losing access to essential services and regional towns in the Bhagirathi–Bhilangna basin (Bhai, 2020; Newton, 2008; Paranjpye, 1988). Displacement is inherently traumatic, and effective rehabilitation requires not only physical relocation but also continuity of services and social networks to mitigate stress and alienation (Cernea, 1997). The gap between policy promises and outcomes in Tehri illustrates systemic failures of accountability—weak monitoring, absent timelines, and ineffective grievance mechanisms—that have allowed corruption and delays to erode the credibility of resettlement frameworks.

As new dams continue to be proposed and constructed in the geologically fragile Himalayan region, the risks of land subsidence and slope destabilisation are becoming increasingly visible. The Himalaya, as young fold mountains, are inherently unstable owing to loose soils, high seismic activity, and steep gradients (Grumbine & Pandit, 2013). These baseline vulnerabilities are frequently amplified by large-scale excavation, tunnelling, and the altered hydrostatic pressures associated with dam construction. Recent cases are illustrative: in the vicinity of the under-construction Vishnugad–Pipalkoti and Tapovan–Vishnugad hydropower projects, the town of Joshimath has experienced alarming subsidence, with cracks appearing in homes, roads, and public infrastructure (Awasthi et al., 2024; Mishra, 1976).

What is striking in such contexts is that many of the displaced residents fall outside official categories of "project-affected persons" and are thus excluded from compensation or rehabilitation frameworks. This reveals a critical oversight in planning, whereby project impacts are narrowly conceived in terms of submergence zones and direct displacement, while indirect but equally severe geophysical consequences remain unacknowledged. Such exclusion not only undermines the legitimacy of resettlement processes but also highlights the inadequacy of existing governance frameworks in accounting for the cascading risks of development in fragile mountain systems.

These cases suggest the urgent need to broaden the definition of displacement to include communities affected by geophysical instability triggered by dam-related interventions. More robust risk assessment protocols—capable of systematically identifying, monitoring, and integrating indirect impacts into planning, mitigation, and compensation mechanisms—are essential if hydropower development in the Himalaya is to be both socially equitable and ecologically responsible.

2.2 Livelihood Loss and Economic Disempowerment

Hydropower development in the Indian Himalayan Region (IHR) produces uneven gains, with women disproportionately bearing its costs. As an often-invisible workforce, women contribute significantly to biodiversity conservation and the maintenance of sustainable livelihoods, yet their labour is routinely undervalued in both statistics and policy (Agarwal, 2009; Elson, 1998). Their dependence on natural resources for food, fodder, fuelwood, manure, building materials, and medicinal herbs ties them closely to fragile ecosystems. This work sustains household economies while preserving ecological knowledge transmitted across generations.

Displacement disrupts these subsistence systems. Women lose direct access to forests and fields, are forced to abandon established livelihood practices, and must adapt to unfamiliar ecological and social environments, often with limited formal education. This results in both economic insecurity and the erosion of cultural knowledge. Although men are perceived as primary wage earners, women historically sustained a degree of economic autonomy through the sale of dairy, fruits, herbs, and fodder (Sidh & Basu, 2011; Singh, 2019). Local barter systems such as *Painchu* in Uttarakhand and *Padyalo* in Himachal Pradesh supplemented these contributions (Bisht, 2009). Displacement dismantles these informal economies, leaving women unable to navigate new cash-dependent markets and thus deepening economic marginalisation.

The outmigration of men in search of work has led women to assume greater responsibility in agriculture, despite its physical demands often finding in it a source of independence and belonging (Ranjan et al., 2020). Many women have gained pride and self-worth from recognising their indispensable role in household survival. The term 'Barkat', often associated with wealth or blessings, is used by women to describe the empowerment they derive from productive work (Bisht, 2009). This increased participation in agriculture and household decision-making has enhanced women's visibility in community life and afforded them a degree of financial independence. For many, survival is accompanied by a desire to redefine identity, reclaim agency and confront challenges with resilience.

Survival strategies developed over generations rooted in strong social bonds, shared experiences and adaptation in the absence of consistent male support have enabled Himalayan women to endure harsh climatic and geographical conditions. Small scale subsistence-based activities such as forest produce collection, dairy production and barter have served not only as economic strategies but also as expressions of community, agency and resilience (Bisht, 2009; Datta & Rustagi, 2012; Sayektiningsih & Hayati, 2021). Displacement from large-scale development projects now threatens these fragile systems. Loss of access to forests, agricultural land and local networks strips women of the very foundations of their autonomy. In resettled areas where traditional skills are often undervalued, uneducated women are pushed back into dependence on male family members for survival. This reversal suppresses nascent entrepreneurship, erasing years of gradual progress.

Post-displacement, women are often forced into a monetised economy where even basic necessities must be purchased, eliminating the safety net provided by barter and reciprocal exchange. Transitioning from a labour-intensive, subsistence-based system to a cash-dependent one has exposed many to economic hardship and social vulnerability. The losses are twofold, the erosion of ancestral lands and livelihood practices and marginalisation within an unfamiliar, often exploitative, economic system where middlemen, profiteers, and even kin may take advantage of their weakened position. Rehabilitation plans frequently overlook these lived experiences, raising concerns over the equity and inclusivity of women in resettlement and resource management strategies. As a result, women remain economically disadvantaged and continue to experience a lower quality of life than men in these regions (Parasuraman, 1993; Asthana, 2012; Datta & Rustagi, 2012; Perera, 2012).

Importantly, displacement is not confined to those whose homes are submerged by reservoirs. It also affects those unable to sustain cultivation in degraded environmental conditions post-project. In some cases, project proponents assert that no new impoundments will be created, claiming instead that water will be sourced from the tailrace tunnel (TRT) of upstream hydropower projects. However, this practice poses significant ecological and hydrological risks. TRTs are designed to return water from run-of-the-river (RoR) projects to the river system, maintaining downstream flows and ecological continuity. Diverting water from TRTs disrupts this process, further depleting river stretches already desiccated by upstream dams (Sahu, 2020; Wang et al., 2024). As a result, both midstream sections often already dry and downstream segments are deprived of critical water flows, severely affecting aquatic ecosystems and dependent communities.

This pattern is evident in several Himalayan projects, viz. the under-construction Rampur HEP drawing water from the TRT of Nathpa Jhakri HEP, the under-construction Surghani-Sundla project drawing from the TRT of Baira Siul HEP and the commissioned Karchham-Wangtoo HEP utilising the head between the tail-waters of Baspa-II HEP and the headwaters of Nathpa–Jhakri HEP (Katoch et al., 2014; Government Audit Report-Comptroller and Auditor General). Alongside these irrigation-related challenges, locals have reported increased noise pollution and dust from construction activities, which disrupt pollination processes and reduce crop yields. Thus, dam-induced displacement is not solely a result of government relocation policies but also of broader ecological and livelihood disruptions that leave affected communities, particularly women, with no viable alternatives.

2.3 *Loss of Access to Natural Resources*

Displaced women consistently report severe impacts from the loss of access to common resource pools (CRPs), including forests, grazing lands, and communal spaces, that historically provided fuelwood, fodder, and essential household materials (Parasuraman, 1993; Bisht, 2009; Datta & Rustagi, 2012; Sayektiningsih & Hayati, 2021; Randell, 2022). These resources have been central to women's survival strategies, supplementing household consumption and generating supplemental income. In the Western Himalaya, women's reliance on CRPs was further tied to tourism and pilgrimage economies, where hotels and resorts created opportunities to sell milk, fuelwood, or other goods sourced directly from communal lands (Sati, 2015; Biswas & Kumar, 2024). Displacement disrupts these contributions, stripping households of critical livelihood buffers and undermining women's role in sustaining food and nutritional security.

Despite their importance, such contributions remain invisible within market-based valuations. Rehabilitation policies rarely account for women's dependence on CRPs, leading to declines in living standards post-displacement. These gaps have been documented across major projects such as the Tehri Dam, Chamera series, Ranjit Sagar HEP, and Nathpa Jhakri HEP, where women's reliance on CRPs was neither recognised nor safeguarded (Asthana, 2012; Bisht, 2009; Dwivedi, 1999; Katoch et al., 2014).

Water insecurity has emerged as a particularly acute challenge. Historically, Himalayan rivers were free-flowing lifelines that ensured fish, fuelwood, drinking water, and irrigation (Erlewein, 2013). Settlements were strategically located near rivers and *nallahs*, enabling women to collect water for household and agricultural needs with relative ease. Post-resettlement, however, colonies are often sil situated far from perennial streams, or the streams themselves have dried due to river diversion. This increases women's physical burden, as they must travel greater distances for basic water needs (Asthana, 2012; Bisht, 2009). Even where piped supply exists, water is intermittent and unreliable. In New Tehri, built at 1500 m, water must be pumped uphill, reaching households only intermittently; tankers and handpumps provide partial relief, but the steep terrain makes carrying water physically exhausting (Asthana, 2012). Similar difficulties are reported in resettled plains, where loss of access to timber, stones, and sand compounds household strain.

Irrigation shortages exacerbate these challenges. In the plains, where compensatory land is often less fertile, irrigation is essential but frequently absent (Jain et al., 2021; Thakkar, 2000). Although policies promise basic amenities before resettlement, irrigation issues persist for years, forcing many families to sell allotted land. With little training or institutional support, women—most of whom lack formal agricultural education—are left to cultivate in unfamiliar conditions (Asthana, 2012; Bisht, 2009). The ecological and cultural distance between Himalayan terrains and the plains deepens their struggle, reflecting governance failures and the absence of gender-responsive adaptation measures.

Traditional Himalayan farming once ensured food security through mixed cropping and rotational systems (*Sar-system*), with diverse staples such as mandua (finger millet), jhangora (barnyard millet), urad (black gram), jowar (sorghum), bajra (pearl millet), tuar (pigeon peas), ramdana (amaranthus), rice, and wheat (Bisht et al. 2007). Fruit cultivation and home-based orchards further supplemented subsistence. Women also foraged wild edibles and medicinal plants, including gucchi mushrooms (*Morchella spp.*), caterpillar fungus (*Ophiocordyceps sinensis*), fiddlehead ferns, burans (*Rhododendron arboreum*), wild berries, nettles, and amaranth leaves (Sharma et al., 2026; Haq et al., 2022; Negi et al., 2020). These practices enriched diets, supported traditional healthcare, and provided supplementary income.

Resettlement severs these intricate socio-ecological linkages. Women are compelled to shift from a subsistence-based to a cash-dependent economy where even basic necessities must be purchased. The consequences are multifold: erosion of crop diversity, loss of traditional ecological knowledge, and diminished opportunities for supplemental income. This transition not only undermines women's economic security but also weakens community resilience, highlighting the structural blind spots of current rehabilitation frameworks.

2.4 Health and Well-Being

Displacement has disrupted women's long-standing social support systems i.e. networks of kin, neighbours and community members that have historically functioned as vital lifelines (Bisht, 2009). These informal systems sustained poor households through the lending of food and money, the sharing of labour and the provision of emotional support during times of crisis. Yet, cost–benefit analyses of large-scale development projects, including hydropower schemes rarely account for such intangible but critical losses.

In pre-displacement contexts, community-based safety nets provided essential assistance in agriculture, childcare, home construction and the exchange of necessities such as food and clothing. Field surveys reveal that, in mountain communities, young widows with children were supported through tightly knit networks built over generations of shared struggle and mutual care (Parasuraman, 1993; Bisht, 2009; Kumar, 2021). These networks offered not only basic material needs but also emotional resilience, enabling women to navigate the challenges of the Himalayan environment. From assistance during childbirth at home to collective responses in times of natural calamity, such communities demonstrated remarkable adaptive capacity. However, resettlement fractures this social fabric, thrusting women into the more individualistic culture of modern resettlement sites. In households where both husbands and young sons have migrated, survival often depends on the cohesion of these community bonds (Asthana, 2012). Women in similar situations consistently report that the absence of these networks, once sources of joy, purpose, and emotional sustenance has led to loneliness, depression and a deep sense of loss. Mental health, once supported by communal living and shared responsibilities, has emerged as an urgent yet largely invisible issue for displaced women (Bisht, 2009). The absence of gender-sensitive mental health frameworks in displacement planning is a critical policy gap, requiring not just clinical care but culturally grounded support that fosters psychosocial recovery and community rebuilding.

The psychological toll of displacement and the collapse of communal life has manifested in physical health issues, including elevated blood pressure and other stress-related illnesses. Daily walks into the forest to gather fuelwood and fodder, once integral to women's routines, were not merely mundane tasks. They were opportunities for social connection, knowledge exchange and mutual encouragement (Asthana, 2012; Bisht, 2009). Such activities reinforced women's autonomy in managing household affairs. In resettlement environments, the disconnection from both nature and social circles has dismantled these routines, eroding emotional well-being and diminishing women's sense of agency. These losses extend beyond the disruption of basic needs, destabilising social, cultural, political, economic and psychological systems that had been established over generations of cohabitation (Chandna et al., 2022).

The health challenges in India's mountainous regions are exacerbated by systemic deficiencies in medical infrastructure. Only a few states, Kerala, Tamil Nadu, Karnataka, Punjab and Delhi meet the WHO benchmark of one doctor per 1000 people while others including states in the Western Himalaya, fall short (Chandna et al., 2022). In these areas, village health centres are scarce and cultural hesitancies toward hospital births mean that many women still give birth at home, often assisted by elderly women or traditional midwives (Chandna et al., 2022). Uttarakhand's maternal mortality rate in 2007–09 was a staggering 359 per 100,000 live births, considerably higher than the national baseline of 212 (National Health Systems Resource Centre). Displacement and community fragmentation further compound this risk by severing access to traditional support systems and medical services. This emphasizes the urgent need for displacement planning to incorporate region-specific health strategies that restore access to maternal care and address structural healthcare gaps.

2.5 *Changing Social Roles and Marginalization*

Displacement has been consistently associated with both the marginalization of women and the reconfiguration of their social roles. In their original mountain villages, women were central to agricultural practices—sowing, weeding, harvesting, and decision-making—often sharing responsibilities on an equal footing with men. In resettlement contexts, however, particularly in the plains where farming depends heavily on hired labour, women's contributions to agriculture are frequently diminished or rendered invisible (Parasuraman, 1993; Bisht, 2009; Asthana, 2012). Men assume the role of primary negotiators for contracts, secure preferential access to resources, and engage with formal institutions, marking a transition from the relatively gender-inclusive farming systems of the Himalaya to more rigidly patriarchal arrangements typical of many urban and semi-urban settings.

This exclusion is not the result of women's unwillingness to engage but of entrenched patriarchal norms that, under the guise of "protection," limit women's participation in public and economic life. Lacking targeted support, many displaced women—often with limited formal education—struggle to adapt to new socio-economic realities. Rehabilitation frameworks, meanwhile, remain narrowly focused on physical relocation and infrastructure provision, failing to address these socio-cultural transitions or to build women's capacity to navigate altered social landscapes.

Even where land has been allocated, structural challenges such as poor irrigation and low soil fertility persist (Asthana, 2012; Bisht, 2009). In such contexts, men frequently seek alternative employment, leaving women with primary responsibility for cultivation and household survival. Non-compensated communities—those excluded from official project-affected zones—face further reductions in agricultural productivity due to diverted streams and dust pollution from construction, intensifying women's dependence on unpaid labour (Goodrich et al., 2019).

Many traditional social roles and community-based customs in the Himalaya have evolved specifically to enable survival in its harsh geographic and climatic conditions. These gendered roles, while deeply interdependent, are highly adapted to the socio-ecological realities of mountain life. The construction of hydropower projects, however, disrupts these systems, destabilising the delicate balance between society and environment. Rehabilitation and resettlement policies typically overlook the intricate web of gendered roles and mutual dependencies, instead operating at a superficial level that fails to account for the deep-rooted socio-ecological interconnections that underpin community resilience.

2.6 Security and Violence Risks

In their native villages, women in the north-western Himalaya often enjoyed a strong sense of safety and mobility, supported by long-standing social trust networks and cultural cohesion. They moved freely during the day and night, participated in community gatherings, and wore traditional jewellery without fear of theft or harassment (Asthana, 2012; Bisht, 2009). Such security was not incidental but rooted in centuries of residence within familiar landscapes where kinship ties and collective vigilance served as protective buffers.

Relocation disrupts these foundations of safety. In resettlement colonies, women frequently report heightened insecurity linked to unfamiliar architectural layouts, altered settlement patterns, linguistic and cultural differences, and at times overt hostility from host communities (Asthana, 2012; Bisht, 2009). What had once been a social environment of belonging is replaced with feelings of constraint, hypervigilance, and anxiety (Mehta & Srinivasan, 2000; Melketo et al., 2023). This erosion of security extends beyond the physical sphere to undermine women's psychological well-being and sense of autonomy.

Coping strategies, such as restricting movement to daylight hours or confining activities to domestic spaces, may mitigate perceived risks but simultaneously deepen dependence on male relatives for routine tasks—farming, marketing, visiting healthcare facilities, or travelling to maternal homes (Asthana, 2012; Bisht, 2009). Such self-imposed restrictions curtail social participation and erode agency, highlighting that the consequences of displacement are not only economic but also profoundly gendered, with direct implications for equity and justice.

Further compounding these vulnerabilities are demographic changes in project areas. Male outmigration for wage labour, combined with the influx of male workers from outside the region, has contributed to rising theft and gender-based crimes (Udas et al., 2019). The weakening of traditional community watch systems intensifies these threats, leaving women—whether resettled or remaining in partially affected villages—with reduced autonomy in their daily lives. The loss is thus multidimensional: diminished physical safety is intertwined with declining mobility, weakened community bonds, and psychological strain, collectively undermining women's capacity to act as independent agents within their new environments.

2.7 Cultural Dislocation, Policy Gaps, and the Need for Inclusive Resettlement Planning

For generations, Himalayan women have navigated harsh climatic and geographical conditions through established social roles rooted in community bonds, spiritual ties, and ecological reciprocity. Displacement, however, disrupts these long-standing patterns, uprooting women from their cultural landscapes and pushing them into unfamiliar socio-environmental contexts that deepen disempowerment (Parasuraman, 1993; Thukral, 1996; Bisht, 2009; Asthana, 2012). Sacred groves, ancestral homes, traditional water mills, agricultural lands, and communal spaces—once central to both livelihoods and identity—are severed by hydropower projects, compounding material loss with cultural and spiritual alienation.

Policy frameworks have largely failed to address these multidimensional impacts. Project proponents frequently abandon initial commitments, altering rehabilitation and resettlement plans mid-course. In the absence of a centrally governed and enforceable legal framework, accountability is weak, and displacement is reduced to the physical relocation of households rather than a holistic process that safeguards cultural, social, and psychological well-being. This governance gap has been repeatedly documented in hydropower development across the region, where families experience disruptions far beyond what resettlement policies acknowledge (Parasuraman, 1993; Judge, 1997; Mehta & Srinivasan, 2000; Hay et al., 2019; Melketo et al., 2023).

As Fig. 1 illustrates, the diversion of land for major hydropower projects in Uttarakhand, Himachal Pradesh, and Jammu & Kashmir has displaced or will displace tens of thousands of families. Such large-scale transformations heighten vulnerabilities for women, who face compounded burdens as primary caregivers, resource managers, and cultural custodians. Yet resettlement policies continue to assume that displaced households will return to farming livelihoods, overlooking aspirations for diversified employment and urban integration. Research shows that resettlement outcomes vary widely across families depending on their socio-economic position and adaptive capacity (Hay et al., 2019). A one-size-fits-all land-for-land approach is thus inadequate. Instead, households should be offered choices that include diversified livelihood options, urban resettlement packages, and capacity-building measures to enable long-term resilience.

Figure 2 provides a synthesis of the gendered vulnerabilities faced by women in the NW Himalaya, underscoring how displacement reshapes their roles, responsibilities, and identities. Integrating flexible, choice-based frameworks into resettlement planning is therefore critical not only for equity but also for strengthening social and ecological resilience. To substantiate these thematic insights, the following section presents empirical case studies of dam-induced displacement in the north-western Himalaya. By examining demographic shifts, literacy patterns, and female workforce participation across affected and control sub-districts, the analysis highlights how structural gender inequalities are reproduced and intensified through resettlement processes.

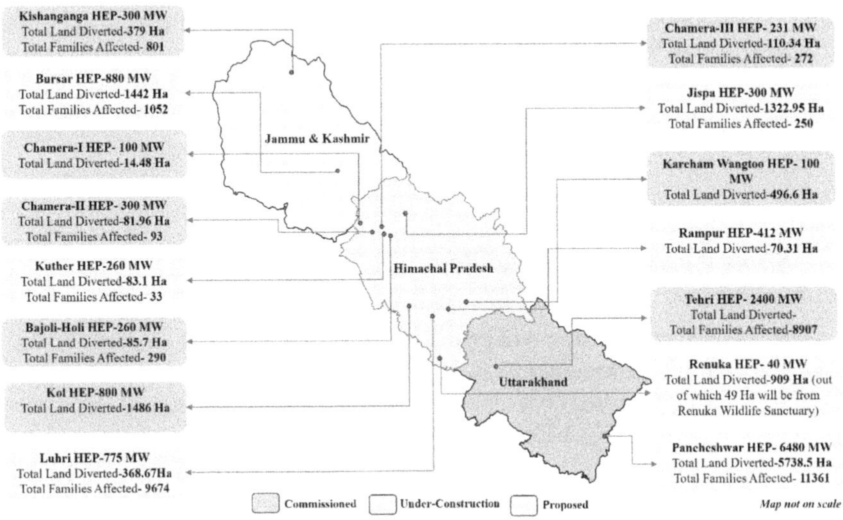

Fig. 1 Map showing geographical locations of major large dams (>100 MW) in the Western Himalayan states of Jammu & Kashmir, Himachal Pradesh, and Uttarakhand that have caused or are projected to cause large-scale displacement, affecting thousands of lives. (*Data source* EJ Atlas, EIA reports, Himdhara)

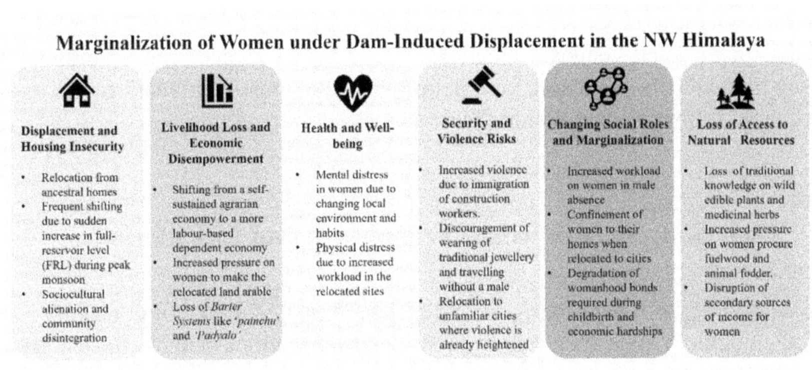

Fig. 2 Gendered vulnerabilities: societal disadvantages and increased risks faced by Himalayan women due to dam-induced displacement

3 Case Studies

Building on the gendered impacts outlined in the preceding section, this part of the review turns to quantitative case analyses of selected dam projects in the north-western Himalaya. While Sect. 2 examined how displacement reshapes women's livelihoods, health, and social roles, the present analysis situates these dynamics within measurable socio-economic patterns—population change, literacy, and workforce participation. By grounding women's experiences in demographic and census-based indicators, the case studies provide empirical evidence that dam development not only exacerbates gendered vulnerabilities but also reshapes community structures in ways that demand inclusive and gender-sensitive policy responses.

3.1 Study Area and Methodology

A quantitative approach was employed to assess the gendered impacts of dam development in sub-districts across Uttarakhand, Himachal Pradesh, and Jammu & Kashmir. The study focuses on female demographic trends, educational attainment, and workforce participation before and after dam construction. Five large dams were selected—Ramganga (1974), Pong (1974), Chamera I (1994), Tehri (2006), and Baglihar (2008)—representing diverse physiographic and socio-cultural contexts (Table 1). Selection was guided by three criteria: (i) large-scale displacement, (ii) prominence in regional hydropower expansion, and (iii) availability of comparable census data.

For each project, "affected sub-districts" were defined as those directly impacted by reservoir submergence or displacement, while "control sub-districts" were chosen for their geographical proximity and socio-environmental similarity but minimal exposure to hydropower development. This pairing allows a comparative assessment of demographic and gendered outcomes. The "pre-dam year" corresponds to the census immediately preceding completion, and the "post-dam year" refers to the first census following completion (Table 1).

3.2 Data Sources, Analytical Framework and Calculations

The analysis draws on Census of India data from 1971 to 2011, enabling the tracing of demographic and socio-economic transformations in dam-affected versus control sub-districts. Census data were selected for their consistent methodology, comprehensive coverage, and gender-disaggregated indicators at sub-district level.

Indicators Extracted

To examine the gendered impacts of hydropower development, a set of demographic and socio-economic indicators was extracted from the Census of India for each study site. These included: (i) total population, (ii) female and male population, (iii) number of workers, (iv) number of literates, and (v) number of Scheduled Caste (SC) and Scheduled Tribe (ST) females. Together, these variables enabled the assessment of

Table 1 Case study dams, affected and control sub-districts, and corresponding census years used for pre- and post-dam analysis

Hydroelectric project	State	District	Affected sub-district	Control sub-district	Pre-dam year	Dam completion year	Post-dam year
Ramganga Dam	Uttarakhand	Pauri Garhwal	Kotdwar	Thalisain	1971	1974	1981
Pong Dam	Himachal Pradesh	Kangra	Fatehpur	Palampur	1971	1974	1981
Chamera I Dam	Himachal Pradesh	Chamba	Bhattiyat	Chaurah	1991	1994	2001
Tehri Dam	Uttarakhand	Tehri	Tehri	Narendranagar	2001	2006	2011
Baglihar Dam	Jammu and Kashmir	Ramban	Ramban	Gool Gulab Garh[1]	1981,[2] 2001	2008	2011

[1] Gool Gulab Garh is in Reasi District, choosen due to data availability.
[2] 1991 census data is not available.

population dynamics, literacy levels, workforce participation, and the representation of marginalized social groups across time.

To capture temporal shifts associated with dam construction, changes in these indicators were quantified by comparing pre-dam and post-dam census years. This approach made it possible to link demographic and socio-economic transformations directly to displacement processes, thereby providing a systematic basis for evaluating gendered vulnerabilities and outcomes.

Female Population Growth Rate (FGR)

The female population growth rate was calculated for both affected and control sub-districts using the standard decadal growth rate formula:

$$Growth\ Rate(\%) = \frac{P_t - P_{t-1}}{P_{t-1}} \times 100$$

where P_t and P_{t-1} denote the total female population in the post- and pre-dam census years, respectively. FGR was used to assess demographic changes, including potential population decline due to displacement.

SC/ST Share Within Female Population (SC/ST%)

The proportion of SC and ST women in the total female population was calculated as:

$$SC\&ST(\%) = \frac{SC_f + ST_f}{Total\ Female\ Population} \times 100$$

where SC_f and ST_f represent the number of ST and SC females, respectively. This metric served as a proxy for assessing the socio-economic vulnerability of marginalized groups in the study areas.

Change in Female Literacy Rate (ΔFLR)

The change in female literacy between pre-dam and post-dam census years was calculated using:

$$\Delta Female\ Literacy(\%) = \frac{FL_{post} - FL_{pre}}{FL_{pre}} \times 100$$

where FL_{pre} and FL_{post} denote the female literacy rates before and after dam completion.

This measure was crucial in evaluating whether hydropower-related development translated into educational improvements for women.

Gender Gap Indicators

Three gender gap measures applied, based on Census 2011 definitions and literature on gendered socio-economic analysis (Kabeer, 2008) include:

- Literacy Rate Gap (LR)

$$LR = Male\ Literacy(\%) - Female\ Literacy(\%)$$

- Work Participation Rate Gap (WPR)

$$WPR = Male\ WPR(\%) - Female\ WPR(\%)$$

  where WPR is proportion of workers to the total population.
- Workforce per Literacy Rate Gap (WLR)

$$WLR = \left(\frac{Female\ Workers}{Female\ Literates} - \frac{Male\ Workers}{Male\ Literates}\right) \times 100$$

The analysis helped to examine the gendered disparities and inequalities in the education and employment outcome across dam-affected sub-districts.

3.3 *Result and Discussion*

Demographic Shifts and Female Population Growth

The analysis of female population growth rates highlights significant demographic divergences between dam-affected and control sub-districts across the five study sites. These trends provide critical insights into the gendered consequences of hydropower development in the NW Himalaya, particularly in relation to displacement and migration.

Figure 3 demonstrates a pronounced decline in female population growth rates across all dam-affected sub-districts in the hilly regions. The sharpest declines are observed in Tehri and Pong, with Tehri recording a negative growth rate of –3.21%, in stark contrast to the gradual decline in its control area. Baglihar shows a moderate decline, while Ramganga and Chamera-I display relatively stable growth rates.

These findings corroborate with broader literature on displacement caused by large dams, where outmigration and social fragmentation disproportionately affect women (Cernea, 2000; World Commission on Dams, 2000). For instance, the Tehri

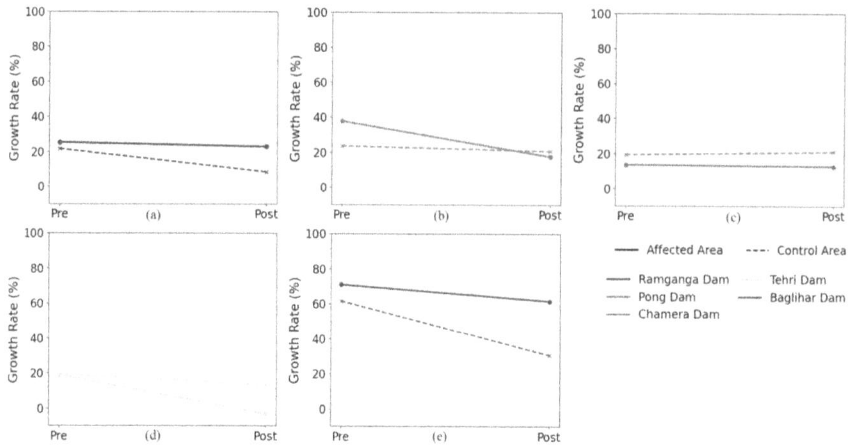

Fig. 3 Slope chart showing the rate of change in female population (%) before and after dam construction in: **a** Ramganga, **b** Pong, **c** Chamera I, **d** Tehri, and **e** Baglihar

Hydropower Project displaced more than 4909 rural and nearly 4000 urban families (Hanumantha Rao Committee [HRC], 1997; Tehri Hydroelectric Development Corporation [THDC]; Bisht, 2009). Similarly, the Pong Dam uprooted between 19,000 and 30,000 people (Judge, 1997). Such large-scale disruptions are strongly reflected in the negative demographic shifts observed in the study.

At the same time, relative demographic stability in certain sub-districts, such as those linked to Ramganga and Chamera-I, may conceal less visible social dislocations. Women's burdens of relocation, livelihood disruption, and loss of community support networks often remain obscured in aggregate statistics, underscoring the necessity of gender-disaggregated data to capture the full developmental impacts (Kothari, 1996; Minocha, 2020).

Caste and Gender Intersections

The intersection of caste and gender reveals layered vulnerabilities in displacement contexts. Figure 4a shows that in Chamera-I, the SC/ST share of the female population remained consistently high, rising only marginally from 38.22% to 38.74% post-dam. This suggests a demography already dominated by historically marginalized groups, coupled with limited outward migration or differential resettlement.

By contrast, Tehri, Ramganga, Pong, and Baglihar registered post-dam increases of 1.26%, 1.49%, 4.01%, and 9.65% respectively, in the proportion of SC/ST women. These patterns may reflect the selective out-migration of dominant groups, leaving behind marginalized communities, or systematic exclusion and undercounting of SC/ST populations in resettlement frameworks. Bisht (2009) highlights how the gendered experiences of SC/ST women are rarely captured in displacement studies, while Bhagat (2017) argues that census data, though limited, remains a crucial tool

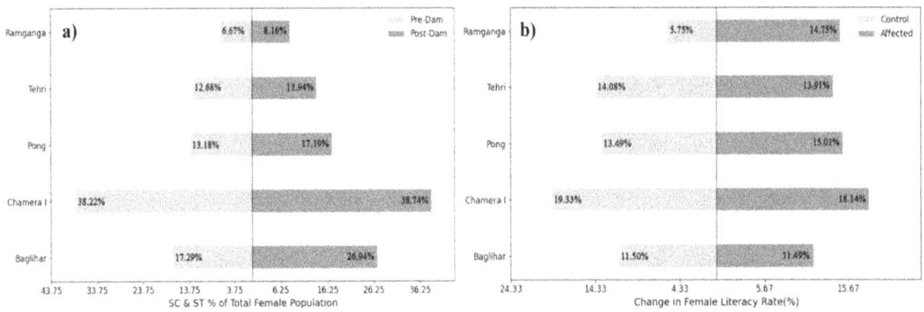

Fig. 4 **a** SC/ST female population (%) in affected areas before and after dam development; **b** Rate of change in female literacy in affected versus control districts post-dam

for revealing such hidden inequities. Taken together, these findings underline how hydropower projects can deepen intersectional disadvantages, disproportionately concentrating displacement risks among marginalized women.

Female Literacy Trends Post-Dam Construction

Despite significant demographic disruptions, female literacy rates show modest but consistent improvements across dam-affected sub-districts (Fig. 4b). In Ramganga and Pong, literacy rose by 9% and 1.52% respectively, even as female population growth declined, suggesting that development projects and associated welfare schemes may create new educational opportunities. Similarly, Chamera-I, Tehri, and Baglihar witnessed incremental literacy gains post-dam.

However, these improvements must be interpreted cautiously. Without parallel increases in women's workforce participation or mobility, rising literacy does not necessarily translate into empowerment. As gender and development literature stresses, education alone cannot dismantle structural barriers unless supported by systemic reforms in employment, resource access, and decision-making spaces (Bisht, 2009; Kabeer, 2010). In this sense, post-dam literacy trends reflect not unqualified progress but rather the tension between formal educational access and persistent socio-economic exclusion.

Gender Gaps in Workforce Participation

The 2011 Census, for the first time, introduced gender-disaggregated data on workforce participation, enabling a more nuanced analysis of women's economic outcomes in dam-affected areas. Figure 5 highlights a persistent male advantage in both literacy and workforce engagement, reflected in the continued literacy rate gap (LR) and work participation rate gap (WPR). This pattern suggests that educational gains among women have not translated into equitable labour force inclusion. Structural barriers—including limited employment opportunities, mobility restrictions

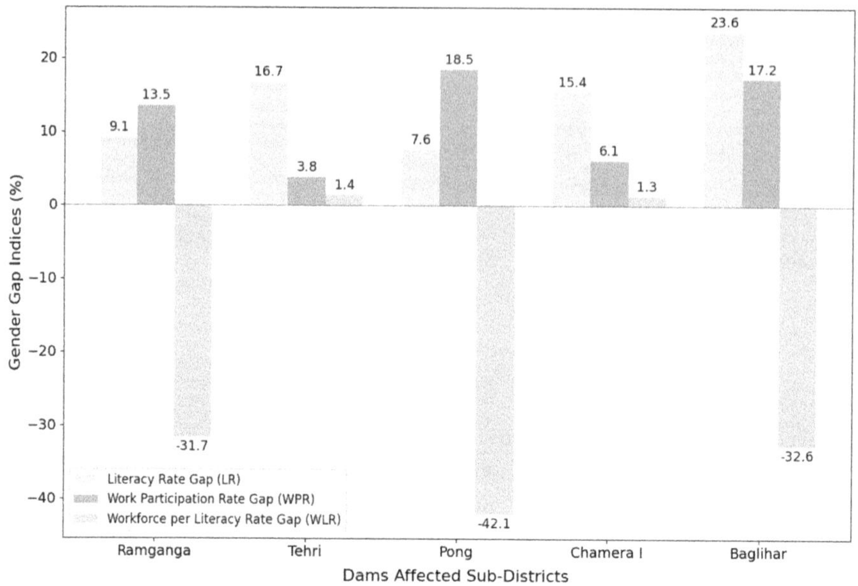

Fig. 5 Gender disparities in dam-affected sub-districts of the NW Himalaya: comparative analysis of female literacy, work participation, and workforce-to-literacy ratio using 2011 Census data

after displacement, and entrenched social norms—continue to constrain women's participation despite rising literacy levels.

Notably, Chamera-I and Tehri display positive Workforce per Literacy Rate (WLR) gaps of 1.3 and 1.4, indicating marginally higher workforce engagement among literate women than literate men. Although counterintuitive, these results may signal adaptive responses in specific contexts, particularly where displacement disrupted traditional male-dominated occupations. Such exceptions underscore the importance of examining localised gendered adaptations rather than assuming uniform outcomes across sites.

Outcome Analyses and Emerging Vulnerabilities

Taken together, the findings reveal a paradoxical outcome. On one hand, hydropower development in the NW Himalaya has generated infrastructure improvements and modest gains in female literacy. On the other, these benefits are overshadowed by sharp demographic decline in affected sub-districts, disproportionate representation of SC/ST women among the displaced, persistently low female workforce participation, and enduring gender disparities.

These dynamics suggest that displacement has not only reinforced pre-existing gender inequalities but also produced new vulnerabilities, particularly for marginalized women. At the same time, instances of increased labour participation among literate women in select contexts (e.g., Tehri and Chamera-I) point to emerging forms of resilience and adaptation that warrant closer examination. Such mixed outcomes highlight the dual role of women in Himalayan development—as populations rendered vulnerable by exclusionary policies, but also as agents of adaptation whose strategies could inform more inclusive and gender-sensitive hydropower planning.

Limitations and Future Directions

This study is constrained by the absence of disaggregated village-level data, migration records, and information on post-displacement infrastructural access such as schools, transport, and healthcare. These gaps restrict a comprehensive assessment of the complex, long-term gendered impacts of dam-induced displacement. Moreover, census data, while valuable for tracking demographic trends, cannot capture the cultural, psychological, and ecological dimensions of women's experiences.

Despite these limitations, the case studies provide important empirical insight into how displacement reshapes women's lives, highlighting not only multidimensional vulnerabilities but also emerging capacities for adaptation. They demonstrate the need for future research that employs intersectional, gender-sensitive methodologies, incorporates qualitative accounts alongside quantitative measures, and builds robust, disaggregated datasets. Such approaches are essential for informing equitable resettlement frameworks and development policies that acknowledge both women's vulnerabilities and their resilience.

4 Women's Agency and Role in Sustainable Development

While women in the north-western Himalaya bear the heaviest costs of displacement and resettlement, their responses reveal significant resilience. Rising literacy, selective workforce participation, and the preservation of social cohesion enable many women to act as more than dependents of development—they emerge as agents who sustain livelihoods, conserve biodiversity, and maintain community resilience under conditions of profound ecological and social change.

Legal reforms in India have begun to address the structural marginalization of women, historically reinforced by patriarchal customs surrounding birth, marriage, and death (Udas et al., 2019). For instance, women have the legal right to inherit parental property, yet in much of the Himalaya this provision remains underutilized. The reluctance to claim inheritance is not simply a reflection of patriarchy but can also be interpreted as a deliberate strategy: in an environment where survival depends on social cohesion, women often prioritize maintaining harmonious relationships with brothers and extended kin over asserting individual property rights (Adger et al., 2003). This demonstrates how survival strategies are negotiated at the intersection of law, culture, and ecology, where women may value relational security above material gain.

Nevertheless, women's participation in public domains such as governance institutions and market-oriented enterprises remains limited, constraining their ability to influence decisions that shape their lives. Their engagement in barter systems and reciprocal labour exchange continues to underpin household and community functioning, but these contributions are often invisible in official economic accounts (Baindur, 2023; Bisht, 2009; Sidh & Basu, 2011). Historical inequalities in land ownership, compounded by low literacy and entrenched caste hierarchies, further curtail women's negotiating power and reinforce their marginalization (Udas et al., 2019).

Despite such structural barriers, women remain central to sustaining adaptive livelihoods. Wild edible plants provide critical nutrition and economic buffers in times of scarcity (Kumari et al., 2025; Mallick et al., 2024), while traditional food systems in the Western Himalaya serve as reservoirs of biodiversity, food security, and cultural identity (Bhatt et al., 2023; Bhattacharyya et al., 2023a, 2023b; Salaria & Kumar, 2025). Women, as custodians and transmitters of this ecological knowledge, rarely receive recognition for their role in sustaining these systems. Their relationship with forests and forest products extends well beyond material needs, being deeply embedded in belief systems, ritual practices, and collective memory.

Historical precedents illustrate the political force of women's ecological agency. The Bishnoi movement in Rajasthan and the Chipko movement in Uttarakhand highlight how women mobilized to defend forests central to their survival and cultural values (Bhatt et al., 2023; Bhattacharyya et al., 2023a, 2023b). Equally significant are women's everyday practices—intercropping, soil conservation, afforestation, and foraging—that safeguard ecological integrity while sustaining households. Such practices underscore that women are not passive victims of environmental change but active custodians of sustainable development whose agency deserves institutional recognition.

Yet, women's ecological knowledge and adaptive strategies remain systematically undervalued in hydropower planning and rehabilitation frameworks. Their exclusion represents not only a gender gap but a missed opportunity for more context-sensitive and ecologically sustainable development. The following section (5.0) builds on these insights to explore how women's agency, intersectional vulnerabilities, and traditional knowledge can be integrated into hydropower governance, ensuring that sustainability agendas in the Himalaya move beyond token participation toward genuine inclusion.

5 Recommendations and Policy Proposals

5.1 Policy Gaps and Limitations

India's key frameworks—including the National Rehabilitation and Resettlement Policy (Ministry of Rural Development, 2007), the Hydro Power Policy (Ministry of Power, 2008), and the National Water Policy (MoEF, 2012)—have guided dam development and resettlement but remain largely gender-blind and inattentive to the socio-cultural and ecological particularities of the north-western Himalaya. This omission has had serious implications in projects such as Tehri, Pong, Bhakra, Chamera, Nathpa-Jhakri, and Baglihar, which together displaced tens of thousands of people, with women, Scheduled Castes (SCs), and Scheduled Tribes (STs) disproportionately bearing the burdens of loss and adaptation (Judge, 1997; THDC, 1998; WCD, 2000; Bisht, 2009).

The relocation of Pong Dam oustees to Rajasthan illustrates the compounded vulnerabilities created by such blind spots. Women not only endured the trauma of leaving ancestral homes but also struggled to adapt to radically different climatic and socio-economic conditions (Judge, 1997). Historically, women in Himalayan communities have been central to the stewardship of forests, water, and medicinal plants—foundations of both household survival and local conservation systems (Gururani, 2002; Sharma et al., 2008). Yet policy frameworks systematically excluded them from decision-making, and literacy improvements have rarely translated into leadership roles or economic autonomy.

Environmental governance mechanisms, particularly the Environmental Impact Assessment (EIA) notification under the Environment (Protection) Act (MoEF, 2006), reveal further weaknesses. The EIA regime does not mandate gender-disaggregated analysis, community-led audits, or systematic social impact monitoring. Implementation has been undermined by weak appraisal quality, mid-project alterations, and superficial public hearings (CAG, 2017). Moreover, the uniform EIA national model ignores the fragile geology and ecology of the Himalaya, where cumulative risks from tunnelling, slope destabilisation, and altered hydrology compound displacement impacts. Exemptions for projects below 25 MW exacerbate this problem by permitting unchecked proliferation of "small" dams in ecologically sensitive catchments (Cernea, 2000; MoEF, 2006).

The 4.5 MW Hul small hydro project in Chamba district starkly demonstrates the consequences of these regulatory gaps. Community submissions reported degradation of approximately 5 km of the Saal River, felling of nearly 1,800 trees and more than 2,000 bushes, and the destruction of traditional watermills (gharats), with estimates ranging from 16 to over 40 mills affected (Bhagat, 2017; Kothari, 1996). These outcomes underscore that even small-scale projects can generate disproportionately high ecological and social costs in mountain environments when robust, locally sensitive assessment and monitoring mechanisms are absent.

5.2 Proposed Reforms

Reforms to India's Environmental Impact Assessment (EIA) and Rehabilitation and Resettlement (R&R) frameworks must account for both the visible and invisible burdens of displacement, including women's unpaid labour, psychological stress, and the erosion of social capital in marginalized groups. Current "one-size-fits-all" assessment models are ill-suited to the diversity and fragility of mountain ecosystems. In the north-western Himalaya—home to multiple Ecologically Sensitive Zones (ESZs) such as the Great Himalayan National Park and Nanda Devi Biosphere Reserve—even small hydropower projects can generate disproportionate ecological and social costs. The exemption of projects below 25 MW from comprehensive appraisal is especially problematic; site-specific EIA norms tailored to fragile mountain ecologies are urgently required.

R&R processes must also move beyond procedural compliance to become participatory, transparent, and trust-based. Affected communities should be actively engaged in all stages of planning and implementation. Compensation strategies should extend beyond one-time cash payments toward benefit-sharing frameworks that guarantee access to project revenues, electricity, healthcare, education, and transport. Equitably designed employment opportunities linked to project activities are essential to restore livelihoods and reduce dependency, with explicit targeting of women and marginalized groups (Schulz & Skinner, 2022).

Finally, where resettlement involves relocation from high-altitude to lowland or urban settings, comprehensive integration support is needed. This includes livelihood training, cultural orientation, and continuous monitoring of assimilation outcomes, ensuring that rehabilitation encompasses social and economic reintegration rather than physical relocation alone.

5.3 Pathways for Inclusive Development

Hydropower development in the Himalaya must aligns economic growth with ecological sustainability and social equity. As illustrated in Fig. 6, planning for new hydropower projects requires moving beyond narrow financial returns to incorporate broader questions: the ecological fragility of mountain environments, health and well-being of displaced communities, livelihood security, greenhouse gas implications, and the viability of alternative renewable energy sources. Integrating these concerns into cost–benefit analyses can help achieve sustainable development and social justice, which has been advocated by serveral workers assessing the cumulative socio-economic and environmental impacts of large dams (Castro-Diaz et al., 2023).

India's policy framework serves as foundation for such inclusivity. For example, the National Policy for Empowerment of Women (2001) promotes mainstreaming gender perspectives in development and enhancing women's participation in

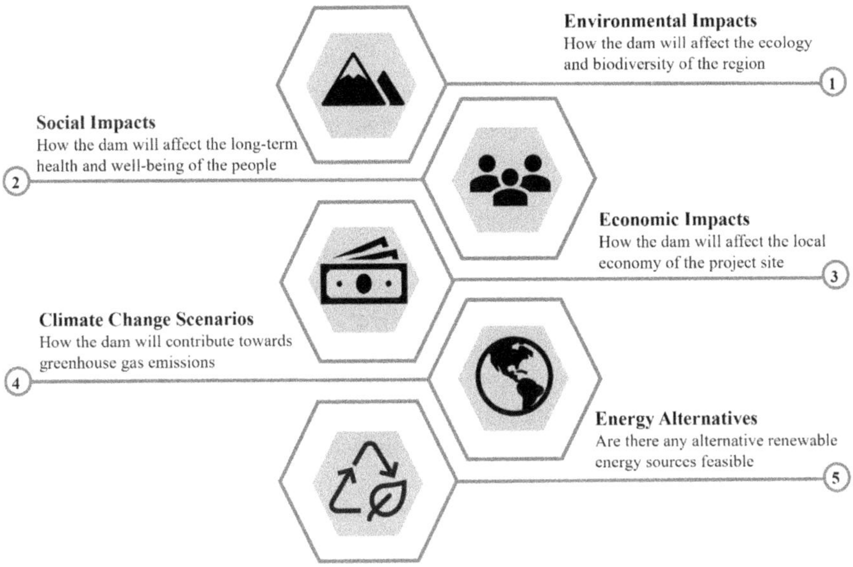

Fig. 6 Key considerations for shaping a new generation of hydropower development in the Himalaya

decision-making (Government of India, 2001). Similarly, the Forest Rights Act (2006) recognises the rights of forest-dependent communities, including women, to protect and sustainably use forest resources, which are also relevant for development and justice in the Himalayan regions (Government of India, 2006).

The redevelopment of New Tehri Town offers an example of more holistic planning, where schools, hospitals, and roads were integrated into resettlement efforts—though not without trade-offs in traditional occupations and cultural practices (Reddy, 2016, 2017). Such examples suggest that if women's ecological knowledge and resource management practices—deeply embedded in local culture—are integrated into planning, they can enhance both community resilience and environmental sustainability. As gender-disaster literature emphasizes, women's knowledge is not a supplementary asset but a critical dimension of adaptive governance (Resurrección, 2019).

Without such context-specific and inclusive approaches, hydropower development risks delivering partial benefits while deepening inequalities, especially for women and marginalized groups.

6 Conclusion

Hydropower development in the north-western Himalaya is not simply an energy project but a socio-ecological transformation that reshapes landscapes, livelihoods, and social relations. Large dams exacerbate existing inequities by disproportionately burdening women, Scheduled Castes, and Tribes through displacement, livelihood loss, and the erosion of social networks. These outcomes emphasize that the costs of development are borne most heavily by those already positioned at the margins.

Women are not passive victims but they act as agents of resilience. Their roles encompass from sustaining households, adapting agricultural practices, innovating barter and reciprocal labour systems, to conserving ecological resources. Such a dual reality of vulnerability and agency demonstrates the limitations of current governance frameworks, which focuses only on physical resettlement and cash compensation rather than the socio-cultural, psychological, and intersectional aspects of displacement.

Our analysis provides a comprehensive understanding of how hydropower reshapes gendered lives in the Himalaya and identifies key directions for more just and sustainable governance. First, it suggests the value of combining demographic indicators with qualitative evidence on livelihoods, health, and social roles. It helps capture both visible and hidden costs of displacement. Second, the intersectional perspective situates women's vulnerabilities at the convergence of gender, caste, and class. It draws attention to their resilience and ecological stewardship rather than as merely passive recipients of development. Third, it argues for governance reforms that should be context-specific and gender-responsive, which should be reflected in site-sensitive EIA norms, gender-disaggregated monitoring, and participatory rehabilitation processes. These reforms will help provide formal recognition of women's ecological knowledge as expertise essential to equitable and sustainable energy transitions.

References

Adger, W. N., Huq, S., Brown, K., Conway, D., & Hulme, M. (2003). Adaptation to climate change in the developing world. *Progress in Development Studies, 3*(3), 179–195. https://doi.org/10.1191/1464993403ps060oa

Agarwal, B. (1994). *A field of one's own: Gender and land rights in South Asia* (No. 58). Cambridge University Press.

Annual Performance Audit Reports, Comptroller and Auditor General (CAG), Government of India. https://cag.gov.in/en/audit-report

Agarwal, B. (2009). Gender and forest conservation: The impact of women's participation in community forest governance. *Ecological Economics, 68*(11), 2785–2799. https://doi.org/10.1016/j.ecolecon.2009.04.025

Asthana, V. (2012). Forced displacement: A gendered analysis of the Tehri Dam project. *Economic and Political Weekly, 47*(7), 96–102. https://www.jstor.org/stable/41720414

Awasthi, S., Jain, K., Sahoo, S., Kumar, R., Goswami, A., Joshi, G. C., & Srivastava, D. C. (2024). Analyzing Joshimath's sinking: Causes, consequences, and future prospects with remote sensing techniques. *Scientific Reports, 14*(1), 10876. https://doi.org/10.1038/s41598-024-60276-3

Baindur, M. (2023). Work, women and landscape in the Himalayas. In *Capital and Ecology* (pp. 254–268). Routledge India.

Bhagat, R. B. (2017). Knowing the social world through census: Reflections on the conditions of Scheduled Castes and Scheduled Tribes. In *Knowing the Social World: Perspectives and Possibilities* (pp. 23–38).

Bhai, V. (2020). The Tehri Dam project: A saga of shattered dreams. In *Water Conflicts in India* (pp. 353–359). Routledge India.

Bhatt, S., Purohit, M. V., Balodi, K. N., Kumar, U., Sharma, A., Shridhar, V., & Arunachalam, K. (2023). The ethnic food system of Jaad Bhotiya tribal community in the Gangotri landscape of western Himalaya: A transition in traditional practices. *AlterNative: An International Journal of Indigenous Peoples, 19*(2), 250–260.

Bhattacharyya, M., Thattantavide, A., & Kumar, A. (2023). Ethnic mountain foods of western and eastern Himalayas, India. In *Wild food plants for zero hunger and resilient agriculture* (pp. 181–205). Springer Nature Singapore.

Bisht, I. S., Mehta, P. S., & Bhandari, D. C. (2007). Traditional crop diversity and its conservation on-farm for sustainable agricultural production in Kumaon Himalaya of Uttaranchal state: A case study. *Genetic Resources and Crop Evolution, 54*(2), 345–357.

Bisht, T. C. (2009). Development-induced displacement and women: The case of the Tehri Dam, India. *The Asia Pacific Journal of Anthropology, 10*(4), 301–317.

Biswas, S. N., & Kumar, P. (Eds.). (2024). *Mountain tourism and ecological impacts: Himalayan region and beyond*. IGI Global.

Castro-Diaz, L., et al. (2023). Impacts of hydropower development on locals' livelihoods: A comparative analysis. *World Development, 167*, Article 105771.

Census of India. (1971–2011). Government of India. https://censusindia.gov.in/census.website/data/population-finder

Cernea, M. M. (1996). Public policy responses to development-induced population displacements. *Economic and Political Weekly, 31*(21), 1515–1523.

Bhattacharyya, M., Thattantavide, A., & Kumar, A. (2023). Ethnic mountain foods of western and eastern Himalayas, India. In *Wild Food Plants for Zero Hunger and Resilient Agriculture* (pp. 181–205). Springer Nature Singapore. https://doi.org/10.1007/978-981-19-6502-9_8 OUCI+1

Cernea, M. M. (1997). Hydropower dams and social impacts: A sociological perspective.

Cernea, M. M. (2000). Risks, safeguards and reconstruction: A model for population displacement and resettlement. In M. M. Cernea & C. McDowell (Eds.), *Risks and Reconstruction: Experiences of Resettlers and Refugees* (pp. 11–55). World Bank.

Chandna, V., Pandey, B. W., Ranjan, O. J., & Jyoti. (2022). Status of reproductive health of women in Himalaya: A case study of selective villages of, Uttarakhand, India. *Journal of the Anthropological Survey of India, 71*(1), 41–57. https://doi.org/10.1177/2277436X221092087

Chopra, R., & Ghosh, D. (2000). Work patterns of rural women in central Himalayas. *Economic and Political Weekly*, 4701–4705. https://www.jstor.org/stable/4410117

Colson, E. (1971). *The social consequences of resettlement: The impact of the Kariba resettlement upon the Gwembe Tonga* (No. 4). Manchester University Press.

Comptroller and Auditor General of India. (2017). *Report No. 39 of 2016: Performance audit of environmental clearance and post-clearance monitoring*. Government of India.

Datta, A., & Rustagi, P. (2012). Status of women in Bihar: Exploring transformation in work and gender relations. International Institute of Social Studies of Erasmus University (ISS) http://hdl.handle.net/1765/34852

Diduck, A. P., Pratap, D., Sinclair, A. J., & Deane, S. (2013). Perceptions of impacts, public participation, and learning in the planning, assessment and mitigation of two hydroelectric projects in

Uttarakhand, India. *Land Use Policy, 33*, 170–182. https://doi.org/10.1016/j.landusepol.2013.01.001
Dwivedi, R. (1999). Displacement, risks and resistance: Local perceptions and actions in the Sardar Sarovar. *Development and Change, 30*(1), 43–78. https://doi.org/10.1111/1467-7660.00107
EJAtlas—Global Atlas of Environmental Justice. https://ejatlas.org/
Elson, D. (1998). Gender and economic growth models. *Feminist Visions of Development: Gender Analysis and Policy*, 156.
Erlewein, A. (2013). Disappearing rivers—The limits of environmental assessment for hydropower in India. *Environmental Impact Assessment Review, 43*, 135–143. https://doi.org/10.1016/j.eiar.2013.07.002
Goodrich, C. G., Prakash, A., & Udas, P. B. (2019). Gendered vulnerability and adaptation in Hindu-Kush Himalayas: Research insights. *Environmental Development, 31*, 1–8. https://doi.org/10.1016/j.envdev.2019.01.001
Government of India. (2001). *National Policy for Empowerment of Women*. Department of Women and Child Development.
Government of India. (2006). *The Scheduled Tribes and Other Traditional Forest Dwellers (Recognition of Forest Rights) Act, 2006*. Ministry of Tribal Affairs.
Grumbine, R. E., & Pandit, M. K. (2013). Threats from India's Himalaya dams. *Science, 339*(6115), 36–37. https://doi.org/10.1126/science.1227211
Gururani, S. (2002). Forests of pleasure and pain: Gendered practices of labor and livelihood in the forests of the Kumaon Himalayas, India. *Gender, Place & Culture, 9*(3), 229–243. https://doi.org/10.1080/0966369022000003842
Hanumantha Rao Committee (HRC). (1997). *Report of Expert Committee on Rehabilitation & Environmental Aspects of Tehri Hydro-Electric Project*. Unpublished report.
Haq, S. M., Hassan, M., Jan, H. A., Al-Ghamdi, A. A., Ahmad, K., & Abbasi, A. M. (2022). Traditions for future cross-national food security—Food and foraging practices among different native communities in the Western Himalayas. *Biology, 11*(3), 455. https://doi.org/10.3390/biology11030455
Hay, M., Skinner, J., & Norton, A. (2019). Dam-induced displacement and resettlement: A literature review. *Available at SSRN*. https://doi.org/10.2139/ssrn.3538211
Huber, A. (2019). Hydropower in the Himalayan hazardscape: Strategic ignorance and the production of unequal risk. *Water, 11*(3), 414.
Indra, D. M. (Ed.). (1999). *Engendering forced migration: Theory and practice* (Vol. 5). Berghahn Books.
Jain, M., Fishman, R., Mondal, P., Galford, G. L., Bhattarai, N., Naeem, S., & DeFries, R. S., et al. (2021). Groundwater depletion will reduce cropping intensity in India. *Science Advances, 7*(9).
Judge, P. S. (1997). Response to dams and displacement in two Indian states. *Asian Survey, 37*(9), 840–851. https://doi.org/10.2307/2645701
Kabeer, N. (2010). Women's empowerment, development interventions and the management of information flows. *IDS Bulletin, 41*(6), 105–113.
Katoch, A., Guleria, J., & Kumar, A. (2014). Impact of Nathpa Jhakri hydroelectric power project on the environment and livelihood in Kinnaur and Shimla Districts of Himachal Pradesh. *Indian Council of Social Science Research, 71*(3), 89–91.
Koenig, D. (2019). Women and resettlement. In *The Women and International Development Annual*, Volume 4 (pp. 21–49). Routledge.
Kothari, S. (1996). Whose nation? The displaced as victims of development. *Economic and Political Weekly*, 1476–1485. https://www.jstor.org/stable/4404269
Kumar, D., & Katoch, S. S. (2015). Small hydropower development in western Himalayas: Strategy for faster implementation. *Renewable Energy, 77*, 571–578. https://doi.org/10.1016/j.renene.2014.12.058
Kumar, S. (2021). Socio-economic impact on displaced women in Jammu Region: A sociological view. *ANTYAJAA: Indian Journal of Women and Social Change*, 6(1–2), 73–88. https://doi.org/10.1177/24556327211068313

Kumari, A., Joshi, R., & Yadav, S. K. (2025). Prospects and potential for sustainable food and nutritional security in North Western Himalayan region. *Indian Journal of Plant Genetic Resources, 38*(2), 145–154. 10.61949/

Maikhuri, R. K., Rao, K. S., & Saxena, K. G. (1996). Traditional crop diversity for sustainable development of Central Himalayan agroecosystems. *The International Journal of Sustainable Development & World Ecology, 3*(3), 8–31. https://doi.org/10.1080/13504509609469926

Mallick, M., Singh, P. K., & Pandey, R. (2024). Harvesting resilience: Tribal home-gardens as socio-ecological solutions for climate change adaptation and sustainable development in a protected area. *Journal of Cleaner Production, 445*, Article 141174. https://doi.org/10.1016/j.jclepro.2024.141174

McCully, P. (1996). *Silenced rivers: The ecology and politics of large Dams.*

Mehta, L., & Srinivasan, B. (2000). Balancing pains and gains: A perspective paper on gender and large dams. Contributing paper for WCD Thematic Review, 1.

Melketo, T. A., Seiber, S., & Bonatti, M. (2023). Gendered analysis of development induced displacement in the Global South: A systematic review. *European Journal of Sustainable Development Research, 7*(3), em0222. https://doi.org/10.29333/ejosdr/13296

Minocha, R. (2020). Development at what cost? A study of migration, loss of livelihood security and development-induced displacement in Himachal Pradesh. In *Displacement, Impoverishment and Exclusion* (pp. 346–378). Routledge.

Ministry of Environment and Forests. (2006). *Environmental Impact Assessment Notification, 2006.* Government of India.

Ministry of Environment and Forests. (2012). *National water policy, 2012.* Government of India.

Ministry of Power. (2008). *National policy on hydro power development.* Government of India.

Ministry of Rural Development. (2007). *National Rehabilitation and Resettlement Policy, 2007.* New Delhi: Government of India.

Mishra, M. (1976). Mishra Committee Report. Government report.

Mishra, V., Prakash, S., Sharma, A., Nautiyal, D., Thakur, A., & Sharma, R. S. (2024a). Limestone Mining in the Himalayas: Advancing Adaptive Governance for Ecological and Socioeconomic Resilience. In *The Himalayas in the Anthropocene: Environment and Development* (pp. 185–201). Cham: Springer Nature Switzerland.

Mishra, V., Thakur, A., Nautiyal, D., Sharma, A., Bhardwaj, S., Prakash, S., ... & Sharma, R. S. (2024b). Himalayan Dam projects in India and their implications: A deep dive into the Dhauliganga hydropower initiative. In *The Himalayas in the Anthropocene: Environment and Development* (pp. 59–92). Cham: Springer Nature Switzerland.

National Health Systems Resource Centre. https://nhsrcindia.org/

Negi, G. C. S., & Punetha, D. (2017). People's perception on impacts of hydro-power projects in Bhagirathi River valley India. *Environmental Monitoring and Assessment, 189*(4), 138. https://doi.org/10.1007/s10661-017-5820-y

Negi, V. S., Rana, S. K., Giri, L., & Rawal, R. S. (2020). Caterpillar fungus in the Himalaya, current understanding and future possibilities. GB Pant National Institute of Himalayan Environment, Kosi-Katarmal, Almora, Uttarakhand, India.

Newton, J. (2008). Displacement and development: The paradoxes of India's Tehri Dam. *The Geographical Bulletin, 49*(1), 2.

Pandey, R. (2021). Male out-migration from the Himalaya: Implications in gender roles and household food (in) security in the Kaligandaki Basin Nepal. *Migration and Development, 10*(3), 313–341. https://doi.org/10.1080/21632324.2019.1634313

Paranjpye, V. (1988). *Evaluating the Tehri Dam: An extended cost benefit appraisal* (No. 1, pp. viii+-139).

Parasuraman, S. (1993). Impact of displacement by development projects on women in India, ISS Working Papers—General Series 18852, International Institute of Social Studies of Erasmus University Rotterdam (ISS), The Hague.

Parasuraman, S. (1993). Impact of displacement by development projects on women in India. International Institute of Social Studies (ISS). ISS working papers. General series No. 159.

Pedraza, S. (1991). Women and migration: The social consequences of gender. *Annual Review of Sociology, 17*, 303–325. https://www.jstor.org/stable/2083345

Perera, J. (2012). The limitations of current approaches to resettlement. In *Resettling Displaced People* (pp. 381–415). Routledge India.

Press Information Bureau: Ministry of Health and Family Welfare, Government of India. https://www.pib.gov.in/newsite/PrintRelease.aspx?relid=98399

Randell, H. (2022). The challenges of dam-induced displacement: Reducing risks and rethinking hydropower. *One Earth, 5*(8), 849–852. https://doi.org/10.1016/j.oneear.2022.07.002

Ranjan, O. J., Anand, S., Pandey, B. W., & Kumria, P. (2020). Spatial analysis of physio-climatic changes and its impact on human adaption in Tawang Valley: A case study of Monpa Tribe, Eastern Himalaya. *Eastern Anthropologist, 73*(1).

Reddy, A. A. (2016). Involuntary resettlement as an opportunity for development: The case of urban resettlers of New Tehri Town. *World Development, 89*, 90–102. https://doi.org/10.1016/j.worlddev.2016.08.011

Reddy, A. A. (2017). *Rehabilitation and resettlement in Tehri Hydro power project*. Administrative Staff College of India.

Resurrección, B. P. (2019). Persistent gender and social exclusion in development. In *The Palgrave Handbook of Gender and Development* (pp. 255–269). Palgrave Macmillan. https://doi.org/10.1007/978-3-319-92288-1_14

Sahu, N., Sayama, T., Saini, A., Panda, A., & Takara, K. (2020). Understanding the hydropower and potential climate change impact on the Himalayan River regimes—A study of local perceptions and responses from Himachal Pradesh India. *Water, 12*(10), 2739. https://doi.org/10.3390/w12102739

Salaria, N., & Kumar, R. (2025). Conservation and Management of Himalayan Plant Biodiversity. In *Biodiversity of the Himalaya: Volume 1: Utilization, Conservation, and Management of Plant Resources*.

Sati, V. P. (2015). Pilgrimage tourism in mountain regions: Socio–economic and environmental implications in the Garhwal Himalaya. *South Asian Journal of Tourism and Heritage, 8*(2), 164–182.

Sati, V. P. (2024). *Farming systems and sustainable agriculture in the Himalaya*. Springer.

Sayektiningsih, T., & Hayati, N. (2021, October). Potential impacts of dam construction on environment, society and economy based on community perceptions. In *IOP Conference Series: Earth and Environmental Science* (Vol. 874, No. 1, p. 012013). IOP Publishing. https://doi.org/10.1088/1755-1315/874/1/012013

Sharma, R., Manjkhola, I., Sharma, A., Sukhralia, M., Chishty, S., & Mishra, V. (2026) Traditional Socio-Cultural Practices in Himalayan Food Systems for Climate Adaptation. *Frontiers in Sustainable Food Systems, 9*, 1718684.

Sharma, R. S., Mishra, V., Singh, R., Seth, N., & Babu, C. R. (2008). Antifungal activity of some Himalayan medicinal plants and cultivated ornamental species. *Fitoterapia, 79*(7–8), 589–591.

Schulz, C., & Skinner, J. (2022). Hydropower benefit-sharing and resettlement: A conceptual review. *Energy Research & Social Science, 87*, Article 102461. https://doi.org/10.1016/j.erss.2021.102461

Scudder, T. (1993). Development-induced relocation and refugee studies: 37 years of change and continuity among Zambia's Gwembe Tonga. *Journal of Refugee Studies, 6*(2), 123–152. https://doi.org/10.1093/jrs/6.2.123

Sidh, S. N., & Basu, S. (2011). Women's contribution to household food and economic security: A study in the Garhwal Himalayas India. *Mountain Research and Development, 31*(2), 102–111. https://doi.org/10.1659/MRD-JOURNAL-D-10-00010.1

Singh, S. (2019). Natural resources—A means of livelihood options for women in Western Himalyas. *IOSR Journal of Economics and Finance (IOSR-JEF)*.

Tehri Hydroelectric Development Corporation (THDC). (1998). *Rehabilitation policy*. THDC, New Delhi.

Thakkar, H. (2000). *Assessment of irrigation in India. World Commission of Dams.*

Thukral, E. G. (1996). Development, displacement and rehabilitation: Locating gender. *Economic and Political Weekly*, 1500–1503. https://www.jstor.org/stable/4404272

Udas, P. B., Tamang, D. D., Unni, A., Hamal, M., Shrestha, K., & Pandit, A. (2019). Basin level gendered vulnerabilities and adaptation: A case of Gandaki River Basin. *Environmental Development, 31*, 43–54. https://doi.org/10.1016/j.envdev.2019.05.002

Wang, X., Wang, P., Deng, Y., Xing, X., Yuan, Q., Du, C., Fang, Y., et al. (2024). Impacts of cascade hydropower development on river ecosystem homeostasis: A review. *Journal of Hydrology, 638*, Article 131484. https://doi.org/10.1016/j.jhydrol.2024.131484

World Commission on Dams. (2000). *Dams and development: A new framework for decision-making: The report of the World Commission on Dams*. Earthscan.

www.ingramcontent.com/pod-product-compliance
Ingram Content Group UK Ltd.
Pitfield, Milton Keynes, MK11 3LW, UK
UKHW021615260926
475829UK00010B/71

9783032240583